Library of Congress Cataloging-in-Publication Data

Rehabilitation / edited by Gerald Goldstein and Sue R. Beers.
    p.   cm. -- (Human brain function)
   Includes bibliographical references and index.
   ISBN 0-306-45662-1
   1. Brain--Diseases--Patients--Rehabilitation. 2. Mental illness-
 -Patients--Rehabilitation. 3. Medical rehabilitation.
 4. Rehabilitation nursing.   I. Goldstein, Gerald, 1931-
 II. Beers, Sue R.   III. Series.
 RC386.2.R44 1998
 616.8'043--dc21   97-50136
                                                               CIP

RC
386.2
.R44
1998

ISBN 0-306-45662-1

© 1998 Plenum Press, New York
A Division of Plenum Publishing Corporation
233 Spring Street, New York, N.Y. 10013

http://www.plenum.com

10 9 8 7 6 5 4 3 2 1

All rights reserved

No part of this book may be reproduced, stored in a retrieval system, or transmitted in any form or by any means, electronic, mechanical, photocopying, microfilming, recording, or otherwise, without written permission from the Publisher

Printed in the United States of America

Contributors

EDMOND AYYAPPA, *Veterans Affairs Medical Center, 5901 East 7th Street, Long Beach, California 90822*

SUE R. BEERS, *Western Psychiatric Institute and Clinic, 3811 O'Hara Street, Pittsburgh, Pennsylvania 15213*

LYNETTE S. CHANDLER, *School of Occupational and Physical Therapy, University of Puget Sound, 1500 North Warner Street, Tacoma, Washington 98416*

ANNE-LISE CHRISTENSEN, *Center for Rehabilitation of Brain Injury, University of Copenhagen, Amager 88 Njalsgade, DK-2300, Copenhagen S, Denmark*

MARTHA CLENDENIN, *Nova Southeastern University, Physical Therapy Program, 3200 S. University Drive, Fort Lauderdale, Florida 33328; and Science Center, PO Box 100154, Gainesville, Florida 32610-0154*

PATRICK J. DOYLE, *VA Pittsburgh Healthcare System, Highland Drive Division, 7180 Highland Drive, Pittsburgh, Pennsylvania 15206-1297*

GERALD GOLDSTEIN, *VA Pittsburgh Healthcare System, Highland Drive Division (151R), 7180 Highland Drive, Pittsburgh, Pennsylvania 15206-1297*

WILLIAM DREW GOUVIER, *Department of Psychology, Louisiana State University, Baton Rouge, Louisiana 70803-5501*

MICHAEL F. GREEN, *UCLA-Neuropsychiatric Institute and Hospital, 760 Westwood Plaza (C9-420), Los Angeles, California 90024-4344*

PAUL D. HANSEN, *School of Occupational and Physical Therapy, University of Puget Sound, 1500 North Warner Street, Tacoma, Washington 98416*

SHIRLEY P. HOEMAN, *Health Systems Consultation, 6 Camp Washington Road, Long Valley, New Jersey 07853; and School of Nursing, Fairfield University, Fairfield, Connecticut 06430*

MARGO B. HOLM, *Occupational Therapy Program, College Misericordia, 301 Lake Street, Dallas, Pennsylvania 18612*

LYNDA J. KATZ, *Landmark College, Rural Route 1, Box 1000, Putney, Vermont 05346*

ROBERT S. KERN, *West Los Angeles VA Medical Center (B116AR), 11301 Wilshire Boulevard, Los Angeles, California 90073*

KATHYE E. LIGHT, *University of Florida, College of Health Professions, Department of Physical Therapy, Health Science Center, PO Box 100154, Gainesville, Florida 32610-0154*

MICHAEL MCCUE, *Center for Applied Neuropsychology, First and Market Building, 100 First Avenue, Suite 900A, Pittsburgh, Pennsylvania 15222*

JUDITH R. O'JILE, *Department of Psychology, Louisiana State University, Baton Rouge, Louisiana 70803-5501*

JACQUELIN PERRY, *Medical Consultant, Pathokinesiology, Rancho Los Amigos Medical Center, 7601 E. Imperial Highway, Downey, California 90242*

MICHAEL PRAMUKA, *James A. Haley Veterans Administration Hospital, 13000 Bruce B. Downs Boulevard, Tampa, Florida 33612*

MARIE A. REILLY, *University of North Carolina at Chapel Hill, Department of Allied Health Sciences, Division of Physical Therapy, CB#7135, Chapel Hill, North Carolina 27599-7135*

JOAN C. ROGERS, *Western Psychiatric Institute and Clinic, 3811 O'Hara Street, Pittsburgh, Pennsylvania 15213*

LAURIE M. RYAN, *Department of Psychology, Louisiana State University, Baton Rouge, Louisiana 70803-5501*

THOMAS W. TEASDALE, *Psychological Laboratory, University of Copenhagen, Amager 88 Njalsgade, DK-2300, Copenhagen S, Denmark*

Preface

This volume is the fourth in a series that is dedicated to the topic of human brain function. The three previous volumes present an in-depth discussion of assessment procedures of the two specialties of neuroimaging and neuropsychology. This book, on the other hand, is focused on assessing brain function as it affects various aspects of the patient's capabilities from the perspectives of the rehabilitation specialties. These specialties generally assess brain function from a broader perspective than neuroimaging or even neuropsychology. That is, they are interested in how the brain interacts with the environment and impacts the patient's ability to manage practical aspects of his or her daily life.

The interest in rehabilitation and habilitation (i.e., treatment efforts directed toward maximizing rather than normalizing function) has grown extensively over the last 20 years, particularly with recent advances in technology and with increased attention directed to the quality of life of patients who present with various neuropsychiatric and medical illnesses. Multidisciplinary assessment procedures and collaborative efforts to enhance rehabilitation outcome are also becoming increasingly important as the various disciplines struggle to provide the most effective services within the economic restraints of the current health care delivery system. In the hopes of fostering future collaboration among the disciplines, Part II of this volume includes descriptions of the various rehabilitation specialties, including the occupational, physical, and speech and language therapies, as well as specialized areas of geriatric rehabilitation and rehabilitation nursing. Part III discusses the assessment methods that are applied to the evaluation of everyday functioning, specific sensory and motor functions, and both internal and external prostheses. Finally, in Part IV, clinical issues are addressed with respect to rehabilitation efforts after psychiatric and developmental disabilities and head trauma.

As the editors, we would like to thank our colleagues in neuropsychology and the rehabilitation specialties for their contributions to this unique multidisciplinary volume. It is hoped that the concepts and ideas presented here will lead to further collaborative efforts among the disciplines, with the ultimate outcome being the enhancement of the quality of care of the patients we share. Special acknowledgment also goes to Drs. Antonio Puente and Erin Bigler, who as series editors helped to conceptualize this collaborative effort across the specialties. Finally, we wish to express our appreciation for the continued support and encouragement of our editors at Plenum Publishing, Eliot Werner and Mariclaire Cloutier.

Sue R. Beers
Gerald Goldstein

Contents

PART I INTRODUCTION

Chapter 1

Introduction to Rehabilitation Assessment 3

Gerald Goldstein and Sue R. Beers

PART II THE REHABILITATION SPECIALTIES

Chapter 2

Occupational Therapy Assessment of Adult Brain Function 9

Margo B. Holm and Joan C. Rogers

    Introduction and Historical Background 9
    Occupational Therapy Research Related to Assessment of
        Brain Function 12
    Functional Assessment of Handicap (Role Performance) 13
        General 13
        Instruments for Assessing Handicap (Role Performance) 14
    Functional Assessment of Disability (Task Performance) 16
        General 16
        Instruments for Assessing Disability (Task Performance) 16
    Technology and Occupational Therapy Assessment 23

Ongoing Research on Occupational Therapy Assessment
of Adult Function 24
Summary 27
References 28

CHAPTER 3

Physical Therapy 33

Kathye E. Light, Marie A. Reilly, and Martha Clendenin

Introduction 33
Historical Background 34
   Early Years 34
   Recent History Marked by Change 35
   Growth and Demand for Services 36
Physical Therapy Evaluation and Assessment 36
   Need for Standardized Assessments 38
   Resistance to Standardized Assessment 38
   Types of Available Assessments 40
Review of the Assessments 40
   Basic Measurements 41
   Pediatric Assessments 41
   Adult Assessments 47
Future Directions 53
Summary 54
References 55

CHAPTER 4

Rehabilitation Assessment and Planning for Neurogenic
Communication Disorders 59

Patrick J. Doyle

Introduction 59
Nomenclature 60
   Dysarthria 60
   Apraxia of Speech 60
   Aphasia 61
The Appraisal Process 61
   Assessment of Dysarthria and Apraxia of Speech 62
   Assessment of Aphasia 65
Summary 67
References 67

Chapter 5

Dynamics of Rehabilitation Nursing 71

Shirley P. Hoeman

- Historical Influences on Rehabilitation Nursing 71
- Rehabilitation as a Nursing Specialty 73
  - Goals of Rehabilitation Nursing 73
  - Functional Health Patterns 73
- The Nursing Process 74
  - Assessment 74
  - Analysis 76
  - Diagnosis 77
  - Planning and Evaluation 77
  - A Continuum of Care 77
- New Subspecialty Directions 78
  - Restorative Care in the Community 79
  - Rehabilitation Nurses as Case Managers 81
  - Resource Management 83
  - Educating and Enabling Clients and Their Families 84
  - Advocacy and Leadership Roles 85
- Future Directions 86
  - Global Trends 86
  - Research Directions 86
- Summary 86
- References 87

Chapter 6

Geriatric Rehabilitation 89

Joan C. Rogers and Margo B. Holm

- Introduction 89
- Assessment Methodologies and Their Limitations 90
- Measurement of Impairment 91
  - Cognitive Impairment 92
  - Affective Impairment 93
  - Sensory-Motor Impairment 94
- Measurement of Disability 95
  - Disability Scaling Approaches 96
  - Disability Instruments 97
- Measurement of Handicap 102
- Future Prospects 103
  - Integrated View of Function 103

Instrumentation 103
Methodology 104
Summary 105
References 105

PART III ASSESSMENT FOR REHABILITATION

CHAPTER 7

Functional Assessment 113
Michael McCue and Michael Pramuka

Overview 113
Functional Assessment Defined 114
Approaches to Functional Assessment for Cognitive Disability 115
Observational Procedures 115
Situational Assessment 116
Functional Interviewing 117
Rating Scales and Questionnaires 118
Ecological Validity Studies 119
Use of Neuropsychological Assessment to Generate Inferences
about Functional Capacities 120
Cognitive Task Analysis 122
Functional Assessment of Executive Abilities: A Simulation
of Everyday Problem-Solving Ability 124
Summary 127
References 127

CHAPTER 8

Assessment of Sensory and Motor Function 131
Paul D. Hansen and Lynette S. Chandler

Introduction and Overview 131
Principles of a Sensory-Motor Evaluation 131
Validity and Its Effect on Diagnostic Accuracy 133
Suggestions for Performing the Neurological Evaluation 135
The Sensory-Motor Exam 135
History 138
Formal Observation 139

    Range of Motion and Muscle Tone Testing 140
    Sensory Testing 141
    Strength Testing 143
    Coordination 149
    Reflexes 149
    Gait 150
Summary 152
References 153

CHAPTER 9

Assessment of External Prostheses 155

Jacquelin Perry and Edmond Ayyappa

Historical Development 155
Scientific Assessment 156
    Berkeley Prosthetic Project 156
    Above-Knee Prosthetic Assessment 158
    Below-Knee Prosthetic Assessment 158
    Modernization of Gait Analysis 159
    Current Scientific Assessment of Below-Knee Prosthetics 161
Clinical Assessment 162
    Socket Interface Shape 163
    Component Design and Selection 164
    Alignment 165
Conclusion 166
References 166

PART IV   CLINICAL CONSIDERATIONS

CHAPTER 10

Rehabilitation Assessment and Planning for Head Trauma 171

Anne-Lise Christensen and Thomas W. Teasdale

Introduction 171
Assessment 173
The Program 174
The Future 179
References 179

Chapter 11

Neuropsychological Assessment for Planning Cognitive Interventions 181

William Drew Gouvier, Judith R. O'Jile, and Laurie M. Ryan

Introduction 181
    Level 1 Assessment: The Criterion Environment of the Real World 182
    Level 2 Assessment: The Neuropsychological Battery 184
    Level 3 Assessment: Fractionating the Deficits and Identifying the Causes of Task Failure 187
    Level 4 Assessment: Daily Monitoring of Treatment Effectiveness 189
    Application of the Levels System of Neuropsychological Assessment for Treatment Planning for Cognitive Disorders 190
    Level 1. Interview the Family and Patient 190
    Level 2. Conduct a Neuropsychological Evaluation to Quantify the Degree of Deficits and the Extent of Residual Strengths in the Following Domains of Neurocognitive Functioning 190
    Level 3. Conduct a Functional Analysis to Examine Why Tests Are Failed and to Fractionate the Causes of the Deficits 191
    Level 4. Plan and Implement Remediation Program 191
Case Example 192
    Selected Assessment Findings 192
    Rehabilitation Strategy for M.L. 194
A Caveat to Clinicians (or Problems with Early Neuropsychological Strategies) 196
Some Final Words 197
References 197

Chapter 12

Rehabilitation Assessment and Planning for Children and Adults with Learning Disabilities 201

Sue R. Beers

Introduction 201
Diagnostic Considerations 203
Heterogeneity of Learning Disability 204
Learning Disability in Adults 206
Assessment 206
    Clinical Interview 207
    Functional Assessment 207
    Neurological Examination 210
    Psychoeducational Assessment 210
    Neuropsychological Testing 216

Application 224
   Demystification 225
   Bypass Strategies 225
   Interventions at Breakdown Points 225
   Strengthening the Strengths 225
Summary 225
References 226

Chapter 13

Assessment and Planning for Memory Retraining 229

Sue R. Beers and Gerald Goldstein

Introduction 229
History of Memory Rehabilitation 230
Theoretical Considerations 230
Current Applications 232
   Assessment 232
   Rehabilitation Procedures following Head Injury 234
   Training Procedures for Patients with Korsakoff's Syndrome 236
Continued Considerations 239
   Outcome Assessment 239
   Instruments 240
Evaluation of Current Rehabilitation Procedures 241
   Alternative Methods of Memory Rehabilitation 241
   Evaluation of External Memory Aids 242
Future Directions 242
Conclusion 243
References 243

Chapter 14

Assessment and Planning for Psychosocial and
Vocational Rehabilitation 247

Lynda J. Katz

Introduction 247
Psychosocial Rehabilitation 249
   Clubhouse Model 249
   Boston University Psychiatric Rehabilitation Model 250
Transitional Employment: A Vocationally Based Psychosocial Model
   for Assessment and Planning 251
Vocational Rehabilitation 252
   Background 252

Sheltered Workshop Model	253
Supported Employment	254
Predicted Validity and Vocational Assessment Practices	255
Integrated Models of Assessment and Planning	256
New Hampshire IPS Model	256
University of Pittsburgh Program in Psychiatric Rehabilitation	257
Future Directions	259
Summary	260
References	260
Appendix	263

Chapter 15

Cognitive Remediation of Psychotic Patients 267

Robert S. Kern and Michael F. Green

Introduction	267
Models of Information Processing	269
Assessment of Cognition and Information Processing	270
Memory—List Learning Measures	270
Abstraction/Problem-Solving Ability—Wisconsin Card Sorting Test	271
Early Visual Processing—Span of Apprehension Task	272
Feasibility Studies of Cognitive Remediation: "Attacking Discrete Deficits"	272
Remediation of Memory Deficits: Modifying Performance on List Learning Tasks	272
Remediation of Abstraction/Problem-Solving Deficits: Modifying Performance on the Wisconsin Card Sorting Test	274
Remediation of Early Visual Processing Deficits: Modifying Performance on the Span of Apprehension Task	277
Comprehensive Programs of Cognitive Remediation	279
Conclusions	280
References	281

Index 285

Introduction I

1

Introduction to Rehabilitation Assessment

GERALD GOLDSTEIN AND SUE R. BEERS

In this volume of *Human Brain Function: Assessment and Rehabilitation,* we look at how rehabilitation specialists conceptualize and practice evaluation of brain function. As we will see, these concepts and practices are quite distinct from those presented in the three volumes on neuroimaging and neuropsychology. The need for an entire volume devoted to this matter will be apparent as one becomes aware of how methods and concepts differ between physicians and clinical neuropsychologists, on the one hand, and rehabilitation specialists on the other. One major distinction is between diagnosis and treatment. Although not entirely true, it is probably fair to say that the neuroradiologist, neurologist, and neuropsychologist are largely concerned with diagnosis, whereas the rehabilitation specialists are primarily concerned with treatment and remediation. Most professional rehabilitation specialists spend their lives engaged in direct treatment efforts, and use assessment primarily to guide and monitor those efforts.

There are two major definitions of rehabilitation, both of which are considered in this book. The more traditional definition is the process of restoring function to normal or near-normal status. The second definition, coming mainly from the field of psychiatric rehabilitation and found in a psychiatric dictionary is an attempt to achieve "maximal function and adjustment, and to prepare the patient physically, mentally, socially, and vocationally, for the fullest possible life compatible with his abilities and disabilities" (Hinsie & Campbell, 1960, p. 639). The first definition is probably more appropriate for sensory and motor disorders, primarily those produced by injury, whereas the second is more relevant to patients with long-term, chronic illnesses. This distinction is of major importance, because it aids in dispelling the notion that rehabilitation is not appropriate for individuals who, by

---

GERALD GOLDSTEIN   VA Pittsburgh Healthcare System, Highland Drive Division (151R), 7180 Highland Drive, Pittsburgh, Pennsylvania 15206-1297.   SUE R. BEERS   Western Psychiatric Institute and Clinic, 3811 O'Hara Street, Pittsburgh, Pennsylvania 15213.

Rehabilitation, edited by Goldstein and Beers. Plenum Press, New York, 1998.

virtue of the nature of their illnesses, will never attain normal or near-normal function. Indeed the term "habilitation" has been introduced to suggest that productive treatment efforts may be devoted to individuals with developmental and early childhood disorders who have never attained normal function in a number of areas.

Definitions related to capacity and disruption of capacity may also be helpful. Rehabilitation specialists make distinctions among terms that are not commonly made by other clinicians. Most clinicians use the terms *function* and *ability* synonymously. We say intellectual functions or intellectual abilities, and mean about the same thing. In rehabilitation parlance, an ability is a capacity within the individual; a function is an exercise of ability in the environment. Motor dexterity is an ability; playing the violin is a function dependent upon numerous abilities. Rehabilitation specialists are primarily concerned with functions; that is, the exercise of various abilities in a natural setting. This book contains a chapter on functional assessment, and one may wonder why it is in the rehabilitation and not the neuropsychology volume, as neuropsychologists traditionally assess functions. The chapter clearly indicates that functional assessment is quite different from neuropsychological assessment with regard to theoretical framework, methodology, and clinical applications.

Correspondingly, when a capacity is diminished or eliminated, we say that the individual has an impairment, disability, or handicap, and frequently use these terms synonymously. To rehabilitation specialists, they are not synonymous. An *impairment* is a disruption of an ability or capacity. The impairment may produce a functional deficiency that is characterized as a *disability*. If the disability prevents the individual from performing some activity in the environment, then he or she has a *handicap*. For example, a hemiparesis from a stroke (impairment) may produce a loss of manual dexterity (disability) that prevents the individual from playing the violin (handicap).

In general, the rehabilitation specialties assess brain function from the perspective of how the brain interacts with the environment. In such assessment, there is almost always an object of reference: a bed to be made, a shirt to be put on, a message to be communicated to or received from another person. Typically, rehabilitation takes place when there is some disruption of this interaction. In some cases, the brain is not directly involved and rehabilitation is devoted to restoration of function of a limb or a sense organ. However, even when the brain is involved, there is the same emphasis on function. In the case of psychiatric rehabilitation, for example, the emphasis is not on "cure" or reduction in the severity of the psychopathology, but rather on assisting the individual with a psychiatric illness in improvement of adaptive function in various behavior settings. Thus, the rehabilitation specialties provide a unique way of looking at the brain.

The volume begins with descriptions of the various rehabilitation specialties, emphasizing those aspects of brain function in which the specialists have particular interest and expertise. We will see that there are specialized conceptual approaches, competencies, and assessment strategies for the areas of task performance, fulfillment of life role expectations, ambulation and motility, self-care, and ability to communicate. In the second section, specific assessment methodologies are discussed including evaluation of function in everyday life, of sensory and motor function, and of functioning of the individual who requires an assistive or prosthetic device.

The third section deals with clinical aspects of rehabilitation in a very broad sense. Extensive research has been done and considerable expertise and technology exist in specialized areas of rehabilitation. These areas of specialization relate to specific disorders, such as head trauma or psychiatric illness, specific aspects of function, such as memory or other cognitive functions, and specific applications, such as vocational planning. Many of these areas have become specialties in themselves, each of them involving interdisciplinary collaboration. Thus, for example, an occupational therapist may utilize one or more forms of functional assessment in the rehabilitation of a child with a developmental disorder. In the rehabilitation process, the occupational therapist may work in collaboration with members of other disciplines, such as neuropsychologists, and with other rehabilitation specialists, such as speech/language pathologists.

Rehabilitation of individuals with neurobehavioral and neuropsychiatric disorders has undergone a revolution during the past several decades. There has been a remarkable growth in the number of rehabilitation hospitals and treatment programs, particularly for patients with traumatic brain damage. New retraining methods have become available in the areas of memory, communication, perception, and problem solving. Assistive devices that support memory and other cognitive functions are being developed and tested. An increasing number of medications are becoming available that have been shown to enhance cognitive function and that may be productively used as parts of comprehensive rehabilitation efforts. Programs are being developed for cognitive rehabilitation of neuropsychiatric patients, especially for individuals with chronic alcoholism and schizophrenia. It is increasingly important to develop multidisciplinary assessment procedures relevant to planning rehabilitation programs, to learn more about how rehabilitation alters brain function and interacts with natural recovery, and to develop and utilize appropriate instruments to assess progress and outcome. Although much remains to be done in these areas, a great deal of work has already been accomplished. This volume provides an overview of the contributions to these endeavors by the various rehabilitation specialties, working in a variety of applications.

Reference

Hinsie, L. E. & Campbell, R. J. (1960). *Psychiatric dictionary* (3rd ed.). New York: Oxford University Press.

II

The Rehabilitation Specialties

2

Occupational Therapy Assessment of Adult Brain Function

MARGO B. HOLM AND JOAN C. ROGERS

INTRODUCTION AND HISTORICAL BACKGROUND

Historically, occupational therapists have established assessment expertise in the direct observation of their patients' functional performance of the everyday tasks that define and bring meaning to their lives (Fisher & Short-DeGraff, 1993; Guralnik, Branch, Cummings, & Curb, 1989; Trombly, 1993). Patients are usually referred to occupational therapy primarily because of problems in everyday *task performance*. The unique contribution of occupational therapy, as a profession, in the study of brain function is at the level of functional task performance, namely, the patient-task-context transaction. In considering function at this level, we bring cognizance of the influence of the factors within the patient (motor, sensory, cognitive, and psychosocial skills and impairments) during the patient's transaction with specific tasks and task contexts. Using a task analysis approach (Faletti, 1984; Fisher & Short-DeGraff, 1993), we also contribute data to the patient and the team about the abilities and behaviors required of a task performer for adequate and safe task performance (task demands). Task analysis also helps us identify the inherent properties and procedures, equipment, and materials involved in critical and representative functional tasks (e.g., dressing, meal preparation, driving) and subtasks (e.g., putting on and taking off a pair of slacks, turning the stovetop burners on and off, and manipulation of adapted steering mechanisms) that the patient wants or needs to perform. In addition, occupational therapists' professional

---

MARGO B. HOLM Occupational Therapy Program, College Misericordia, 301 Lake Street, Dallas, Pennsylvania 18612.   JOAN C. ROGERS Western Psychiatric Institute and Clinic, 3811 O'Hara Street, Pittsburgh, Pennsylvania 15213.

Rehabilitation, edited by Goldstein and Beers. Plenum Press, New York, 1998.

education enables them to appraise a task context (clinic or naturalistic setting) for its potential to support a patient's abilities, provide appropriate levels of stimulation, and promote safe task performance. Therefore, task performance is pivotal to all levels of occupational therapy functional assessment and intervention (Holm & Rogers, 1989) and this focus is the occupational therapy profession's particular contribution to the study of brain function.

Occupational therapy functional assessment and intervention at the levels of person, task, and context might best be depicted using an adaptation of the World Health Organization (WHO) model (WHO, 1980). Rehabilitation theorists (Nagi, 1977; Wood, 1980; WHO, 1980) attempted to clarify the construct of function by viewing traumatic and chronic health conditions as a multidimensional experience. The conceptual model endorsed by WHO, for example, distinguishes among four concepts: pathology, impairment, disability, and handicap (see Figure 1, left).

WHO defines *pathology* as a disruption of normal body systems or processes. It includes acquired medical diagnoses such as stroke, traumatic brain injury, multiple sclerosis, arthritis, repetitive motion disorders and spinal cord injury; developmental disorders such as cerebral palsy and developmental delay; and psychiatric disorders such as schizophrenia, depression, and dementia. If a pathology is active or cannot be cured, but only controlled, impairment may result.

Impairment refers to loss or abnormality of an anatomic, physiologic, cognitive, or emotional structure or function. Inability to think in the abstract, memory

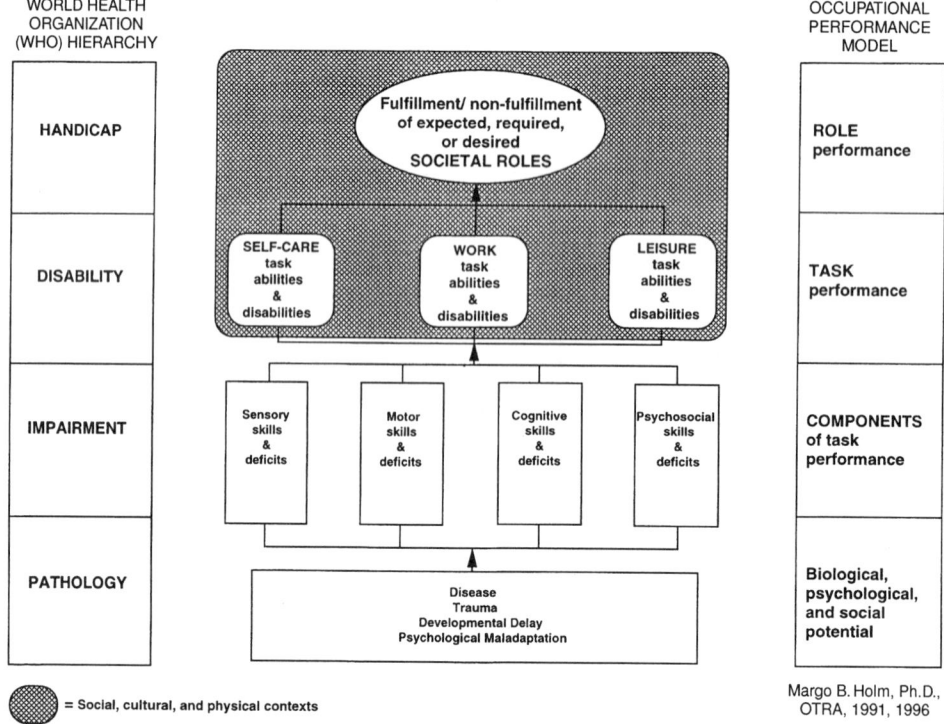

Figure 1. A model of occupational therapy practice based on the World Health Organization model.

deficits, visual-perceptual dysfunction, parasthesia, muscle weakness, restricted range of motion, low self-esteem, and apathy are examples of impairments. Impairment, then, is concerned with the function of body parts, systems, or processes at their basic levels, in isolation.

Disability, however, is concerned with the integrated functioning of the entire person that is required to accomplish tasks or to interact meaningfully in society. Examples of disabilities are problems in eating, dressing, purchasing groceries, finding a theater, driving, operating a drill press, changing a diaper, or taking notes during a class.

Everyday tasks, combined in different ways, form the basis of social roles (e.g., mechanic, homemaker, student). If task disabilities are severe, social role performance can become dysfunctional and the person will no longer be able to enact roles satisfactorily. The experience of pathology at this level is referred to as *handicap.* Handicap involves a social disadvantage such that the individual is limited or prevented from fulfilling a social role that is regarded as appropriate based on age, sex, and social and cultural norms.

The WHO model focuses only on the negative consequences that can result from pathology. The model, however, can also be adapted to represent the multiple levels of functional performance that are the focus of occupational therapy assessment and intervention, and which include the patient's strengths as well as deficits (see Figure 1, center). Although the WHO model is multidirectional, it highlights the consequences of pathology up the hierarchy through the categories of impairment, disability, and handicap. Occupational therapists, however, prefer an approach to assessment that could be labeled top-down (see Figure 1, right; Trombly, 1993; Holm & Rogers, 1989). Beginning with the WHO concept of handicap, or social role performance, occupational therapists must first ascertain which roles and tasks patients performed prior to illness or trauma, the level of role competence, and the meaningfulness of those roles (Trombly, 1993; Holm & Rogers, 1989).

When the current level of functioning is different from what patients desire to do or are required to do to carry out meaningful role performance, occupational therapists directly observe the performance of the tasks critical to role functioning to identify patients' abilities and disabilities. These task abilities are then used to enhance role performance. Task disabilities are also identified, and rehabilitation interventions are designed to restore performance, prevent further deterioration of performance, or compensate for deficits in performance with adaptive strategies or assistive technology devices (Holm & Rogers, 1991; Rogers & Holm, 1991; Trombly, 1993). Abilities and disabilities do not occur in isolation, however, nor can they be assessed as such. Therefore, occupational therapists prefer to assess task performance in the naturalistic context when possible, or to simulate the naturalistic context in the clinic setting. If the occupational therapist's assessment establishes that the task context fails to support or perhaps even obstructs task performance, then environmental adaptations are identified and recommended.

If task disability is based on impairment at the level of the anatomic, physiologic, cognitive, or affective functioning of a patient, however, occupational therapists assess these components, in isolation, to identify which components are adequate to support task performance, and what specific impairments exist. If the component is adequate, then its contribution to task performance is exploited; if the component is impaired, then rehabilitation intervention is designed to restore its specific function or identify strategies to adapt for its loss so as to prevent further disability.

OCCUPATIONAL THERAPY RESEARCH RELATED TO ASSESSMENT OF BRAIN FUNCTION

Instead of an exhaustive review of current research for each level of occupational therapy assessment, we have chosen to highlight assessment research that represents the primary approaches to functional assessment currently being used by occupational therapists. Using the top-down approach, we identify primary methods of assessment for the level of handicap and disability and highlight ongoing research into these methods. Because assessment of impairment is specific to each system being evaluated and focuses on function in the narrowest sense, we have chosen to address the more complex levels of functional assessment.

In general, interview, direct observation of performance, and testing (standardized and nonstandardized) are the primary methods used by occupational therapists to gather information about patients' functional performance. Based on the level of function being assessed, the assessment methods vary, however, in their reliability and validity, usefulness as measures of functional status, and ability to measure change in functional status. For example, in the assessment of meal preparation, an interview item may yield excellent interrater and test–retest reliability, because the patient perceives meal preparation as heating a frozen entree and consistently gives the response "no difficulty." For direct observation of performance, the setup, tasks and subtasks, and directions must be determined in advance and operationally defined. In addition, rating scales should have mutually exclusive categories. If these conditions are not present, then the chance of acceptable levels of interrater reliability and test–retest reliability decrease rapidly as task difficulty (i.e., number of meal preparation criteria to be rated) increases. When test items are standardized, reliability is easier to achieve, but if the item has been simplified to achieve reliability, it may not be valid. For example, when the only meal preparation item of a functional assessment includes heating water in an automatic teapot to make a cup of tea, the reliability of the item may be excellent, but its validity as an assessment of everyday meal preparation could hardly be considered valid.

The complexity of functional assessment increases as the therapist is required to observe, measure, and interpret

> units of behavior nested within progressively larger chunks of a person's ongoing stream of life activities. Behavioral units can range from the simple movement of a thumb to the grasping of a cup, the drinking of coffee, and the consumption of breakfast. Each of these progressively larger units of behavior happens within an environmental surround that itself broadens as one assembles the units into meaningful patterns. These meaningful patterns or chunks of behavior are often referred to as tasks and roles. (Christiansen, 1993, p. 258)

When the top-down approach to assessment is used, occupational therapists are faced with unnesting the functional units to identify the level at which performance deficits are affecting functional performance (e.g., disability or impairment); this analysis of performance deficits is a necessary prerequisite for intervention (see Figure 1, center).

In 1991, the American Occupational Therapy Foundation, the American Occupational Therapy Association, and the Occupational Therapy Center for Research and Measurement at the University of Illinois cosponsored the *Symposium on Measurement and Assessment: Direction for Research in Occupational Therapy*. One of the purposes of the symposium was to identify for which "levels of function" occupational therapists still need to construct quality assessment tools. Another purpose of the symposium was to identify assessment methods and measurement models that are appropriate for functional assessment (Fisher & Short-DeGraff, 1993). Recommendations emerging from the symposium included increased emphasis on assessment of the social and physical context of task performance and the relationship between the components of task performance (i.e., sensory, motor, cognitive, and psychosocial) and the task performance. Also emanating from the symposium were a series of papers that appeared in two issues of the *American Journal of Occupational Therapy* that focused on the current status of functional assessment in the profession, methods of measurement, and alternative strategies for functional assessment (Fisher & Short-DeGraff, 1993; Short-DeGraff & Fisher, 1993). Because of increased research interest in occupational therapy regarding the person-task-context interaction, this chapter is limited to a discussion of assessment research at the WHO levels of handicap (i.e., role function) and disability, where this interaction is most evident in functional performance assessment.

FUNCTIONAL ASSESSMENT OF HANDICAP (ROLE PERFORMANCE)

GENERAL

The three primary methods used by occupational therapists to assess role performance include interview, job analysis, and performance-based observation of the numerous tasks that are combined in role functions. When the interview method is used, depending on the role being assessed (e.g., self-caregiver, parent, laboratory technician, caregiver) those being interviewed can include the patient, family members, employers, or caregivers. An evolutionary process has been under way since the 1960s in an attempt to standardize the interview process, obtain relevant historical information, and develop scoring criteria and measures that yield a concise narrative or quantitative summary (Kielhofner & Henry, 1988b).

Job analysis has its roots in vocational rehabilitation, and occupational therapy has been included in federal legislation related to vocational rehabilitation for more than 75 years (Bitter, 1979; Public Law 65-178; Public Law 66-236; Public Law 74-271; Public Law 78-346; Public Law 89-97; Public Law 98-527; Reed, 1992). In addition to the focus on vocational rehabilitation, legislation such as Public Law 98-527 included training in independent living and integration into the community for individuals with developmental disabilities. The most recent legislation to impact the role of occupational therapists in the areas of job analysis and community integration is the Americans With Disabilities Act (ADA) (Public Law 101-336), which requires the specification of essential, marginal, and peripheral job functions. Individuals with disabilities must be able to perform the essential functions of a job, but not necessarily the marginal or peripheral functions (Jacobs, 1992). A job analysis, therefore, delineates the essential functions that a patient must be able to perform to fulfill a work role. Those essential functions then serve as the performance criteria to be rated during a performance-based post-job-offer assessment, functional capacity evaluation, or work hardening rehabilitation intervention (AOTA, 1986; Blankenship, 1989; Ellexson & Kornblau, 1992; Jacobs, 1992; U.S. Department of Labor, 1986; Wyrick, Niemeyer, Ellexson, Jacobs, & Taylor, 1991). Although an expanded review of job analysis is not within the purview of this chapter, a job's essential functions, as well as the necessary activities of daily living (ADL) and instrumental activities of daily living (IADL) required to prepare for a work role, can also guide the therapist's choice of a more global performance-based observation.

A performance-based observation (PBO) requires the patient to perform a specific task in a specific environment while the occupational therapist observes and rates the performance according to pre-established criteria. PBO is conducted in the clinic if necessary, but preferably in the naturalistic setting (e.g., home, school, job site) because of its ecological validity (Dunn, Brown, & McGuigan, 1994). The focus of the observation is usually performance problems identified during the interview with the patient or patient advocate. Task performance is rated based on predetermined criteria from a job analysis or a task analysis of the skills, sequences, equipment, materials, and environment necessary to accomplish role behavior (U.S. Department of Labor, 1986). Examples of instruments used by occupational therapists to gather data about role performance are included below, and the intended population, functional use, content, format and scoring, and reliability and validity information are described.

INSTRUMENTS FOR ASSESSING HANDICAP (ROLE PERFORMANCE)

The Occupational History (Moorhead, 1969) is a semistructured interview consisting of 76 possible questions, and was designed to be used with various patient populations. The Occupational Role History (ORH; Florey & Michelman, 1982), based on Moorhead's work, is designed for use with a mental health population, and focuses on five critical areas related to a patient's role performance (sequence of roles and role requirements, satisfaction derived from roles, ability to balance several roles, perceived areas of skill and deficits, and balance among work, chores, and leisure pursuits). A modified version of the ORH was later developed for use with the physically disabled population (Kielhofner, Harlan, Bauer, & Maurer, 1986). The ORH also served as the prototype for the Occupational Performance History Interview.

THE OCCUPATIONAL PERFORMANCE HISTORY INTERVIEW (OPHI) (KIELHOFNER, HENRY, & WAYLENS, 1989). (NB: The OPHI emanated from the previously described interview tools, circa 1969, 1982).

Population. The OPHI was designed to be used with all populations.

Functional Use. The OPHI is designed to help the therapist better understand the forces that affect a patient's performance in life roles, and thus assists with planning for intervention and discharge.

Content. The OPHI consists of 39 recommended questions that cover five content areas: (1) contextual resources and barriers; (2) life roles; (3) organization of daily living routines; (4) perceptions of ability and responsibility; and (5) interests, values, and goals.

Format and Scoring. The interview uses a semistructured format, and for each of the five content areas an ordinal rating scale is used to quantify a patient's ability to adapt: A 1 indicates completely maladaptive and a 5 indicates completely adaptive. Items address both past and present patient behaviors and the context in which roles are enacted. Since the OPHI is a guide, and therapist discretion dictates which items will be used, there is no meaningful summative score.

Reliability and Validity. Test–retest reliability studies were conducted on 153 interviews of patients aged 13 to 93 whose diagnoses included both psychiatric and physical pathologies. Test–retest reliabilities ranged from $r = .31$ to $r = .49$ for present behaviors and $r = .55$ to $r = .68$ for past behaviors. Interrater reliability coefficients for 129 of the 153 subjects, from therapists with similar frames of reference, for present items ranged from $r = -.12$ to $r = .63$, and for past items ranged from $r = .23$ to $r = .55$. Flaws in the study designs negatively impacted the reliabilities. For example, different raters and different time spans were used for comparisons (Kielhofner & Henry, 1988).

THE ROLE CHECKLIST (OAKLEY, 1982; OAKLEY, KIELHOFNER, BARRIS, & REICHLER, 1986)

Population. Although initially designed for use with a mental health population, the checklist is appropriate for adolescents, adults, and older adults with mental health or physical dysfunction (Barris, Oakley, & Kielhofner, 1988).

Functional Use. The Role Checklist is meant to be used as a screening tool to ascertain a patient's major life roles, their continuity over time, and the value that the individual places on each role. If the patient is not performing those roles that are expected or valued, a more in-depth evaluation of role functioning is carried out using other tools and methods.

Content. The Role Checklist consists of two parts. Part I includes 10 roles, namely, student, worker, volunteer, caregiver, home maintainer, friend, family member, religious participant, hobbyist/amateur, and participant in organizations. A role is checked if the patient has fulfilled it (past), currently fulfills it (present) or intends to fulfill it (future). In Part II, the patient identifies the degree to which each role is valued.

Format and Scoring. Each role and time period (i.e., past, present, future), as well as the term *valuable* are operationally defined. The patient self-administers the checklist and the therapist remains with the patient to answer questions and provide assistance if needed. For Part I of the Role Checklist a dichotomous scale is used for each role and each time frame. If a role and time frame are checked, a 1 is assigned; if they are not checked, a 0 is assigned. In Part II, for each role the value checked (i.e., not valuable, somewhat valuable, and very valuable) is assigned a 1 and the other values are assigned a 0. Nine summary scores can be calculated: continuous roles, disrupted roles, role changes, past roles, present roles, future roles, not valuable, somewhat valuable, and very valuable (Asher, 1989).

Reliability and Validity. Test–retest reliability percent agreements ranged from 77% to 93% and averaged 87%, with little difference between 1–4-week and 5–7-week intervals. Content validity of the roles selected was established based on a review of the social psychology, sociology, and occupational therapy literature. Construct validity studies have indicated that hospitalized patients attached significantly less value to roles than did nonhospitalized patients. Another validity study found that noninstitutionalized older adults, who had checked greater numbers of roles, experienced significantly greater life satisfaction than did those who fulfilled fewer roles (Barris et al., 1988).

ROUTINE TASK INVENTORY-2 (RTI-2) (ALLEN, KEHRBERG, & BURNS, 1992; HEIMANN, ALLEN, & YERXA, 1989). See the following section for a description of the RTI-2 items on role performance.

FUNCTIONAL ASSESSMENT OF DISABILITY (TASK PERFORMANCE)

GENERAL

Occupational therapy research on disability assessment has focused on the development of assessment tools that can be used across populations as well as on the identification of those tasks critical for independent living. Research at this level has also focused on the measures used (i.e., ordinal, interval), how the tools are scored (i.e., additive, nonadditive), and the impact of environment (i.e., clinic, home) on performance. The differential difficulty of mobility, ADL and IADL tasks, the impact of impairments on task difficulty level, and the number and types of assists necessary for safe and successful task performance are other avenues of current disability assessment research. Examples of instruments used by occupational therapists to gather data about task disability are listed below based on order of approximate initial publication or description in the literature. For each instrument, the intended population, functional use, content, format and scoring, and reliability and validity information are described.

INSTRUMENTS FOR ASSESSING DISABILITY (TASK PERFORMANCE)

KOHLMAN EVALUATION OF LIVING SKILLS (KELS) (McGOURTY, 1979, 1988)

Population. Initially designed for use with mental health populations, it has also been used with geriatric populations.

Functional Use. The KELS was designed to assist with discharge planning.

Content. The KELS includes 18 living skills in five content areas: self-care, safety and health, money management, transportation and telephone, and work and leisure.

Format and Scoring. The KELS is a knowledge-based assessment wherein problems, including stimulus materials, are presented to a patient for resolution. The KELS is administered, using a standardized format, to a patient seated at a table. A dichotomous scoring system is used, with "independent" scored as 0 and "needs assistance" scored as 1. Independent is defined as "the level of competency required to perform the basic living skill in a manner that maintains the safety and health of the individual without direct assistance of other people" (McGourty, 1988, p. 138). A summed score ≤6 indicates that the patient needs assistance to live in the community. The therapist is also asked to write a summary note and include information about a patient's reality orientation, judgment, and time use.

Reliability and Validity. Interrater reliability has been reported as percent agreements ranging from 74% to 94% (Ilika & Hoffman, 1981b). Pilot predictive validity studies, using percent agreement of predicted and actual outcome ranged in accuracy from 50% to 100%, but design flaws in the studies have made the results inconclusive (McGourty, 1988; Morrow, as cited in McGourty, 1988). Concurrent validity with the Bay Area Functional Performance Evaluation was established at $r = -.84$ ($p < .0001$) and between $r = .78$ and $r = .89$ ($p < .001$) with the Global Assessment Scale in samples of inpatient mental health patients (Ilika & Hoffman, 1981a; Kaufman, 1982). Construct validity with independent living status (lower scores) was also established in a study comparing independent and sheltered living environment subjects (Tateichi, as cited in McGourty, 1988).

PERFORMANCE ASSESSMENT OF SELF-CARE SKILLS, VERSION 3.1 (PASS) (ROGERS, 1984; ROGERS & HOLM, 1989A, 1994)

Population. Initially designed for use with inpatient and outpatient geriatric populations, the PASS has been used with healthy adults as well as those from the following diagnostic populations: osteoarthritis, dementia, depression, cardiopulmonary disease, schizophrenia, bipolar disorder, and mental retardation. The PASS is designed for use with adults.

Functional Use. Both versions of the PASS (Clinic and Home) are designed to establish functional status at baseline, and to measure change at discharge. In addition, the assist hierarchy provides data for treatment planning and discharge disposition by identifying the number and types of assists that are necessary for safe and successful task performance.

Content. The instrument is designed to provide data about the performance of 26 tasks: 5 functional mobility (MOB) tasks (e.g., bed mobility, stairs, bathtub mobility), 3 personal self-care (ADL) tasks (e.g., dressing, oral hygiene, trimming toenails), and 18 instrumental activities of daily living (IADL) tasks (e.g., oven cooking, stovetop cooking, medication management, small repairs, environmental safety awareness, emergency procedures).

Format and Scoring. The PASS is a criterion-referenced, performance-based observational tool with standardized procedures. There are two versions of the PASS: one to be administered in a clinic context (PASS-Clinic) and one to be administered in the home (PASS-Home). Both versions contain the same items and the same criteria for performance. The PASS rating protocol identifies the conditions (context and setup) for each item and the verbal instructions to be given to the patient. The instructions to the patient in the PASS are global, but the rating protocol divides tasks into critical subtasks for rating purposes. This enables the therapist to identify the specific point or points in the task sequence at which breakdown occurs, and to codify the type and number of task or contextual modifications, or human assists, necessary for safe and successful task completion. The PASS (version 3.1) yields three types of summary scores: independence, safety, and outcome. Since the PASS is criterion-referenced, it may be given in total, or target items may be used alone or in combination. The total PASS takes 1 1/2 to 3 hours to administer, depending on the patient's impairments and disability level.

Reliability and Validity. Interrater reliability of the PASS 2.0–Clinic, based on 30 females from 74 to 85 years, 2 observers, and 1,965 observations yielded percent agreements as follows: 96%, Performance Score; 99%, Independence Score; 99%, Person Assists; 99%, Task/Environment Assists. Interrater reliability of the PASS 2.0–Home, for the same subjects, with 2 observers, yielded the same percent agreements. Three observers, rating 5 subjects in both settings, reached mean percent agreement scores of 93%, 98%, 99%, and 99%, respectively, for the same categories. Subjects in the reliability study included the well elderly as well as individuals who carried primary diagnoses of arthritis, dementia, depression, and cardiopulmonary disease in addition to stroke, diabetes, bursitis, macular degeneration, glaucoma, and incontinence (Holm & Rogers, 1990a). Mini-Mental State Examination scores ranged from 17 to 30 (mean = 26), and all subjects resided independently in the community. Reliability studies, including test–retest studies, are currently under way for the PASS 3.1.

Content validity of the PASS is based on the interview schedules of the OARS Multidimensional Functional Assessment Questionnaire—Activities of Daily Living Scale (Pfeiffer, 1976), the Comprehensive Assessment and Referral Evaluation (Gurland et al., 1977–78), the rating scales for Physical Self-Maintenance and Instrumental Self-Maintenance (Lawton 1972), and the Functional Assessment Questionnaire (Pfeffer, 1987). ADL/IADL categories, tasks, and task specifications explored in these instruments were delineated. Performance test items were then developed for each category (excluding transportation), incorporating as many of the task specifications as possible. Evidence of the *construct validity* of the PASS is gleaned from two investigations. In the first study, the IADL scores of patients with primary degenerative dementia were found to be proportionally lower, and to have more variability, than ADL scores. In the second study, healthy control subjects exhibited the highest level of performance, followed by patients with depression, and then those with dementia. These results are consistent with the hierarchy of ADL and IADL and expected differences in level of ADL/IADL disability based on diagnosis and impairment severity.

Milwaukee Evaluation of Daily Living Skills (MEDLS) (Leonardelli, 1988a)

Population. The MEDLS was designed for use with chronic mental illness populations.

Functional Use. The MEDLS was designed to establish baseline behaviors necessary for developing treatment objectives and guiding intervention.

Content. The MEDLS consists of a screening form to assist with the record review in order to determine which subtests need to be administered. The 21 subtests assess the areas of communication, personal self-care, safety, money management, medication management, and use of transportation.

Format and Scoring. The MEDLS is primarily a measure of discrete performance skills which a patient performs either in the naturalistic setting (e.g., dressing), simulates (e.g., bathing), or describes how performance is usually carried out (e.g., transportation). Items are administered using standardized procedures; however, based on the screening the therapist selects the items needing further evaluation. The items selected, therefore, determine the context for observation of the performance skills: the naturalistic setting, a simulated setting (the clinic), or seated at a table with stimulus items presented according to protocol (Leonardelli, 1988b).

Each subtest has its own score based on the tasks performed. There is no summative score for the MEDLS, as not all items are always administered. For each subtest, if a patient's performance skill is deemed deficient, the therapist uses preselected "key words" to identify the skill that is deficient (e.g., cleans). If items cannot be completed in the time specified, the therapist must make a clinical judgment as to the cause (e.g., comprehension deficit, motivational deficit).

Reliability and Validity. Interrater reliability (N = 18) ranged from $r = .40$ to $r = 1.00$, with most subtests above $r = .80$. There was no significant difference between experienced and nonexperienced raters (T = 47.5, $p = .10$). Content validity is based on a review of the literature and other evaluation tools. Specific tools or articles are identified as content validity references for each subtest (Leonardelli, 1988a).

ARNADÓTTIR OT-ADL NEUROBEHAVIORAL EVALUATION (A-ONE)
(ARNADÓTTIR, 1988)

Population. The A-One is designed for use with individuals who have central nervous system (CNS) dysfunctions of cortical origin. It is suggested that the evaluation is most appropriate for people over 16 years of age who have developed neurological processing strategies for the ADLs tested.

Functional Use. The purpose of the A-One is twofold: (1) to assess the functional independence level and types of assistance needed for several ADL, and (2) to identify the types and severity of neurobehavioral impairment. Functional independence and neurobehavioral dysfunction are assessed to assist with setting goals and planning treatment for patients with neurological dysfunction.

Content. The A-One consists of two parts: Part I, the Functional Independence Scale (IP) and Neurobehavioral (NB) Specific Impairment Scale, and Part II, the Neurobehavioral Scale Summary Sheet. Part I includes five primary ADL activities, namely, dressing, grooming and hygiene, transfer and mobility, feeding, and communication, as well as an NB impairment scale for each ADL. Part II is an optional form to assist therapists with identifying the possible lesion site associated with the observed neurobehavioral deficits (Arnadóttir, 1990).

Format and Scoring. Part I is a performance-based evaluation designed to be administered in a hospital room or occupational therapy clinic that includes a sink, toilet, and tub, and should be scheduled at the time a patient normally does morning care routines. Part I does not use standardized procedures or directions, but relies on observation of ADL task independence and the therapist's clinical judgments about the degree of neurobehavioral dysfunction for specified NB impairments (e.g., motor apraxia, unilateral spatial neglect) associated with each ADL. Each ADL is scored on a 5-point ordinal scale that ranges from a 4 (independent and able to transfer activity to other environmental situations) to a 0 (unable to perform; totally dependent on assistance). For each ADL and its associated NB impairments, the NB scale ranges from 0 (no neurobehavioral impairments observed) to 4 (patient is unable to perform due to neurobehavioral impairment; needs maximum physical assistance). The scores for each ADL and associated NB impairments are meant to stand alone and are not meant to be summed; nor are the ADL designed to be weighted equally.

Part II is a grid that can be used to match NB impairment scores with a chart of possible lesion sites for several types of pathology. Part II is not meant to be a diagnostic tool, but rather to help the therapist understand the underlying central nervous system (CNS) dysfunction so as to plan intervention appropriately.

Reliability and Validity. Interrater reliability of the A-One, Part I, achieved a kappa of .84. For Part II, an average kappa of .76 for all raters was achieved, although for the temporal lobe, a criterion kappa of .70 was not met, indicating the need for further refinement of that item. Test–retest reliability within 1 week for Part I was reported at $r_s = .85$. Content validity of the A-One is based on literature reviews and expert opinion. Normative and concurrent validity studies have also been conducted and reported (Arnadóttir, 1990).

ASSESSMENT OF MOTOR AND PROCESS SKILLS (AMPS) (FISHER, 1994). (NB: Initial version of the AMPS research edition was published in 1989.)

Population. Adolescents, adults, and older adults with multiple sclerosis, stroke, arthritis, and dementia and other psychiatric diagnoses have been tested with the AMPS in North America, and several other countries. The AMPS database also includes healthy subjects.

Functional Use. The AMPS is used to establish functional competence to plan appropriate interventions.

Content. The AMPS consists of more than 50 calibrated tasks (e.g., making a tossed salad, ironing a shirt, repotting a plant, making a bed, handwashing dishes) and two categories of items to be rated: 16 motor skills (e.g., reaching, lifting, calibrating, pacing) and 20 process skills (e.g., pacing, initiating, handling, searching, adjusting, benefiting). The AMPS must be administered by a therapist who has received training in the administration of the AMPS and has also achieved acceptable levels of intrarater and interrater reliability (Doble, Fisk, Fisher, Ritvo, & Murray, 1994; Park, Fisher, & Velozo, 1994).

Format and Scoring. Based on a pre-evaluation interview in which a patient's normal routines are identified, the therapist offers five or six tasks to the patient, from which the patient chooses between one and three tasks to perform. Each motor and process skill item is rated on a 4-point ordinal scale ranging from a 1 (i.e., deficit is severe enough to result in damage, danger, or task breakdown) to a 4 (i.e., competent—no evidence of a deficit that impacts performance) (Doble et al., 1994). AMPS item scores are then transformed from ordinal measures to interval measures with a many-faceted Rasch analysis program (Linacre 1989) that allows the measures to be adjusted based on rater leniency and task challenge. It also allows prediction of performance on the calibrated tasks not performed (Doble et al., 1994).

Reliability and Validity. Of the 300 trained raters, 95% have achieved Rasch goodness-of-fit statistics that indicate adequate interrater reliability (Doble et al., 1994). For tasks to be considered valid, they must fit a pre-established Rasch measurement model. Numerous studies have been reported on issues related to the reliability and validity of the AMPS, including between settings (Nygård, Bernspång, Fisher, & Winblad, 1994; Park et al., 1994), and across cultures (Fisher, Liu, Velozo, & Pan, 1992).

ROUTINE TASK INVENTORY–2 (RTI-2) (ALLEN, KEHRBERG, & BURNS, 1992; HEIMANN ET AL., 1989)

Population. The RTI-2 is designed for use with adolescents and adults who have cognitive deficits secondary to developmental, traumatic, undiagnosed, or psychiatric pathologies, and who may also have physical impairments.

Functional Use. The RTI-2 was designed to establish the level of functional status and document change based on the Allen Cognitive Levels (ACL; Allen, 1985, 1990) and assist with making comparisons among methods of gathering data (i.e., self-report, caregiver report, and observation of performance).

Content. The RTI-2 content is based on WHO categories and items. It includes 28 behaviors to be rated, 7 under each of the following categories: self-awareness disability (i.e., grooming, dressing, bathing), situational awareness disability (i.e., housekeeping, spending money, shopping), occupational role disability (e.g., planning/doing major role activities, pacing and timing actions, speaking), and social role disability (e.g., communicating meaning, following instructions, caring for dependents).

Format and Scoring. The RTI-2 can be administered using three methods: self-report (not recommended for individuals at ACL levels 1–4), caregiver report, and observation of performance. Although it is possible to administer the RTI-2 in the naturalistic setting (home or job site), the emphasis is on the clinical setting.

Each of the 28 behaviors to be rated and the six possible ACL levels for each behavior are operationally defined. The RTI-2 does not use standardized procedures or directions, but relies on a semistructured interview (self-report) or sustained observation of performance, and the therapist's clinical judgments to score the ACL level reported or observed for each behavior. A 3- to 6-point ordinal scoring system is provided for each behavior, with lower scores indicating lower abilities and higher scores indicating higher abilities. For example, under the category of self-awareness disability—grooming, scores range from 1 (ignores personal appearance; may not hold head up or keep head turned to the side to cooperate with the caregiver) to 5 (initiates and completes grooming without assistance). For the less complex behaviors in the self-awareness disability and for the situational awareness disability categories, the ACL levels defined for scoring ranged from 1 to 6 or 2 to 5; for the more complex behaviors in the occupational role disability and social role disability categories, the lowest ACL level defined for scoring is level 3 and the highest is level 6.

Reliability and Validity. Interrater reliability was established on the original version of the RTI ($r = .98$) as was test–retest reliability ($r = .91$). Internal consistency was also established on the original RTI ($\alpha = .94$). One concurrent validity study established correlations between the RTI behaviors and the ACL levels at $r_s = .54$ to $r_s = .56$ (Heimann et al., 1989; Wilson, Allen, McCormack, & Burton, 1989). A second study reported a correlation of $r_s = .61$ between the RTI and the Mini-Mental State Examination, which supported the relationship between functional decline and cognitive impairment (Allen, Kehrberg, & Burns, 1992).

CANADIAN OCCUPATIONAL PERFORMANCE MEASURE (COPM) (LAW ET AL., 1991)

Population. The COPM is designed for used across the lifespan with all diagnostic and disability groups. Because the COPM is an interview tool, for younger patients or nonverbal patients a caregiver/proxy may need to be the respondent.

Functional Use. The COPM is designed to identify problems appropriate for intervention and document change following intervention.

Content. The COPM evaluates three areas of performance: self-care, productivity, and leisure. For each area, the therapist identifies several examples of activities and then probes further as to whether the individual needs, wants, or is expected to perform any of the activities. The therapist continues to identify all relevant activities for each category.

Format and Scoring. The COPM is a semistructured interview that uses a self-report or proxy report format. If during the interview the patient identifies a problem in any of the three areas of performance, three more aspects of performance are then scored: importance of the activity (I), ability to perform the activity (P), and satisfaction with the way the activity is performed (S). Scores for I, P, and S scales range from 1 to 10. The criteria for each extreme on the three scales is as follows: I scale, 1 = not important at all and 10 = extremely important; P scale, 1 = not able to do it and 10 = able to do it extremely well; S scale, 1 = not satisfied at all and 10 = extremely satisfied. All P and S scores are then multiplied by the I score for decision making prior to intervention. All patients identify problems unique to themselves and therefore the number of problems may differ; the COPM has the therapist sum the P and S scores and divide them by the number of identified problems. Change is measured by using the same scoring system, except that at follow-up the original I score is used. The degree of change measured by the COPM is derived by subtracting the initial P and S scores from the follow-up scores.

Reliability and Validity. The COPM has been pilot tested with more than 256 patients and a total of 1,084 performance problems were identified. Patients in the pilot sample ranged in age from <5 years to >90 years who were seen at inpatient, outpatient, and community-based pediatric, geriatric, physical dysfunction, and mental health settings (Law et al., 1994). Validity of the COPM is based on a review of 136 instruments, of which 39 were selected as models (Pollock et al., 1990). Reliability studies are currently under way.

TECHNOLOGY AND OCCUPATIONAL THERAPY ASSESSMENT

For over two decades, computer-assisted assessments have been used and developed by occupational therapists for evaluating and treating motor and cognitive impairments (e.g., perceptual-motor skills, motor accuracy, executive functions), but more recently, occupational therapists have used technology (Bakeman & Quera, 1992; Repp, Karsh, van Acker, Felce, & Harman, 1989) to enhance assessment of disability, as well as the task context (Christianson, 1995). Another area of technology research has been computerized data management and reporting.

A system that makes real-time recording of patient task performance possible is the Portable Computer System for Observational Use (Observe) developed by Communitech International. The Observe program allows for keys on a laptop computer to be assigned to different behavior codes (e.g., patient dresses upper body, caregiver gives verbal prompt). Keys can also be assigned to record the frequency of an event or its frequency and duration. The real-time data collection enabled by a system such as Observe allows an observer to record the interactions between the therapist and the patient so that the probability of patient behaviors occurring with and without therapist intervention can be calculated. The Observe program is currently being used in a collaborative study among occupational therapy, nursing, and behavioral medicine (Rogers, Holm, Burgio & McDowell, 1993).

EASE™ (Enhancements Adapting Senior Environments) is another computerized assessment system designed by and for occupational therapists (Lifease, Inc., New Brighton, MN). It is designed to be used as an in-home assessment tool, but can also be used by therapists prior to discharge to ease the transition from the rehabilitation setting to the home. EASE is a multilevel tool that supports clinical decision making. First, EASE assists the therapist and the patient in the collaborative identification of the patient's priorities for independence in the home. Second, EASE includes an impairment screen to establish impairment status or include previously gathered data about impairment status to be entered. Third, based on the patient's home environment and everyday activities, EASE uses dynamic questioning and a set of decision rules to assist the therapist with recommendations about alternative methods of task performance, products to enhance independence, safety during task performance, and services to support independent living.

Another computerized assessment tool, OT-FACT™, is a functional assessment system that uses dynamic questioning to help the therapist expeditiously document levels of function at the impairment and disability levels. With its graphic displays, OT-FACT is able to provide quick reports on variables of interest, documenting patient progress during rehabilitation from assessment to discharge (Smith, 1993).

Ongoing Research on Occupational Therapy Assessment of Adult Function

The focus of occupational therapy assessment research is currently driven by many factors including shortened hospital stays, increased community-based care, a new emphasis on function and functional outcomes, legislative mandates that affect the profession, and the use of technology. Table 1 is a synopsis of ongoing research on, or using, the identified assessment tools, and the following summary is thus broader in scope.

As the focus of occupational therapy practice in adult assessment and rehabilitation is on the person-task-context transaction, the assessment tools used and the measures they include must be appropriate for activities of daily living, work, leisure tasks, and environments. Norm-referenced tools are inappropriate for the most part, because there are no norms for adult dressing, toileting, bathing, money management, and so on, except the broader societal norms regarding what is acceptable. Therefore, occupational therapy research has focused on the development of appropriate measurement instruments for activities of daily living. Research on instrument development has centered on criterion-referenced tools, measurement models such as Rasch analysis that can differentiate between person and task difficulty, and dynamic assessment. More basic still is the ongoing research into the identification of those tasks and roles that are critical to independent living in the community and that should therefore be included in assessments of function. Reliability and validity studies are common as the instruments are used with new diagnostic populations and as critical tasks and items are added, refined, or deleted when the results of ongoing studies become available. Two instruments have recently been included in large federally funded studies that include objectives related to instrument development, reliability, validity, and identification of critical tasks and levels of task performance necessary for independent living (AMPS, NIMH; PASS, NINR, NIA, AHCPR).

Methodologic and measurement issues are also the focus of ongoing research in occupational therapy. For example, the relationship between subjective (self-report, proxy report) and objective (PBO, standardized instruments) indicators of function has been receiving more attention, because the subjective instruments

TABLE 1. SYNOPSIS OF ONGOING RESEARCH ON OCCUPATIONAL THERAPY ASSESSMENT OF ADULT FUNCTION

WHO LEVEL/TECHNOLOGY Functional assessment tool	Reliability	Validity	Critical roles/tasks for independent living	Methods of measurement	Subjective vs. objective measures	Task vs. person difficulty	Dynamic assessment	Relationship of impairment & disability	Task context	Rehabilitation outcomes	Documentation of function/outcomes
Handicap (role function)											
Occupational Performance History Interview (OPHI)			X	X							
Routine Task Inventory - 2 (RTI-2)			X	X							
The Role Checklist			X								
Disability (task function)											
Kohlman Evaluation of Living Skills (KELS)	X		X							X	X
Performance Assessment of Self-Care Skills (PASS)	X	X	X	X	X		X	X	X	X	X
Milwaukee Evaluation of Daily Living Skills (MEDLS)	X	X	X					X		X	X
Arnadóttir OT-ADL Neurobehavioral Evaluation (A-ONE)				X				X		X	X
Assessment of Motor and Process Skills (AMPS)	X	X	X	X		X		X	X		X
Routine Task Inventory - 2 (RTI-2)			X	X	X						X
Canadian Occupational Performance Measure (COPM)	X	X	X	X	X						X
Technology and assessment											
OT FACT™			X					X		X	X
Observe			X						X	X	
EASE™			X					X	X		X

take less time and skill to administer than the PBO administered by occupational therapists. Patients, however, both overestimate and underestimate their abilities, based on diagnosis and degree of cognitive impairment (Ferrucci et al., 1992; Rogers, Holm, Goldstein, & Nussbaum, 1994; Rubenstein, Schairer, Wieland, & Kane, 1984; Sager et al., 1992; Zimmerman, Magaziner, Fox, & Hebel, 1992). Ongoing research, therefore, attempts to establish whether there is a systematic bias between subjective and objective data about MOB, ADL, or IADL, and if so, whether it is systematic for any particular types of tasks (e.g., IADL versus MOB), impairments, or pathologies. A related area of subjective assessment research is aimed at ascertaining the value of everyday activities and their meaning and importance to patients. If an activity is not important to a patient, and the patient does not wish to perform it and is not required to perform it, then rehabilitation focused on accomplishing the activity is not beneficial or functional for the patient.

Another methodological aspect of research is the use of dynamic assessment (Vygotsky, 1978), an approach to assessment and intervention that actively includes the observer-therapist in the patient-task-context transaction. Whereas psychometric models advocate an objective and detached observer who intervenes only after all baseline data are collected, the dynamic assessment model allows the observer to intervene during the assessment to ascertain whether or not an intervention furthers task performance. The intervention, however, must be recorded, as well as the response, and both pieces of data become part of the baseline data. Dynamic assessment allows the therapist to ascertain the specific point of task breakdown as well as the type of intervention that furthers safe task performance. The PASS is designed based on the dynamic assessment approach, and provides usable data for immediate planning of the rehabilitation program, thus shortening the time needed between baseline data collection and intervention implementation.

The specific aspect of function that is being rated is another focal point of occupational therapy research. For example, when the therapist uses the AMPS to rate patient performance, the therapist rates the motor skills (e.g., reaching, lifting, calibrating, pacing) and process skills (e.g., pacing, initiating, handling, searching, adjusting, benefiting) embedded within each task. When therapists use the PASS to rate patient performance, they record each of nine possible types of assists given to further task performance (e.g., encouragement, verbal nondirective guidance, verbal direction, gesture, task/environment rearrangement, demonstration, physical guidance, physical support, or total assist), and then use these data to make a judgment about task independence. In addition, therapists rate the safety and outcome of the performance (process and quality). Research that examines the specific aspect of deficit performance (A-ONE, AMPS, PASS) or the types of assists that are needed to safely perform a task (PASS) provide critical information for planning the rehabilitation program. Research using the AMPS, PASS, MEDLS, A-ONE, OT-FACT™, and EASE™, among others, has also examined the relationship between pathology, impairment, and disability. For example, research studies have been conducted to calibrate items on instruments for various diagnostic groups, compare performance of tasks and types of tasks (e.g., MOB, ADL, IADL) among diagnostic groups, and identify how various impairments (e.g., cognitive, motor, sensory) affect safe and independent task performance.

Measurement versus scoring is another research thrust in occupational therapy, and involves the application of the Rasch probabilistic model to observation data in order to calibrate assessment item measures. Rasch analysis is used to convert ordinal rating scores of patient performance into difficulty measures of the rating items (Linacre, 1989; Linacre & Wright, 1994). The Rasch procedure establishes the differential difficulty of seemingly alike items on an assessment. In addition, Rasch can also map out how individual patients perform on each item in relation to one another. Once assessment items have been calibrated for a specific diagnostic group, a new patient's score can be entered, and the patient's performance can be compared with that of other patients whose rehabilitation potential and outcomes are already known. Rasch analysis studies have been used with both the AMPS and the PASS (Doble et al., 1994; Fisher et al., 1992; Hohman, Holm, Rogers, Chandler & Stone, 1995; Nygard et al., 1994; Park et al., 1994).

Research on the impact of context on task performance has been receiving more attention by occupational therapy researchers in recent years (Christianson, 1995; Hohman et al., 1995; Holm & Rogers, 1990a, b; Letts et al., 1994; Nygard et al., 1994; Park et al., 1994). Although therapists have always known intuitively that clinic conditions cannot adequately replicate home conditions during an assessment, it is only recently that the data have confirmed their clinical judgments. Research comparing task performance in clinical and community contexts helps therapists to understand the true differences between settings and the impact on performance, so that inferences, interventions, and recommendations can be appropriate and accurate.

Occupational therapy research on rehabilitation outcomes and on the use of technology for expeditious documentation of rehabilitation status, progress, and outcomes is increasing in response to consumer demands and the changing health care environment (Fuhrer, 1987; King, 1993; Letts et al., 1994; Niemeyer, Jacobs, Reynolds-Lynch, Bettencourt, & Lang, 1994; Rogers & Holm, 1994; Smith, 1993). A testament to this is a recent issue of the *American Journal of Occupational Therapy* that was devoted to the topic of functional outcomes (October, 1994). The issue included several examples of outcomes studies, methods of documenting outcomes, and recommendations for further research on outcomes of occupational therapy intervention.

SUMMARY

As we stated earlier, patients are usually referred to occupational therapy because of problems in everyday *task performance*. We believe that the unique contribution of occupational therapy as a profession, in the study of brain function, is at the level of functional task performance, namely, the patient-task-context transaction. We discussed several functional assessments aimed at assessing the consequences of brain pathology at two WHO model levels: handicap (role function) and disability.

We then summarized the focus of ongoing occupational therapy assessment research, which is being driven by numerous factors, including shortened hospital stays, increased community-based care, a new emphasis on function and functional outcomes, legislative mandates that impact the profession, and the use of technology. Table 1 is a synopsis of ongoing research on, or using, the assessment tools previously discussed.

References

Allen, C. K. (1985). *Occupational therapy for psychiatric diseases: Measurement and management of cognitive disabilities.* Boston: Little, Brown.

Allen, C. K. (1990). *Allen cognitive level test manual.* Colchester, CT: S & S/Worldwide.

Allen, C. K., Kehrberg, K., & Burns, T. (1992). Evaluation instruments. In C. K. Allen, C. A. Earhart, & T. Blue (Eds.), *Occupational therapy treatment goals for the physically and cognitively challenged* (pp. 31-68). Rockville, MD: American Occupational Therapy Association.

American Occupational Therapy Association (AOTA). (1986). Work hardening guidelines. *American Journal of Occupational Therapy 40,* 841-843.

Americans With Disabilities Act of 1990. (Public Law 101-336), 42 U.S.C. § 12101.

Arnadóttir, G. (1988). *A-ONE.* Reykjavik Iceland: Arnason.

Arnadóttir, G. (1990). *The brain and behavior: Assessing cortical dysfunction through activities of daily living.* St. Louis, MO: Mosby.

Asher, I. A. (1989). *An annotated index of occupational therapy evaluation tools.* Rockville, MD: American Occupational Therapy Association.

Bakeman, R., & Quera, V. (1992). SDIS: A sequential data interchange standard. *Behavior Research Methods, Instruments, & Computers,* 554-559.

Barris, R., Oakley, F., & Kielhofner, G. (1988). The role checklist. In B. J. Hemphill (Ed.), *Mental health assessment in occupational therapy* (pp. 75-91). Thorofare, NJ: SLACK.

Bitter, J. A. (1979). History of the rehabilitation program. In J. A. Bitter (Ed.), *Introduction to rehabilitation* (pp. 15-36). St. Louis, MO: Mosby.

Blankenship, E. L. (1989). *Functional capacity evaluation and work hardening* [Seminar syllabus]. Macon, GA: American Therapeutics.

Christiansen, C. (1993). Continuing challenges of functional assessment in rehabilitation: Recommended changes. *American Journal of Occupational Therapy, 47,* 258-259.

Christianson, M. (1995). Assessing an elder's need for assistance: One technological tool. *Generations, 19*(1), 54-55.

Developmental Disabilities Assistance and Bill of Rights Act (Public Law 98-527). (1984).

Doble, S. E., Fisk, J. D., Fisher, A. G., Ritvo, P. G., & Murray, T. J. (1994). Functional competence of community-dwelling persons with multiple sclerosis using the Assessment of Motor and Process Skills. *Archives of Physical Medicine and Rehabilitation, 74,* 843-851.

Dunn, W., Brown, C., & McGuigan, A. (1994). The ecology of human performance: A framework for considering the effect of context. *American Journal of Occupational Therapy, 48,* 595-607.

Ellexson, M., & Kornblau, B. (1992). *Job analysis: What every rehabilitation professional in the U.S.A. should know about the ADA.* Miami, FL: ADA Consultants.

Faletti, M. V. (1984). Human factors research and functional environments for the aged. In I. Altman, M. P. Lawton, & J. F. Wohlwill (Eds.), *Human behavior and environment: Vol 7. Elderly people and the environment* (pp. 191-237). New York: Plenum.

Ferrucci, J. M., Guralnik, J. M., Marchionni, S., Costanzo, M., Lamponi, M., Salani, B., & Baroni, A. (1992). *Self-report and performance-assessment of functioning in Italian community dwellers.* Paper presented at the 45th annual scientific meeting, Gerontological Society of America, Washington, DC.

Fisher, A. G. (1994). *Assessment of Motor and Process Skills manual (research ed. 7.0)* [Unpublished test manual]. Fort Collins, CO: Colorado State University.

Fisher, A. G., & Short-DeGraff, M. (1993). Improving functional assessment in occupational therapy: Recommendations and philosophy for change. *American Journal of Occupational Therapy, 47,* 199-201.

Fisher, A. G., Liu, Y., Velozo, C. A., & Pan, A. W. (1992). Cross-cultural assessment of process skills. *American Journal of Occupational Therapy, 46,* 876-885.

Florey, L., & Michelman, S. (1982). Occupational role history: A screening tool for psychiatric occupational therapy. *American Journal of Occupational Therapy, 36,* 301-308.

Fuhrer, M. J. (Ed.), (1987). *Rehabilitation outcomes: Analysis and measurement.* Baltimore, MD: Brookes.

Guralnik, J. M., Branch, L. G., Cummings, S. R., & Curb, J. D. (1989). Physical performance measures in aging research. *Journal of Gerontology, 44,* 141-146.

Guriand, B., Kuriansky, J., Sharpe, L., Simon, R., Stiller, P., & Birkett, P. (1977-78). The Comprehensive Assessment and Referral Evaluation (CARE). *International Journal of Aging and Human Development, 8,* 9-42.

Heimann, N. E., Allen, C. K., & Yerxa, E. J. (1989). The routine task inventory: A tool for describing the functional behavior of the cognitively disabled. *Occupational Therapy Practice, 1,* 67-74.

Hohman, M. J., Holm, M. B., Rogers, J. C., Chandler, L. S., & Stone, R. G. Task performance in clinic versus home contexts: A Rasch analysis approach. Unpublished manuscript, University of Puget Sound, Tacoma, WA.

Holm, M. B. (1990, October). The role of allied health in effectiveness research: A response to the case study on low back pain. In *Proceedings: Allied health MEDTEP research issues conference*. Key West, FL. Washington, DC: Agency for Health Care Policy and Research.

Holm, M. B., & Rogers, J. C. (1989). The therapist's thinking behind functional assessment, II. In C. Royeen (Ed.), *Assessment of function: An action guide*. Rockville, MD: American Occupational Therapy Association.

Holm, M. B., & Rogers, J. C. (1990a). Functional assessment outcomes: differences between settings. *Archives of Physical Medicine and Rehabilitation, 71*, 761.

Holm, M. B., & Rogers, J. C. (1990b). Functional performance differences between the health care setting and the home. *Gerontologist, 30*, 327A.

Holm, M. B., & Rogers, J. C. (1991). High, low, or no assistive technology devices for older adults undergoing rehabilitation. *International Journal of Technology and Aging, 4*, 153-162.

Ilika, J., & Hoffman, N. G. (1981a). *Concurrent validity study on the Kohlman Evaluation of Living Skills and the Global Assessment Scale*. Unpublished manuscript.

Ilika, J., & Hoffman, N. G. (1981b). *Reliability study on the Kohlman Evaluation of Living Skills*. Unpublished manuscript.

Jacobs, K. (1992). Integrating the Americans With Disabilities Act of 1990 into client intervention. *American Journal of Occupational Therapy, 46*, 445-449.

Kaufman, L. (1982). *Concurrent validity study on the Kohlman Evaluation of Living Skills and the Bay Area Functional Performance Evaluation*. Unpublished master's thesis, University of Florida, Gainesville.

Kielhofner, G., & Henry, A. (1988a). Development and investigation of the Occupational Performance History Interview. *American Journal of Occupational Therapy, 42*, 489-498.

Kielhofner, G., & Henry, A. (1988b). Use of an occupational history interview in occupational therapy. In B. J. Hemphill (Ed.), *Mental health assessment in occupational therapy* (pp. 61-71). Thorofare, NJ: SLACK.

Kielhofner, G., Harlan, B., Bauer, D., & Mauer, P. (1986). The reliability of a historical interview with physically disabled respondents. *American Journal of Occupational Therapy, 40*, 551-556.

Kielhofner, G., Henry, A., & Waylens, D. (1989). *A users Guide to the Occupational Performance History Interview*. Rockville, MD: American Occupational Therapy Association.

King, P. M. (1993). Outcome analysis of work-hardening programs. *American Journal of Occupational Therapy, 47*, 595-603.

Law, M., Baptiste, S., Carswell-Opzoomer, A., McColl, M. A., Polatajko, H., & Pollack, N. (1991). *Canadian Occupational Performance Measure*. Toronto, Ontario, Canada: Canadian Association of Occupational Therapists.

Law, M., Polatajko, H., Pollock, N., McColl, M. A., Carswell, A., & Baptiste, S. (1994). Pilot testing of the Canadian Occupational Performance Measure: Clinical and measurement issues. *Canadian Journal of Occupational Therapy, 61*, 191-197.

Lawton, M. P. (1972). Assessing the competence of older people. In D. P. Kent, R. Kastenbaum, & S. Sherwood (Eds.), *Research planning and action for the elderly*. New York: Behavioral Publications.

Leonardelli, C. A. (1988a). *The Milwaukee Evaluation of Daily Living Skills*. Thorofare, NJ: SLACK.

Leonardelli, C. A. (1988b). The Milwaukee Evaluation of Daily Living Skills (MEDLS). In B. J. Hemphill (Ed.), *Mental health assessment in occupational therapy* (pp. 151-162). Thorofare, NJ: SLACK.

Letts, L., Law, M., Rigby, P., Cooper, B., Stewart, D., & Strong, S. (1994). Person-environment assessments in occupational therapy. *American Journal of Occupational Therapy, 48*, 608-618.

Linacre, J. M. (1989). *Many-faceted Rasch measurement*. Chicago: MESA.

Linacre, J. M., & Wright, B. D. (1994). *A user's guide to BIGSTEPS*. Chicago: MESA.

McGourty, L. K. (1979). *Kohlman Evaluation of Living Skills*. Seattle, WA: KELS Research.

McGourty, L. K. (1988). Kohlman Evaluation of Living Skills (KELS). In B. J. Hemphill (Ed.), *Mental health assessment in occupational therapy* (pp. 133-146). Thorofare, NJ: SLACK.

Medicare and Medicaid Amendments (Public Law 89-97). (1965).

Moorhead, L. (1969). The occupational history. *American Journal of Occupational Therapy, 23*, 329-334.

Nagi, S. Z. (1977). The disabled and rehabilitation services: A national overview. *American Rehabilitation, 2*, 26-33.

Niemeyer, L. O., Jacobs, K., Reynolds-Lynch, K., Bettencourt, C., & Lang, S. (1994). Work hardening: Past, present, and future—the work programs special interest section national work-hardening outcome study. *American Journal of Occupational Therapy, 48*, 327-339.

Nygård, L., Bernspång, B., Fisher, A. G., & Winblad, B. (1994). Comparing motor and process ability of persons with suspected dementia in home and clinic settings. *American Journal of Occupational Therapy, 48,* 689-696.

Oakley, F. (1982). *The model of human occupation in psychiatry.* Unpublished master's project, Medical College of Virginia, Virginia Commonwealth University, Richmond.

Oakley, F., Kielhofner, G., Barris, R., & Reichler, R. K. (1986). The role checklist: Development and empirical assessment of reliability. *Occupational Therapy Journal of Research, 6,* 157-169.

Park, S., Fisher, A. G., & Velozo, C. A. (1994). Using the Assessment of Motor and Process Skills to compare occupational performance between clinic and home settings. *American Journal of Occupational Therapy, 48,* 697-709.

Pfeffer, R. I. (1987). The Functional Activities Questionnaire. In I. McDowell & C. Newell (Eds.), *Measuring health: A guide to rating scales and questionnaires.* New York: Oxford University Press.

Pfeiffer, E. (1976). *Multidimensional functional assessment: The OARS methodology.* Durham, NC: Center for the Study of Aging and Human Development.

Pollock, N., Baptiste, S., Law, M., McColl, M. A., Opzoomer, A., & Polatajko, H. (1990). Occupational performance measures: A review based on the guidelines for the client-centered practice of occupational therapy. *Canadian Journal of Occupational Therapy, 57,* 77-81.

Reed, K. L. (1992). History of federal legislation for persons with disabilities. *American Journal of Occupational Therapy, 46,* 397-408.

Repp, A. C., Karsh, K. G., van Acker, R., Felce, D., & Harman, M. (1989). A computer-based system for collecting and analyzing observational data. *Journal of Special Education Technology, 9,* 207-216.

Rogers, J. C. (1984). *Performance Assessment of Self-Care Skills (PASS) (version 1.0).* Unpublished functional performance test. Pittsburgh, PA: University of Pittsburgh.

Rogers, J. C., & Holm, M. B. (1989a). *Performance Assessment of Self-Care Skills-Revised (PASS-R).* Unpublished functional performance test. Pittsburgh, PA: University of Pittsburgh.

Rogers, J. C., & Holm, M. B. (1989b). The therapist's thinking behind functional assessment, I. In C. Royeen (Ed.), *Assessment of function: An action guide,* Rockville, MD: American Occupational Therapy Association.

Rogers, J. C., & Holm, M. B. (1991). Task performance of older adults and low assistive technology devices. *International Journal of Technology and Aging, 4,* 93-106.

Rogers, J. C., & Holm, M. B. (1994). *(Performance Assessment of Self-Care Skills (PASS).(version 3.1).* Unpublished functional performance test. Pittsburgh, PA: University of Pittsburgh.

Rogers, J., Holm, M. B., Burgio, & McDowell (1993). ADL intervention for nursing home residents with DAT. Unpublished grant application.

Rogers, J., Holm, M. B., Goldstein, G., & Nussbaum, P. D. (1994). Stability and change in functional assessment of patients with geropsychiatric disorders. *American Journal of Occupational Therapy, 48,* 914-918.

Rubenstein, L. A., Schairer, C., Wieland, G. D., & Kane, R. (1984). Systematic biases in functional status assessment of elderly adults. *Journal of Gerontology, 39,* 686-691.

Sager, M. A., Dunham, N. C., Schwantes, A., Mecum, L., Halverson, K., & Harlowe, D. (1992). Measurement of activities of daily living in hospitalized elderly: A comparison of self-report and performance-based methods. *Journal of the American Geriatrics Society, 40,* 457-462.

Serviceman's Readjustment Act (Public Law 78-346). (1944).

Short-DeGraff, M., & Fisher, A. G. (1993). A proposal for diverse research methods and a common research language. *American Journal of Occupational Therapy, 47,* 295-297.

Smith, R. O. (1993). Computer-assisted functional assessment and documentation. *American Journal of Occupational Therapy, 47,* 988-992.

Smith-Fess Act (Public Law 66-236). (1920).

Social Security Act of 1935 (Public Law 74-271).

Soldiers Rehabilitation Act (Public Law 65-178). (1918).

Trombly, C. (1993). Anticipating the future: Assessment of occupational function. *American Journal of Occupational Therapy, 47,* 253-257.

U.S. Department of Labor (1986). *Dictionary of occupational titles* (4th ed.). Washington, DC: Author.

Vygotsky, L. S. (1978). Interaction between learning and development. In M. Cole, V. John-Steiner, S. Scribner, & E. Souberman (Eds.), *Mind in society: The development of higher psychological processes* (pp. 79-91). Cambridge, MA: Harvard University Press.

Wilson, D. S., Allen, C. K., McCormack, G., & Burton, G. (1989). Cognitive disability and routine task behaviors in a community based population with senile dementia. *Occupational Therapy Practice, 1,* 58-66.

Wood, P. H. N. (1980). Appreciating the consequences of disease—The classification of impairments, disabilities, and handicaps. *The WHO Chronicle, 34,* 376-380.

World Health Organization (WHO). (1980). *International classification of impairments, disabilities, and handicaps: A manual of classification relating to the consequences of disease.* Geneva: author.

Wyrick, J. M., Niemeyer, L. O., Ellexson, M., Jacobs, K., Taylor, S. (1991). Occupational therapy work hardening programs: A demographic study. *American Journal of Occupational Therapy, 45,* 109-112.

Zimmerman, S. I., Magaziner, J., Fox, K., & Hebel, J. R. (1992). *Activities of daily living: A comparison of residence-based performance and self-report.* Paper presented at the 45th annual scientific meeting, Gerontological Society of America, Washington, DC.

3

Physical Therapy

KATHYE E. LIGHT, MARIE A. REILLY, AND
MARTHA CLENDENIN

INTRODUCTION

Physical therapists are licensed health-care professionals who are involved actively in the assessment and treatment of clients with neurologic disorders, as well as clients with other types of movement dysfunction. The physical therapy profession is characterized by high academic standards for both admission and completion of the educational program and licensure. The profession is guided by the American Physical Therapy Association's defined Code of Ethics and Standards of Practice, as well as by state licensure laws outlining the scope of practice. The national association that represents physical therapists in the United States is the American Physical Therapy Association (APTA), which is also supported by state and district physical therapy associations. The Neurology Section is a specialty component of the APTA with the particular focus on neurologic dysfunction. The mission statement of the Neurology Section reads, "Our mission is to provide a forum for physical therapists, physical therapy assistants, and students having a common interest in neurologic injury and disease. We are committed to prevention and treatment of impairment and disability and minimizing societal barriers to function" (APTA Neurology Section, 1994). The Neurology Section of the APTA offers board certification, via credentialing and testing, for physical therapists to become Neurology Certified Specialists. Other specialty sections, including Pediatrics, Geriatrics, and Electrotherapy, are also relevant to the practice of physical therapy with clients having neurologic dysfunction.

KATHYE E. LIGHT   University of Florida, College of Health Professions, Department of Physical Therapy, Health Science Center, PO Box 100154, Gainesville, Florida 32610-0154.   MARIE A. REILLY   University of North Carolina at Chapel Hill, Department of Allied Health Sciences, Division of Physical Therapy, CB#7135, Chapel Hill, North Carolina 27599-7135.   MARTHA CLENDENIN   Nova Southeastern University, Physical Therapy Program, 3200 S. University Drive, Fort Lauderdale, Florida 33328; and Science Center, PO Box 100154, Gainesville, Florida 32610-0154.

Rehabilitation, edited by Goldstein and Beers. Plenum Press, New York, 1998.

As defined by the APTA, the primary purpose of the physical therapy profession is "the promotion of optimal human health and function through the application of scientific principles to prevent, identify, assess, correct, or alleviate acute or prolonged movement dysfunction" (APTA House of Delegates, 1983). The emphasis of evaluation and treatment is on movement dysfunction and all the factors that cause movement problems. Using a scientific educational background based in neuroscience, kinesiology, biomechanics, anatomy, physiology, exercise physiology, motor control, motor learning, pathology, and clinical medicine, the physical therapist evaluates and develops interventions to address the overall movement abilities of the client. A partial listing of neurologic diagnostic groups in which physical therapists actively address movement dysfunction includes cerebral palsy, developmental delay, Down's syndrome, preterm infants, spina bifida, stroke, head trauma, multiple sclerosis, Parkinson's disease, spinal cord injury, Guillain-Barré syndrome, Alzheimer's disease, vestibular disorders, peripheral neuropathies, and nervous system tumors and infections.

Historical Background

Early Years

In the United States physical therapy, a relatively young profession, has developed out of critical service needs during times of historical crises. All of the early physical therapists were female and were recruited by orthopedic surgeons to serve the rehabilitation needs of soldiers injured during World War I. These therapists were initially recruited from the fields of nursing and physical education, were provided only six to eight weeks of training in rehabilitation, and were called reconstruction aides (Beard, 1961; Pinkston, 1989). The service origin of the physical therapy profession, based on the practical restorative needs of those injured and ill, was critical to the future pathway of professional development. Although those early reconstruction aides had very little training or sophisticated technology to assist them in their rehabilitation efforts, their backgrounds in nursing and physical education melded into the focus that would evolve into the highly specialized field that physical therapy is today.

After World War I, the first physical therapists organized schools of physical therapy and developed the original framework for the professional organization that eventually became the American Physical Therapy Association (Beard, 1961; Pinkston, 1989). Two other historical crises directed the growth of physical therapy toward practical service needs: the polio epidemic and World War II.

During the polio epidemic of the 1930s, the curriculum in approved schools of physical therapy included coursework on exercise conditioning, muscle testing, muscle reeducation, flexibility measurement and training, hydrotherapy, and the use of heat and cold modalities. Following the polio epidemic, the experiences of rehabilitating many patients with residual dysfunction led to increased sophistication in theory and application of neurophysiology to therapeutic exercise. Large rehabilitation hospitals were built to manage the care of these clients, and physical therapy played an integral role in that care. By 1940, physical therapists were employed in hospitals, schools, physicians' offices, private physical therapy practices, crippled children's organizations, public health clinics, veterans administration facilities, and industry (Beard, 1961; Pinkston, 1989).

World War II created another surge in the need for and growth of the physical therapy profession. Soldiers returning home with head injuries, spinal cord injuries, peripheral nerve injuries, and amputations required assistance to regain functional and productive lives. Educational programs increased coursework to accommodate the updated information that would be useful to physical therapists working with injured soldiers. Coursework with titles such as "Movement Disorders: Measurement, Assessment and Intervention" appeared in schools throughout the country (Beard, 1961; Pinkston, 1989).

By the 1950s, hospital-based certificate programs in physical therapy were declining and the minimum level of approved educational programs in physical therapy were those offering baccalaureate degrees. In 1960 there were 42 approved schools of physical therapy, and the driving forces of practice were the emergence of medicare legislation and a growing geriatric population (Beard, 1961).

Recent History Marked by Change

Today, entry-level practitioners of physical therapy are educated in academic programs in accredited colleges and universities that provide professional education at the baccalaureate and postbaccalaureate levels. The number of accredited physical therapy programs in the United States is 138 at the time of this writing, and many others are awaiting accreditation. The APTA's House of Delegates voted in 1979 to raise the professional degree to the postbaccalaureate level by 1990 (APTA, 1984).

Enormous change has occurred in the physical therapy profession in a short period of time. Over the past ten years the predominant entry level of the profession has shifted to postbaccalaureate and represents the development of the physical therapy practitioner from technician to specialized professional. As in many professional fields, physical therapy is progressing toward gender neutrality. Educational credentials of those teaching in physical therapy have also shifted from master's degrees with a primary focus on teaching and service, to doctoral degrees with emphasis on research as well. Physical therapy research is reported commonly in the journal of the American Physical Therapy Association, *Physical Therapy,* as well as many other scientific journals.

Physical therapists use the services of other assistive personnel, including physical therapy assistants, technicians, and aides. Physical therapy assistants have an educational background at the associate degree level; technicians and aides are trained on the job. As physical therapy professional education programs have increased to meet the expanding need for services, physical therapy assistant programs have also grown. Currently, 140 accredited physical therapy assistants programs exist in the United States. These programs offer the associate degree and are located typically in community colleges. Physical therapy assistants are also licensed by the state in which they practice, and must practice under the supervision of a licensed physical therapist.

Over the history of the physical therapy profession, not only have the educational backgrounds of therapists changed, but also the relationships between therapist and physician. Initially, the practice of physical therapy was under the direct and prescriptive supervision of a physician, most frequently an orthopedic physician. Because of the initial status of the physical therapist as technician, this relationship was originally by design; the roles were along typical gender lines—female therapists and male physicians. As the profession has grown in numbers, level of education, and stature, and as society has changed toward accepting women as having equal minds, the therapist/physician relationship has changed to one of collegial interaction as team members rather than prescriber/technician. The evolution of this therapist/physician relationship has produced dramatic legislative changes in state licensure laws, such that currently physical therapists in most states are licensed to practice autonomously without the prescriptive referral of a physician. Thus, physical therapy has become a possible entryway into the health-care system (APTA, 1984). These changes in practice do not preclude client referral. Today physical therapists continue to welcome referrals of clients by physicians and dentists along Medicare or other third-party payer guidelines, or by the recommendation of other health-care professionals.

Growth and Demand for Services

As can be determined from this brief review of physical therapy history, the physical therapy profession has changed dramatically over a short period. The focus of the profession was originally, and continues to be, restoring functional movement and improving the quality of life for those clients who have movement dysfunction. Today physical therapy is one of the top ranked professions in the United States in job growth and job satisfaction.

Physical Therapy Evaluation and Assessment

Within the last 25 to 30 years there has been an explosion of information and technology concerning neuroscience. This has resulted in the United States Congress designating the 1990s as the "Decade of the Brain." The neurological basis for movement has been reevaluated since the early work of such nobel laureates as Sherrington, who provided the original framework for our understanding of movement and movement dysfunction. This early work presented hierarchical or reflex models of the central nervous system (CNS) as the basis for movement control. For more than 30 years popular treatment approaches for clients with neurologic disorders have been based on reflex and hierarchical models of the CNS. The assumption behind these approaches has been that sensory input is necessary for motor output and that the development of motor control proceeds from lower to higher levels of the nervous system (Horak, 1992). Movement dysfunction was thought to occur after damage to higher CNS centers by the loss of inhibitory control, which resulted in the release of primitive reflexes and abnormal muscle tone. Treatment intervention was directed toward the inhibition of these released reflexes and the normalizing of muscle tone, while facilitating more normal movement patterns and postures. Intervention was based on the assumption that all subsequent motor phenomena post–CNS damage were caused by neurophysiological mechanisms gone awry (Gordon, 1987). This work is being challenged by current neurological modeling and clinical studies that suggest a systems or dynamic organizational model in the CNS that is operational for the motor planning, control, and execution of movement. In the dynamical model of organization, each functioning system within the CNS is weighted in importance, based on the movement control necessary at a particular time, but overall all systems are equal and interactive.

The facilitation approach to physical therapy intervention had certain limitations, including failure to provide adequate motor learning environments. Client roles have been too passive with the facilitated movement approach and, therefore, not adequately challenged to find motor learning strategies to solve motor problems. Furthermore, the facilitation treatment approaches did not address the nonneural and secondary effects of a CNS lesion, such as the alteration of the musculoskeletal and cardiopulmonary systems, or the limitation of environmental restrictions. Certainly, all physiological systems interact to make functional movement possible, and the continuous interaction with a changing environment provides the need and drive for a human to move functionally.

In recent years there has been a shift in therapeutic interventions toward incorporating current theories and knowledge of CNS control. Physical therapy approaches have broadened as the science of motor control and learning expands and research in child development, biomechanics, and pathokinesiology pro-

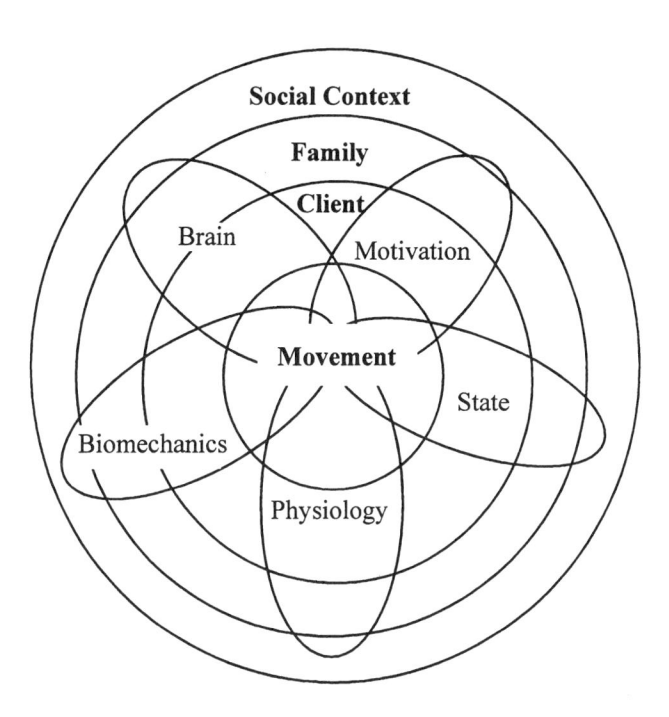

Figure 1. Movement develops from multiple interacting systems.

gresses. A multidimensional, systems oriented, and task specific therapeutic model for neurologic rehabilitation is developing. This model is based on the assumption that movement develops from dynamic interactions of many systems that are organized around goal oriented functional behaviors rather than on isolated normal movement patterns (Figure 1).

From the first teams of reconstruction aides organized and trained in muscle training and corrective exercise by Dr. Robert Lovett in the 1920s to the curricula of today with entire courses in exercise science, exercise physiology, therapeutic exercise, and neurological assessment, the specialists in movement dysfunction have been physical therapists (Pinkston, 1989). The struggle continues for the best methods of assessing and solving the movement intervention needs of clients with neurologic disorders.

Need for Standardized Assessments

The methods used for physical therapy assessment of neurologic clients are currently under close scrutiny by the physical therapy profession. With an increased base in scientific literature, a growing emphasis on a scientifically based evaluation process has evolved. In addition to the rapidly growing science and training of physical therapists in the scientific approach, health-care reform has begun. Health-care reform means many different things to different people, but the clear message to all health-care providers is the need for functional outcomes and accountability. Therapists are being encouraged to use standardized assessments in their evaluations and to use these assessments to direct the treatment interventions toward the quickest, if not always the best, outcomes. Health insurance providers have added pressure to this process of standardization. The physical therapy community, as a whole, is in a state of continuous debate over issues of evaluation and intervention planning.

Physical therapists recognize the importance of standardizing evaluations, as well as the associated need for validity and reliability. Numerous evaluation batteries have been developed, and research on evaluation instruments continues at a feverish pitch. The APTA has issued standards for tests and measurements in physical therapy practice (Rothstein & Echternach, 1993). Recent books devoted to the issues of measurement, standardized testing, and description of reliability and validity of available tests have impacted the profession of physical therapy (Cole, Finch, Gowland & Mayo, 1994; Rothstein, 1985; Rothstein & Echternach, 1993; Wade, 1992). Continuing education workshops for teaching standard tests for outcome measures are ongoing across the country.

Resistance to Standardized Assessment

When evaluating clients with neurologic lesions, the physical therapist is concerned with capturing the problems that the client has concerning movement. To understand the methods used by physical therapists to perform this evaluation, one must consider the therapist's educational background in kinesiology, biomechanics, exercise physiology, neuroscience, pathology, and clinical medicine. Physical therapy's history of finding practical solutions to practical problems also guides the assessment. Although many evaluation batteries have been developed, none serves all the needs of the physical therapy assessment. When balancing the time allowed for evaluation and the limitations of information acquired with the standardized tests, many physical therapists avoid the standardized tests and continue to use a descriptive approach to client assessment. The descriptive approach allows the physical therapy assessment to be oriented toward the problem list and goals of the individual client. Physical therapists analyze movement control via the method of task analysis, exploring the types of movements and tasks that are important to the particular client. Emphasis on the individual's needs according to age, gender, motivation, goals, family support, and living environment is considered to be important to the physical therapy assessment. The physical therapist's treatment interventions are goal oriented, with the goals being developed by the client, client's family, and therapist. The goals are also developed with relevant input from other members of the health-care team when appropriate.

In the present clinical environment, physical therapists have not resolved the conflict over issues of the needs of the individual client versus those for objectivity in standardized testing. Generally, physical therapists do not yet demonstrate a preference for standardized tests, and are not convinced that the standardized tests currently available are an improvement over the more personalized evaluations that have traditionally been performed. To provide an example of how a physical therapy clinician addresses evaluation, two cases will be presented. Consider the physical therapy evaluations required to develop appropriate treatment interventions for two 80-year-old women, both of whom have suffered right middle cerebral artery strokes. Before the stroke, one of these elderly women lived at home alone on a farm, with a few cattle to feed, two pet dogs, and several cats. The woman is energetic and motivated, loves to garden and had walked to her neighbor's house, one-half mile away, each day for socializing. The other elderly woman lived in a small town, in a house with her sister. She has been somewhat reclusive, enjoys the company of her sister, and has never wanted to go out much. She has never participated in exercise or outdoor activity. This woman enjoys keeping her little house clean and entertaining a few friends and neighbors when they come to visit. Without belaboring the differences further, it is clear that these two women do not have the same motivation, goals, family support, economic level, living environment, or interest in active rehabilitation. Most physical therapists would choose to help these individuals meet their own goals, for which there is no standardized assessment. Indeed, one of the most important operating principles in motor learning is that movement training must be specifically directed toward the task that is to be learned. Practicing general types of tasks will not transfer to the learning of a specific task; evaluation must be based on the types of tasks that are important to each individual. Using activities of daily living (ADLs), extended ADLs, balance tests, strength tests, coordination batteries, and so on does not address the most important purpose of the physical therapy evaluation. Standardized tests, therefore, will not become the total method of physical therapy assessment.

In the future, third-party payers will require greater standardization in the evaluations performed by physical therapists to document treatment efficacy. Standardization guidelines are needed to develop adequate and appropriate care in these days of rationing health-care dollars. Clearly, confusion and upset continues in the physical therapy clinics throughout the country. Although therapists recognize the need for standardization, they are more likely to have a greater appreciation of the individuality of their clients and will reflect this in their evaluations.

Types of Available Assessments

Although resistance to standard tests exists among physical therapy clinicians, a multitude of tests and test batteries are available. Part of the resistance and confusion among physical therapists regarding the usefulness of these tests is related to their sheer number and their lack of appropriate fit to particular client needs. Therapists are concerned about the inadequate categories within each standard evaluation, failure of the assessments to address movement quality, and the insensitivity of the assessment tools to the gradual change typical in rehabilitation.

Assessments are classified in a variety of ways, which adds to the confusion concerning their use in full physical therapy evaluation schemes. Pediatric and adult assessments are usually categorized separately. In both the pediatric and adult areas of practice, screening tools and assessment tools are available. The adult assessments vary more in methods of classification. Many of the assessments are categorized by physiological system testing, such as with strength testing by dynamometry or range-of-motion testing by goniometry. Some test batteries are diagnosis specific, such as the Fugl-Meyer Assessment of Sensorimotor Recovery after Stroke (Duncan, Propst, & Nelson, 1983; Fugl-Meyer, 1980) and many are appropriate for assessment of function regardless of diagnosis. Tables 2, 3, and 4 list many of these tests by categories.

Many of the standardized tests have addressed various criteria for reliability and validity; however, even the best tests have not been evaluated for all of the criteria reported in testing and measurement literature as important (Rothstein & Echternach, 1993). One very important test criterion that few tests or test batteries have addressed is that of measurement sensitivity for the dependent variables. Measurement sensitivity concerns the level of detectable change addressed by the measurement system. The importance of measurement sensitivity cannot be overemphasized. Many tests exist, with adequate reliability and validity, that have extremely poor sensitivity for detecting a meaningful small change in movement dysfunction. If these tests are to be used to judge the effectiveness of rehabilitation interventions, establish the monies appropriated for health care, and limit the time for allowed treatment intervention, we must all insist that the tests detect gradual meaningful change. Physical therapists are not alone in this concern, as can be seen in the recent NIH publication (Public Health Service, 1993) concerning the focus of current need in functional assessment. Currently, research monies are being made available to improve the functionality and sensitivity of rehabilitation assessment instruments (Public Health Service, 1993).

Review of the Assessments

The topic of assessment and evaluation in physical therapy would require many books for adequate representation. The following section on physical therapy assessment and evaluation has been developed to provide the reader with an overall sense of the breadth and detail required to understand the methods used in physical therapy. We list most of the recognized standard tests but do not discuss all evaluation instruments. Rather, a variety of evaluations is presented as an example of the state of the art in evaluation. Evaluation instruments are described and the issue of reliability and validity is addressed for those particular assessments. We address basic measurements of a general nature first; pediatric and adult assessments are then presented separately.

BASIC MEASUREMENTS

In addition to analyzing relevant functional movements, the typical physical therapy evaluation will include a basic neurologic examination; testing of passive and active range-of-motion, strength, balance, coordination, endurance, bed mobility, and transfer ability; and a gait or wheelchair usage analysis. The physical therapist's basic neurologic examination consists of those items appropriate to the client's movement function, a brief mental status assessment, testing of cranial nerves, simple motor and sensory testing, reflex testing, and simple coordination testing such as the finger-to-nose test or the rapid alternating movement test (Goldberg, 1992).

Measurements of strength, flexibility, coordination, balance, endurance, speed, and other performance criteria reflecting physical capacity are standard in physical therapy assessments for clients with neurologic dysfunction. Physical capacity issues are addressed differently for adult versus pediatric populations. For the purpose of basic understanding, these measurement issues are addressed in the section on adult evaluation.

PEDIATRIC ASSESSMENTS

PEDIATRIC EVALUATION AND TESTS. Along with alteration of treatment approaches in recent years, evaluation procedures and tests used for pediatric motor assessments have proliferated. No one instrument is responsive to all the needs of pediatric physical therapy assessment; therefore, consideration of the purpose, cost, reliability, and validity of the instrument in determining the most appropriate test is necessary. Clinical measures and tests are used in physical therapy to discriminate among children functioning normally and those who are motor delayed, to predict future outcome, and to evaluate change in function (Campbell, 1991). Various tests have been developed for each of these purposes. The pediatric physical therapist often works in a team of professionals including speech pathologists, nurses, occupational therapists, and teachers. The assessments either are performed together, or one individual on the team will complete the standardized assessment. For these reasons, the tests presented here are not unique to physical therapy, but are common to pediatric evaluation.

Many of the older norm-referenced developmental scales are administered as part of a diagnostic assessment in which the goal is to identify children meeting eligibility criteria for special services. Another use is monitoring development. In contrast, criterion-referenced scales are designed to compare performance to external criteria or standards. A list of pediatric assessment batteries is reported in

Table 1. Examples of the better-known and better-constructed tests are described below. Presentation of the pediatric assessments is according to the following classification categories: (1) developmental tests, (2) diagnostic- or population-specific tests, (3) tests of movement quality, (4) newborn assessments, and (5) new measures of disability.

DEVELOPMENTAL TESTS. The Bayley Scales of Infant Development (BSID) were the first nationally normed infant developmental assessment (Bayley, 1993). This assessment package was originally published in 1969 and has been recently revised and renormed on a national, stratified random sample of 1,700 children whose characteristics and demographics were reflected in the 1988 U.S. Census of population for infants and toddlers from 1 month to 42 months of age. The authors of the BSID II suggested that it could be used to diagnose exceptional performance and to plan intervention strategies. Like the previous version, the BSID II consists of three parts: a Mental scale, a Motor scale, and a Behavior Rating scale. For the purpose of this review only the Motor scale or Psychomotor Developmental Index (PDI) is discussed The new items added to this scale are those deemed to assess sensory integration and perceptual-motor integration. Reliability coefficients for the Motor scale have been reported, ranging from $r = .75$ to $r = .91$. The test–retest stability coefficient, combined for all ages, was .78 and tester–observer reliability was .75.

The BSID II instructional manual reports evidence of validity for both the original BSID and its revision. When each version was administered in counterbalanced order to a sample of 200 children there was a moderate correlation ($r =$

TABLE 1. LIST OF DEVELOPMENTAL OR PEDIATRIC ASSESSMENTS

Alberta Infant Motor Scale (AIMS)
Assessment of Preterm Infant Behavior (APIB)
Basic Gross Motor Assessment (BGMA)
Batelle Developmental Inventory Screening Test
Bayley Scales of Infant Development (Psychomotor Scale)
Brazelton Neonatal Behavioral Assessment Scale
Bruininks–Oseretsky Test of Motor Proficiency (BOTMP)
Gross Motor Function Measure (GMFM)
Gross Motor Performance Measure (GMPM)
Movement Assessment of Infants (MAI)
Neurological Assessment of the Full and Preterm Infant
Neurological Examination of the Full-Term Infant
Peabody Developmental Motor Scales
Pediatric Evaluation of Disability Inventory (PEDI)
Posture and Fine Motor Assessment of Infants (PFMAI)
Test of Infant Motor Performance (TIMP)
Test of Motor Impairment
Toddler and Infant Motor Evaluation (TIME)

References: Physical Rehabilitation Outcome Measures, by B. Cole, E. Finch, C. Gowland, and N. Mayo, 1994, Toronto, Ontario, Canada: Canadian Psychotherapy Association; "Physical Therapy Assessment in early Infancy," in *Clinics in Physical Therapy*, by I. J. Wilhelm, 1993, New York: Churchill Livingstone.

.63) between the two PDI scores. Correlations between the BSID II and another developmental assessment, the McCarthy Scales, were high for the mental status scales but relatively low for the PDI, with the highest correlation ($r = .59$) for the motor subscale. To examine predictive validity, several small studies were completed with children in certain diagnostic categories, including children who were premature, HIV seropositive, exposed to drugs in utero, or oxygen deprived at birth, as well as children with Down's syndrome, autism, developmental delay and chronic otitis media. Each group demonstrated performance differences when compared to the normative sample, indicating increased risk for developmental delays. The original BSID Motor scale was not found to be predictive of development at later ages (Coryell, Provost, Wilhelm, & Campbell, 1989; Crowe, Deitz, & Bennett, 1987). Future studies are needed to determine the responsiveness of the revised BSID Motor scale to children with motor dysfunction. Although the BSID and other developmental scales are helpful in determining eligibility for physical therapy intervention, they are of little use for the physical therapist in planning developmental treatment programs.

The Peabody Developmental Motor Scales (PDMS) were developed by educators, to identify children with delays, assess change in performance, compare gross and fine motor function, and serve as a curriculum-based assessment when used with the activity cards provided in the test kit (Folio & Fewell, 1983). This test was normed on a large stratified sample of children from birth to 83 months of age, primarily from middle-class fanulies. The PDMS consists of separate Gross and Fine Motor scales which are divided into separate skill categories. Interrater reliability was reported to be $r = .97$ and $r = .94$ for Gross and Fine Motor scales respectively. The PDMS was also reported to be stable over time (r of .95 and .80). Several researchers have reported moderate to high concurrent validity with the BSID and the Gessell Schedules (Campbell, 1990; Hinderer, Richardson, & Atwater, 1989). The PDMS addresses normal motor development, an important concern for the physical therapist; however, no measures of movement quality are addressed.

In the past, the majority of motor assessments were based on the rate and sequence of normal motor development. These types of scales have limited sensitivity to the changes in movement abilities for children with moderate to severe motor impairments.

DIAGNOSTIC- OR POPULATION-SPECIFIC TESTS. The Bruininks–Oseretsky Test of Motor Proficiency (BOTMP) was developed to identify children with motor dysfunction and developmental handicaps, to assess fine and gross motor skills in individual children, and to assist in developing and evaluating motor training programs for children from 4.5 years to 14.5 years of age (Bruininks, 1978). This test was normed on 765 children representing the 1970 U.S. Census for sex, race, community size, and geographic location. The BOTMP consists of 46 items in eight subtests: running speed and agility, balance, bilateral coordination, strength, upper-extremity coordination, response speed and agility, visual-motor control, and upper-extremity speed and dexterity. The complete battery provides three standard scores including the Gross Motor Composite, Fine Motor Composite, and Total Battery Composite. Test–retest reliability coefficients were obtained for two age groups, second graders and sixth graders. Coefficients for the three composites ranged from $r = .68$ to $r = .89$. Subtest coefficients ranged from $r = .29$ to $r = .89$. The standard error of the measurement was also determined for both composite scores (4–5 standard score points) and subtests (2–3 standard score points). Interrater reliability coefficients for the visual-motor control subtest were reported to be $r = .98$ and $r = .90$ for two groups of raters.

Construct validity of the BOTMP was determined by demonstrating that scores for each subtest significantly correlated with chronological age. The authors tested normal children and compared scores to children with mental retardation and learning disabilities. Comparison of groups demonstrated that the scores of children in the normal group were significantly different from the scores of the other groups on all composites and subtests.

The Movement Assessment of Infants (MAI), a criterion-referenced test, was developed by physical therapists as a clinical tool for localizing motor problems in high-risk infants and to predict the risk of cerebral palsy (Chandler, Andrews, & Swanson, 1980). The MAI has not been empirically normed, but profiles have been developed for children four and eight months of age. It is a systematic appraisal of motor behaviors in the first year of life. Categorization of assessment procedures enables the examiner to obtain an estimate of motor function based on muscle tone, primitive reflexes, automatic reactions, and volitional movement. Harris, Haley, Tada, and Swanson (1984), using the MAI, studied preterm and full-term infants at four months of age and found fair reliability for the total risk score (interobserver $r = .72$ and test–retest $r = .79$). Predictive validity was assessed by Harris (1987), who compared the sensitivity and specificity of the MAI and the Bayley Motor Scale to predict cerebral palsy in a group of infants who were followed to ages 3 to 8 years. Sensitivity was 73.5% for the MAI and 35.5 % for the Bayley. The MAI has potential as a diagnostic tool (Schnieder, Lee, & Chasnoff, 1988), and further test construction is currently under way to improve reliability and validity.

TESTS OF MOVEMENT QUALITY. Many of the tests developed by nontherapists are concerned with attainment of motor milestones. Although recent tools have been developed to examine functional abilities, the tests of motor milestones and the tests of functional abilities have not considered many of the qualitative aspects of movement control such as posture, movement transitions, and coordination. Therapists have responded to this dearth in measurement by designing several new tests which analyze movement quality. Some of these tests are described briefly in this section.

The Alberta Infant Motor Scale (AIM) focuses on the development of postural control in children from birth to 18 months of age. Its 58 items are scored from observations of the child in four different positions: supine, prone, sitting and standing (Piper & Darrah, 1994). Intrarater reliability ($r = .99$) and interrater reliability ($r = .99$) were both high, and content, construct, and concurrent validity were reported to be high (Cole et al., 1994, pp. 134–135).

The Test of Infant Motor Performance (TIMP) is currently being completed by Campbell (1993). The TIMP assesses postural alignment and control during functional movement activities of infants from 32 weeks gestational age to 3.5 months adjusted age. The processes tested will include the ability to orient to various stimuli and stabilize the head in space, to perform this head orientation while in various body positions and during movement transitions, and control of body alignment, distal limb control, and antigravity control of extremity movements.

The Toddler and Infant Motor Evaluation (TIME), developed by Miller and Roid (1993), assesses quality and sequence of movement in children from birth to 42 months. This test allows the examiner to observe the child's spontaneous movements from various positions while the necessary handling or positioning is done by the parents.

NEWBORN ASSESSMENTS. A review of pediatric assessments would not be complete without mention of the numerous newborn assessments performed by physical therapists. For a thorough review of the many infants tests available, the reader is referred to Wilhelm (1993). The purposes of these tests are for diagnosis of an evident neurological problem, evaluation of changes in behavior, prognosis, and parental education. Often serial examinations can be administered to assess the progress of the infant over time and to determine frequency of follow-up.

The Neurological Examination of the Full-Term Infant (Prechtl, 1977) was developed to diagnose neurological abnormalities and identify infants at risk. The examination is appropriate for both full-term and preterm infants who have reached term age. Items include the evaluation of physical features, tone, reflexes, automatic reactions, posture, and spontaneous movement.

The Neurological Assessment of the Full and Preterm Infant (Dubowitz & Dubowitz, 1981) includes assessment of habituation, reflexes, orientation, state, muscle tone, and movement. This test was developed to capture the functional state of the infant's nervous system and to document change with maturation or recovery.

The Brazelton Neonatal Behavioral Assessment Scale (BNBAS) (Brazelton, 1984) was originally developed to study individual differences in newborns, but it can also be used to examine transactional processes between parents and infants. The BNBAS consists of 18 elicited responses including assessment of tone, vestibular and reflex responses, and 28 behavioral items that score response decrement, orientation, behavioral state, behavioral control, and motor abilities. The BNBAS can be used to assess full-term infants from birth to one month. It can also be used with the preterm infant after the infant reaches 38 weeks gestational age.

The Assessment of Preterm Infant Behavior (APIB; Als, Lester, Tronick, & Brazelton, 1983) is a behavioral test that is a refinement and extension of the BNBAS. Maneuvers from the BNBAS are grouped by increasing levels of stimulation, while assessing infant behavior in five dimensions including physiology, motor, state, interactional, and self-regulatory. This assessment can be used with both preterm and full-term infants up to 44 weeks gestational age.

NEW MEASURES OF DISABILITY. With ongoing changes in health-care reform, and following the passage of the Individuals with Disability Education Act (IDEA), a greater emphasis is now placed on accountability, functional outcome, and family participation in pediatric therapeutic intervention. The World Health Organization (1980) has developed a classification system that categorizes motor problems according to their pathology, impairment, disability, and handicap. Campbell (1992) has suggested that, since the consequences of motor problems range from the level of the organ system to societal level, the World Health Organization framework would provide a consistent approach to the development of a comprehensive assessment strategy for children with neuromotor dysfunction. Historically, physical therapy assessments have been focused on the level of impairment, at which measures such as strength and range of motion are assessed. Tests that demonstrate the child's ability to function within his or her environment are also needed. In line with this perspective, a new assessment, the Pediatric Evaluation of Disability Inventory, has been developed to evaluate motor behaviors at the WHO levels of disability and handicap.

The Pediatric Evaluation of Disability Inventory (PEDI) measures both basic functional skills and the performance of complex functional activities in children from 6 months to 7.5 years of age (Haley & Baryza, 1990; Haley, Baryza, & Webster, 1992; Haley, Coster, Ludlow, Haltiwanger, & Andrellos, 1990). The PEDI considers change in the three levels: functional skill level, caregiver assistance, and the number and type of modifications needed to successfully complete an activity. Items encompass self-care, mobility, transfers, communication, and social function. The PEDI is an assessment that is administered by interview and parent report. The test was standardized on large groups of children with and without disabilities. Inter-interviewer reliabilities for both the normative sample and a clinical sample demonstrated interclass correlations ranging from .79 to 1.0 and intercorrespondent reliabilities were reported to range from .74 to .96. Feldman, Haley, and Coryell (1990) assessed children with and without disabilities on the PEDI and the Batelle Developmental Inventory Screening Test (BDIST), a test that assesses function in five domains. Concurrent validity was supported by correlations in summary scores ranging from .70 to .80. PEDI scores were significantly different for the two groups and were found to be better group discriminators than the BDIST. The authors also report that the PEDI was responsive to change in children recovering from head injury and children with severe disabilities.

The Gross Motor Function Measure (GMFM) was developed by a multidisciplinary group from Canada (Russell et al., 1993). The intent was to construct an instrument responsive to clinically important changes in disability over time. The GMFM is a criterion-referenced test for use with children with cerebral palsy. It has been useful for therapists in describing current level of function, determining treatment goals, and documenting progress. The test consists of 88 items divided into five domains: (1) lying and rolling, (2) sitting, (3) crawling and kneeling, (4) standing and walking, and (5) running and jumping. The GMFM focuses on ability and quality of performance. Items represent motor functions that are typically performed by 5-year-old children without motor dysfunction.

Intrarater and interrater reliabilities were determined through test–retest conditions. For all domains, interrater ICCs for reliability ranged from .87 to .99 and intrarater reliability ranged from .92 to .99. Interclass correlations range from .73 for scores on lying and rolling to .97 for scores on standing. The authors report evidence of validity and responsiveness when the GMFM was used to assess children receiving various interventions including intensive physical therapy, dorsal rhizotomies, and baclofen injections. As hypothesized, the GMFM also demonstrated greater change scores in children recovering from acute head injury than children with cerebral palsy. The test developers have responded to the need in physical therapy to have valid and reliable assessments that can be used to evaluate treatment effects and also to plan treatment progression. The same group is currently developing another evaluative measure, the Gross Motor Performance Measure (GMPM). The purpose of the new measure is to assess the more qualitative aspects of motor performance, such as postural alignment and disassociated movements.

SUMMARY. In recent years, new pediatric assessments have proliferated, yet no single assessment meets the measurement needs for analyzing all levels of motor dysfunction. To develop such an approach, and to analyze the total picture of the child's motor competencies, the physical therapist must use a battery of tests.

ADULT ASSESSMENTS

The multiplicity of adult assessments used in the field of physical therapy makes the task of description overwhelming. For the purpose of organization and representation, the evaluations for adult clients with neurological problems are categorized in three areas: (1) basic measurements, (2) functional assessments, and (3) diagnosis-specific assessments. These categories are not mutually exclusive of each other, and not every evaluation available is listed Reports on the reliability and validity are presented briefly for a representative sample of tests. The reader is referred to two recent books on measurement that include more inclusive commentaries on assessment descriptions, reliability, and validity: *Measurement in Neurological Rehabilitation* (1992) by Wade and *Physical Rehabilitation Outcome Measures* (1994) by Cole et al.

BASIC MEASUREMENTS. When considering basic measurement for human movement dysfunction, we must consider the physiological systems that affect movement. Movement is, in fact, affected by all systems of the body. The systems most directly involved with movement abilities are the neurologic, musculoskeletal, cardiovascular, and pulmonary systems. Other systems are, however, also involved. Could anyone disagree that the immune, gastrointestinal, and urinary systems also affect movement abilities? The physiological systems are well integrated and balanced for function, and function implies controlled movement. When physical therapists evaluate movement abilities in clients with neurologic dysfunction, the musculoskeletal, pulmonary, cardiovascular, and neurologic systems are evaluated for impairment. Problems with other systems are considered for precautions in treatment planning but are not evaluated directly. Some of the basic measurements of physiological systems will be presented by categories.

Musculoskeletal System. Two common measurement areas of the musculoskeletal system are those of strength and range of motion. The most common method of measuring strength is by manual muscle testing (MMT), but the use of dynamometers and isokinetic testing is also prevalent. MMT is a general method of strength testing that is easily administered in any clinical setting, and it requires no specialized equipment. The MMT is an ordinal scale assessment with grades from 0 to 5 and half-step grades indicated by adding pluses and minuses to each number (e.g., 3^+ or 5^-). MNT requires carefully standardized methods for positioning, stabilizing, and application of resistive force. Therapists require training and extensive practice with MMT methods before becoming adequately skilled or reliable in this assessment. Intrarater reliability of experienced therapists has been reported to be in 96%–98% agreement within full muscle-grade levels, and the interrater reliability in 95%–97% agreement for full-grade measures. When pluses or minuses were included in the scaling, the exact interrater agreement fell to only 60–67%. Generally, the MMT is reported to have good reliability if the scale is expanded to full scales. The MMT has face validity and content validity because it is a direct measure of the torque generated by muscle groups. The MMT does not measure all types of muscle contractions or the rate of muscle tension development. Predictive and construct validity are not established for MMT. Evidence has not been presented in the literature for the ability of MMT to predict future functional outcomes, and the degree to which MMT can be generalized to present functional behaviors is also not clear (Lamb, 1985).

Handheld dynamometry is replacing MMT for some clinicians. The reliability of handheld dynamometry is reported to be better than for MMT when the dynamometry is performed carefully and correctly. The technique requirements for handheld dynamometer are much more precise than for MMT. The same issues of validity apply to handheld dynamometer as for MMT (Bohannon, 1986; Bohannon & Andrews, 1987).

Manual goniometry is the most traditional method of measuring either passive or active range of motion (ROM). The universal goniometer is a protractor with two movable arms. The protractor is placed over the center of the joint axis, and the arms are aligned with the body segments on either side of the joint. This method of joint ROM measurement is the most practical for general usage. The method is easy to learn, relatively quick, and involves a nontechnical, portable instrument. Miller (1985) reviewed studies of intrarater and interrater reliability and reported that it was generally good to high. Experienced testers performed more reliably than less skilled testers. Regarding the validity of manual goniometry, little information is reported in the literature. Comparisons of manual goniometry to trigonometric measurements or radiographic measurements have been reported to be in close agreement. Goniometry does not provide information on the cause of ROM dysfunction, nor does it provide predictive validity of future outcomes (Miller, 1985). Other methods of flexibility testing exist. For example, the sit and reach test is a standardized method of evaluating hamstring and low back flexibility. Electrogoniometry is a method of measuring joint angles while an individual is moving. Other devices used to measure ROM include flexible rulers, fluid goniometers, and pendulum goniometers (Miller, 1985).

Cardiopulmonary System. When evaluating the cardiopulmonary systems, simple measures of vital signs are taken during rest and varying types and intensity of exercise. Reliability is problematic even for the simple vital sign measurements of heart rate, respiration rate, and blood pressure. Blood pressure readings are the least reliable (Eilertsen & Humerfelt, 1968; Stolt, 1990). General exercise endurance is assessed by a standardized exercise test such as the 6-minute walk or 12-minute walk. The Borg Perceived Exertion Scale is used frequently to analyze the effort involved in the activity (Borg, 1982). Another easily performed standardized exercise test of endurance is the Self-Paced Walking Test to Predict V_{O_2}

(Bassey, Fenton, & MacDonald, 1976). Although used with clients having neurologic dysfunction, these tests have been standardized more typically with nondisabled populations.

As examples of these testing methods, the 6-minute and 12-minute walk test are discussed briefly. The 12-minute walk test was developed originally by Cooper (1968) to assess V_{O_2} max of athletes, but was modified later for disabled individuals. The 6-minute walk test was a further modification to allow the evaluation of cardiovascular endurance for clients with dysfunction. To assess endurance the individual performance is measured by the distance walked in a 6-minute time interval (Cole et al., 1994, p. 122). The test–retest reliability for one individual with obstructive lung disease was $r = .97$, but the reliability was worse for group mean comparisons between repeated tests. The coefficient of variation for 13 subjects with chronic bronchitis was ±8.2%. Regarding construct validity, the 12-minute walk-run test was reported to correlate with V_{O_2} max of other exercise tests in the laboratory ($r = .90$). Numerous studies of concurrent validity were reported for pulmonary function tests and cycle ergometry. The correlations of the walk test to pulmonary tests were conflicting but were significant for cycle ergometry ($r = .60$). The concurrent validity between the 6-minute and 12-minute walk tests was also reported to be high (Cole et al., 1994, p. 123).

Neurological System. When assessing clients with neurologic dysfunction, neural motor control is most carefully examined. This includes simple mental tests of memory and reason; measurements for balance, coordination, and spasticity; and measurements of movement timing, sequencing, and force gradation during the movement. Examples of a few basic tests follow.

Common balance tests used include the Functional Reach Test (Duncan, Weiner, Chandler, & Studenski, 1990), SensoriInteraction Test (Shumway-Cook & Korak, 1986), Berg Balance Test (Berg, Williams, Wood-Dauphine, & Maki, 1992), Tinetti Balance Assessment (Tinetti, 1986) and the Wolfson Balance Test (Wolfson, Whipple, Amerman, & Kleinberg, 1986). Of these tests, the Functional Reach Test has probably been evaluated most carefully for reliability and validity; it scores high in these areas (Duncan, Studenski. Chandler, & Prescott, 1992; Duncan et al., 1990; Weiner, Duncan, Chandler, & Studenski, 1990). The Berg Balance Scale is another well-constructed and easily administered objective measure of balance performance (Berg, Maki, Williams, Holliday, & Wood-Dauphine, 1992; Berg, Williams, et al., 1992). The test consists of 14 everyday tasks that challenge the subject's abilities to maintain off-balanced positions or to maneuver with a progressively decreasing base of support. The scoring is based on a five-point ordinal scale. Reliability measures of internal consistency (Cronbach's alpha = .96), intrarater reliability (ICC = .99), and interrater reliability (ICC = .99) were all excellent. Construct validity for the Berg Balance Scale with the Barthel Index ($r = .80$ to $r = .94$) and the Fugl-Meyer Exam ($r = .62$ to $r = .94$) were acceptable. Concurrent validity was established between the Berg Balance Scale and the Tinetti balance subscale ($r = .91$), the Barthel Mobility subscale ($r = .67$), and the Timed Up and Go test ($r = -.76$). One of the most impressive features of the Berg Balance Scale is that it has been partially tested for predictive validity and functional sensitivity. These issues are seldom addressed in test development and are two of the key issues in today's health-care environment. Scores of <45 on the Berg

Balance Scale were found to be predictive of multiple falls, and the Berg Balance Scale was found to be discriminative for mobility characteristics and outcome groups for clients poststroke (Cole et al., 1994, p. 51).

Spasticity or tone is an assessment factor of concern to most physical therapists. Methods to evaluate spasticity or tone include the following: Ashworth Scale, Modified Ashworth Scale, Pendulum Drop Test, H Reflexes, or Resistance to Passive Movement (Bajd & Vodovnik, 1984; Bohannon & Smith, 1987). The assessment of tone is most easily addressed with the Ashworth Scale or Modified Ashworth Scale. The intrarater and interrater reliability is generally good for these scales (Bohannon & Smith, 1987; Wade, 1992, p. 54). Validity is questionable for all measurements of tone or spasticity. No convincing evidence exists that tone or spasticity correlate with disability or function (Wade, 1992, p. 54).

Summary. The tests performed under the basic measurement section have varying degrees of reliability and validity. Many publications exist on these issues for many of the basic measurements.

FUNCTIONAL ASSESSMENTS. In today's health-care environment, functional assessments are receiving the greatest support and attention. These are promoted as outcome assessments. Table 2 presents a list of these functional assessments.

Many ADL and extended ADL tests have been developed over the years. These generally have good intrarater and interrater reliability, and the practical and obvious importance of each item on the tests speaks for content validity. The ADL scales are similar. The Barthel ADL Index is one of the most familiar scales and is reported to have good reliability and inherent content validity because of its direct look at basic functional movements (Cole et al., 1994, p. 53). The Barthel

TABLE 2. LIST OF FUNCTIONAL ACTIVITY ASSESSMENTS

Activity Index
Barthel Index
Canadian Neurological Scale (CNS)
Clinical Outcome Variable Scale (COVS)
Frenchay Activities Index
Functional Independence Measure (FIM)
Functional Autonomy Measurement System (SMAF)
Katz Index of Activities of Daily Living
Kenny Self-Care Evaluation
Klein-Bell Activities of Daily Living Scale
Lawton and Brody Instrumental Activities of Daily Living Scale (IADLs) (Lawton & Brody, 1969)
Level of Rehabilitation Scale (LORS-II)
Northwestern University Disability Scale
Nottingham Extended ADL Index
Rivermead ADL Assessment
Rivermead Motor Assessment (RMA)
The PULSES Profile
Tinetti Balance and Gait Evaluation (Tinetti, 1986)

References: Cole, Finch, Gowland, and Mayo, 1994; *Measurement in Neurological Rehabilitation*, by D. T. Wade, 1992, New York: Oxford University Press.

ADL Index has been reported to be predictive of clients' living arrangements after leaving a rehabilitation center (Dejong & Branch, 1982) and predictive of functional improvement capacity (Wylie & White, 1964). Problems with the Barthel Index include its failure to detect gradual changes in function and its low ceiling effect, which allows the test to be worthwhile only for low-functioning individuals.

The Functional Independence Measure (FIM) provides a general overview of functional status and is scored by a multidisciplinary team of rehabilitation professionals (Keith, Granger, Hamilton, & Sherwins, 1987; Cole et al., 1994, p. 54). A national data base has been established to collect data from rehabilitation centers throughout the country to test the FIM's prediction of outcomes. The FIM consists of 23 items addressing seven basic functions: self-care, sphincter control, mobility, locomotion, communication, social adjustment/cooperation, and cognition/problem solving. Test-retest and intrarater reliability have not been reported for the FIM, but the interrater reliability was reported to be high (.83-.96). The reliability and validity testing were performed primarily with spinal-cord-injured clients. Content and concurrent validity are reported to be good. The modified Barthel and modified FIM were highly correlated. The FIM was found to be predictive of life satisfaction for a group of clients with multiple sclerosis (Cole et al., 1994, p. 55). Physical therapists are concerned with the gross measurement nature of the FIM as an outcome measure. The lack of sensitivity in the instrument limits its usefulness to very general outcome categories. This lack of sensitivity to the small and gradual changes typical of most rehabilitation clients could make the FIM a detrimental evaluation tool for the client. Therapists across the country are concerned that the welfare of their clients will not be served if the FIM is used to establish health insurance coverage and cutoff times for rehabilitation intervention.

The extended ADLs are more important than the ADLs in evaluating functional abilities of individuals in their homes and in the wider community. The Nottingham Extended ADL Index and the Frenchay Activities Index are two scales that have adequate reliability and validity and are very simple and general in assessment. The Lawton IADLs are more detailed and specific (Lawton & Brody, 1969). An instrument developed as a joint measure of ADL and extended ADLs is the Functional Autonomy Measurement System (SMAF). The SMAF was an attempt to integrate key items from the Katz ADI, Psychiatric Behavioral Scales, and Lawton's Instrumental Activities of Daily Living to assess the level of independence of elderly clients with disability and handicap (Cole et al., 1994, p. 72). The SMAF has greater sensitivity than the FIM and was designed according to the World Health Organization classification of impairment, disability, and handicap. Test-retest reliability of the SMAF was reported to be .68 to .78, interrater reliability has not been reported. Concurrent validity when the SMAF was validated to a measure of nursing care time was rated poor to good for various subscales (ADL .89, Mobility .83, Communication .58, and Mental Functions .63). Predictive validity has not been reported (Cole et al., 1994, p. 73).

DIAGNOSTIC SPECIFIC ASSESSMENTS. Many rehabilitation assessments have been developed specifically for diagnostic groups of clients. Tables 3 and 4 denote many of these assessments. One assessment for each diagnostic group will be discussed briefly.

Evaluation After Stroke. The Fugl-Meyer Assessment of Sensorimotor Recovery after Stroke is an evaluation of impairment after stroke. Items on motor recovery, balance, sensation, and joint ROM are evaluated. The evaluators require extensive training and must be well practiced to be considered competent and reliable testers. Test–retest reliability is excellent for the total assessment ($r = .98$ to $r = .99$), and for the subtests ($r = .87$ to $r = 1.00$). Interrater reliability is good for the lower- and upper-extremity total scores, but the coordination and reflex scores have less reliability. Construct validity is adequate. Concurrent validity was established by correlating the Fugl-Meyer Assessment with the DeSouza methods ($r = .97$), the Fugl-Meyer Assessment to the Barthel Index ($r = .67$ to $r = .76$), and the Fugl-Meyer Assessment to the Action Research Arm test ($r = -.91$ to $r = .94$) (Duncan et al., 1983; Cole et al., 1994, p. 57). The predictive validity and sensitivity have not been reported.

TABLE 3. REHABILITATION ASSESSMENTS SPECIFIC TO THE DIAGNOSES OF STROKE AND PARKINSON'S DISEASE

Diagnostic category	Name of the assessment
Stroke	Motor Assessment Scale
	Chedoke–McMaster Stroke Assessment
	Fugl-Meyer Assessment of Sensorimotor Recovery after Stroke
	Brunnstrom Hemiplegia Evaluation (Sawner & Lavigne, 1992)
	Bobath Hemiplegia Evaluation (Bobath, 1978)
Parkinson's disease	Columbian Rating Scale
	Hoehn and Yahr Grades
	New York Rating Scale
	Parkinson's Disease Impairment Index
	Parkinson's Disease Disability Index
	Parkinson's Disease: Lieberman Index
	Self-Assessment Parkinson's Disease Disability Scales
	Webster Rating Scale
	Unified Parkinson's Disease Rating

References: Cole, Finch, Gowland, and Mayo, 1994; Wade, 1992.

TABLE 4. REHABILITATION ASSESSMENTS SPECIFIC TO THE DIAGNOSES OF SPINAL CORD INJURY, TRAUMATIC BRAIN INJURY, AND MULTIPLE SCLEROSIS

Diagnostic category	Name of the assessment
Spinal cord injury	Spinal Injury: Frankel Scale
	Spinal Cord Injury Motor Index and Sensory Indices
Traumatic brain injury	Glasgow Coma Scale
	Glasgow Assessment Schedule
	Rapaport Disability Rating Scale
	Neurobehavioral Rating Scale
Multiple sclerosis	Minimal Record of Disability
	Kurtzke Expanded Disability Status Scale

References: Cole, Finch, Gowland, and Mayo, 1994; Wade, 1992.

Evaluating Clients with Parkinson's Disease. As can be seen by the list of assessments for Parkinson's disease, many instruments have been developed. Wade (1992, p. 343) has suggested that none of these assessments are of particular value, and recommends the use of general measures of disability rather than these specialized assessments. The Unified Parkinson's Disease Rating Scale is probably the best of the Parkinson's scales and has adequate reliability and validity.

Spinal Cord Injury Assessments. Two of the more common scales used in spinal cord rehabilitation are the Spinal Injury: Frankel Scale and the Spinal Cord Injury Motor Index and Sensory Indices. Although used often to classify a spinal cord injury, the Frankel Scale has not been tested for reliability and validity. The Spinal Cord Injury Motor Index and Sensory Indices is useful in acute rehabilitation to classify level of injury, but it also has not been adequately tested for reliability and validity. The predictive validity has been established for high cervical lesions (Wade, 1992, pp. 354–355).

Traumatic Brain Injury Evaluation. The Rappaport Disability Rating Scale has been developed to monitor clients post head injury from the initial evaluation until discharge from rehabilitation. The scale is reported to be adequately reliable and valid for use in large multicenter studies. Extensive testing of the scale has not been reported in the literature (Wade, 1992, pp. 310–311).

Scales for Multiple Sclerosis. The Kurtzke Scales were developed specifically to assess clients with multiple sclerosis. The scales are criticized as being clinically impractical and insensitive to changes in disability. Validity and reliability measures have been reported to be acceptable (Wade, 1992, pp. 285–290).

FUTURE DIRECTIONS

Physical therapy will continue to change and respond to new scientific information and the alteration of our health-care systems, but standardized tests will never replace individualized assessments. A more flexible, comprehensive, problem-oriented, and individualized plan will always be necessary; however, the use of standardized tests as a component of each physical therapy evaluation is becoming a reality and a necessity. Evaluation standardization is important to allow examination of the effectiveness of intervention strategies.

None of the existing assessment batteries address the potential for or actual amount of motor learning, and all lack sensitivity in following the processes by which movement dysfunction is altered. The sciences with an emphasis on movement control are increasing research and training efforts toward this goal. Doctoral programs in movement science oriented toward the issues of movement dysfunction are developing to serve the needs of these research and training efforts. The sciences of motor control, motor learning, biomechanics, and exercise physiology are influencing the types of assessments and interventions developed for clients.

Physical therapy evaluations are often directed by time restraints and the level of intervention that is possible for the client. For these and other reasons, a levels of analysis approach to evaluation is needed. The World Health Organization classification of pathology, impairment, disability, and handicap is currently being promoted as one type of approach to the need for levels (1980). Physical therapy assessments have been criticized for being too focused on impairment as opposed to disability and handicap. Wade (1992, pp. 27–96) has addressed the issues of measurement classification for impairment, disability, and handicap. Future standardized tests will be greatly improved by following the World Health Organization model. Some of the newer assessments, for example, the Functional Autonomy Measurement System for adult evaluation (Cole et al., 1994, p. 72) and the Pediatric Evaluation of Disability Inventory (Haley et al., 1992), have considered this model.

The future direction of physical therapy assessment is evidenced by a recent special edition physical therapy journal devoted to the topic of functional assessment, outcome measures, and the World Health Organization classification model (Rothstein, 1994).

In addition to the need to consider levels of disability and handicap, physical therapy assessments must be developed that are sensitive to the gradual changes in movement dysfunction typical of clients in rehabilitation. A recent publication by NIH encourages the research agenda of improving functional assessments and increasing sensitivity in the assessments. A particular need to improve assessments for mild to moderate disabilities was noted (Public Health Service, 1993).

Perhaps the most important change in physical therapy and health care in general is the emphasis on improving interdisciplinary interactions to guide practice and assessment. With the client as the focus, the health-care team needs to collaborate and develop efficient, cost-effective, and goal-directed assessments that consider the client as a whole person. The fragmented and isolated practices of the various professions must change to meet the new health-care objectives. This change is long overdue.

Summary

Physical therapists are licensed health-care professionals concerned with movement dysfunction. From simple and practical beginnings, the physical therapy profession has grown and changed significantly throughout its history. This change has been driven by flourishing advances in science and the pressures of economic accountability within health care.

Physical therapy evaluations are problem oriented, goal directed, and individualized by design. Physical therapists recognize the need for standardized measurements, and the current scientific and health-care accountability requirements will insure improvement in standardization. Numerous tests and evaluation batteries are available for use in physical therapy assessments; these tests vary in reliability, validity, and sensitivity. Research on these issues is currently intense and increasing. Physical therapists are concerned that the instruments used have the sensitivity to evaluate incremental changes in movement dysfunction. This issue is of extreme importance for the welfare of our clients if health insurance coverage is to be based on improvement in function.

Physical therapy will continue to grow and change. Physical therapists are involved actively in the improvement of standardized assessments toward the World

Health Organization model of pathology, impairment, disability, and handicap. The sciences of motor control, motor learning, biomechanics, and exercise physiology influence the types of assessments and interventions developed for clients. Physical therapists are devoted to the concept of professional team building and the interdisciplinary approach to assessment and client intervention. The welfare of our clients demands that we integrate our efforts and focus on efficient and meaningful intervention strategies.

REFERENCES

Als, H., Lester, B. M., Tronick, E. Z., & Brazelton, T. B. (1983). Manual for the assessment of preterm infants' behavior. In H. E. Fitzgerald, B. M. Lester, & M. W. Yogman (Eds.), *Theory and Research in Behavioral Pediatrics*. New York: Plenum.

American Physical Therapy Association (APTA). (1983). House of Delegates fifty-ninth annual conference. Kansas City, MO.

American Physical Therapy Association (APTA). (1984). *Physical therapy education and societal needs: Guidelines for physical therapy education*. Alexandria, VA: Author.

American Physical Therapy Association (APTA) Neurology Section. (1994). Neurology Section membership pamphlet. Alexandria, VA: American Physical Therapy Association.

Bajd, T., & Vodovilik, L. (1984). Pendulum testing of spasticity. *Journal of Biomedical Engineering, 6*, 9–16.

Bassey, E. J., Fenton, P. H., & MacDonald, I. C. (1976). Self-paced walking as a method for exercise testing in elderly and young men. *Clinical Science and Molecular Medicine, 51*, 609–612.

Bayley, N. (1993). *Bayley Scales of Infant Development* (2nd ed.). San Antonio, TX: Psychological Corporation.

Beard, G. (1961). Foundations for growth. A review of the First forty years in terms of education, practice, and research. *Physical Therapy Review, 41*, 843–861.

Berg, K. O., Maki, B. E., Williams, J. I., Holliday, P. J., & Wood-Dauphine, S. L. (1992). Measuring balance in the elderly: Validation of an instrument. *Canadian Journal of Public Health, 83*, 7–11.

Berg, K. O., Williams, J. I., Wood-Dauphine, S. L., & Maki, B. E. (1992). Clinical and laboratory measures of postural balance in an elderly population. *Archives of Physical Medicine and Rehabilitation, 73*, 1073–1080.

Bobath, B. (1978). *Adult Hemiplegia: Evaluation and Treatment* (2nd ed.). London: William Heinemann Medical Books.

Bohannon, R. W. (1986). Test-retest reliability of hand-held dynamometry during a single session of strength assessment. *Physical Therapy, 66*, 206–209.

Bohannon, R. W., & Andrews, A. W. (1987). Interrater reliability of hand-held dynamometry. *Physical Therapy, 67*, 931–933.

Bohannon, R. W., & Smith, M. B. (1987). Interrater reliability of a Modified Ashworth Scale of muscle spasticity. *Physical Therapy, 67*, 206–207.

Borg G. (1982). Psychophysical basis of perceived exertion. *Medicine and Science in Sports and Exercise, 14*, 377–381.

Brazelton, T. B. (1984). Neonatal Behavioral Assessment Scale. *Clinics in Developmental Medicine*. Philadelphia: Lippincott.

Bruininks, R. H. (1978). *The Bruininks-Oseretsky test of motor proficiency: Examiner's manual*. Circle Pines, MN: American Guidance Service.

Campbell, S. K. (1990). Consensus conference on efficacy of physical therapy in the management of cerebral palsy. *Pediatric Physical Therapy, 2*, 123–125.

Campbell, S. K. (1991). Framework for the measurement of neurological impairment and disability. In M. J. Lister (Ed.), *Contemporary management of control problems: Proceedings of the II Step Conference* (pp. 143–153). Alexandria, VA: Foundation for Physical Therapy.

Campbell, S. K. (1992). Measurement of motor performance in cerebral palsy. In H. Frossberg & H. Hirschfield (Eds.), *Movement disorders in children* (pp. 264–271). Basel, Switzerland: Karger.

Campbell, S. K. (1993). Future directions for physical therapy assessment in early infancy. In I. J. Wilhelm (Ed.), *Physical therapy assessment in early infancy* (pp. 293–308). New York: Churchill Livingstone.

Chandler, L., Andrews, M., & Swanson, M. (1980). *Movement assessment of infants: A manual.* Rolling Bay, WA: Authors.

Cole, B., Finch, E., Gowland, C., & Mayo, N. (1994). *Physical rehabilitation outcome measures.* Toronto, Ontario, Canada: Canadian Physiotherapy Association.

Cooper, K. H. (1968). A means of assessing maximal oxygen intake. *Journal of the American Medical Association, 203,* 135-138.

Coryell, J., Provost, B., Wilhelm, I. J., & Campbell, S. K. (1989). Stability of the Bayley Motor Scales in the first year of life. *Physical Therapy, 69,* 834-844.

Crowe, T. K., Deitz, J. C., & Bennett, F. C. (1987). The relationship between the Bayley Scales of Infant Motor Development and preschool gross motor and cognitive performance. *American Journal of Occupational Therapy, 41,* 374-379.

Dejong, G., & Branch, L G. (1982). Predicting the stroke patient's ability to live independently. *Stroke, 13,* 648-655.

Dubowitz, L., & Dubowitz, V. (1981). The neurological assessment of the full and preterm infant. *Clinics in Developmental Medicine, 79,* 10-44.

Duncan, P. W., Propst, M., & Nelson, S. G. (1983). Reliability of the Fugl-Meyer assessment of sensorimotor recovery following cerebrovascular accident. *Physical Therapy, 63,* 1606-1610.

Duncan, P. W., Weiner, D., Chandler, J., & Studenski, S. (1990). Functional reach: A new clinical measure of balance. *Journal of Gerontology, 45,* M192-197.

Duncan, P. W., Studenski, S., Chandler, J., & Prescott, B. (1992). Functional reach: Predictive validity in a sample of elderly male veterans. *Journal of Gerontology, 47,* M93-98.

Eilertsen, E., & Humerfelt, S. (1968). The observer variation in the measurement of arterial blood pressure. *Acta Medica Scandinavica, 184,* 145-157.

Feldman, A. B., Haley, S. M., & Coryell, J. (1990). Concurrent and construct validity of the Pediatric Evaluation of Disability Inventory. *Physical Therapy, 70,* 602-610.

Folio, R. M., & Fewell, R. R. (1983). *Peabody Developmental Motor Scales and Activity Cards.* Allen, TX: DIM Teaching Resources.

Fugl-Meyer, A. R. (1980). Post-stroke hemiplegia assessment of physical properties. *Scandinavian Journal of Rehabilitation Medicine, 7,* 85-93.

Goldberg, S. (1992). *The Four-Minute Neurologic Exam.* Miami, FL: Medmaster.

Gordon, J. (1987). Assumptions underlying physical therapy intervention: Theoretical and historical perspectives. In J. H. Carr, R. B. Shepherd, J. Gordon, A. M. Gentile, & J. M. Held (Eds.), *Movement science: Foundations for physical therapy in rehabilitation* (pp. 1-30). Rockville, MD: Aspen.

Haley, S. M., & Beryza, M. J. (1990). A hierarchy of motor outcome assessment: Self-initiated movements through adaptive motor function. *Infants and Young Children, 3,* 1-14.

Haley, S. M., Coster, W. J., Ludlow, L. H., Haltiwanger, J. T., & Andrellos, P. J. (1990). *Pediatric Evaluation of Disability Inventory (PEDI): Development, standardization and administration manual.* Boston: New England Medical Center.

Haley, S. M., Baryza, M. J., & Webster, H. C. (1992). Pediatric rehabilitation and recovery of children with traumatic injuries. *Pediatric Physical Therapy, 4,* 24-30.

Harris, S. R. (1987). Early detection of cerebral palsy: Sensitivity and specificity of two motor assessment tools. *Journal of Perinatology, 7,* 11-15.

Harris, S. R., Haley, S. M., Tada, W. L., & Swanson, M. W. (1984). Reliability of observational measures of the movement assessment of infants. *Physical Therapy, 64,* 471-476.

Hinderer, K. A., Richardson, P. K., & Atwater, S. W. (1989). Clinical implications of the Peabody developmental motor scales: A constructive review. *Physical and Occupational Therapy in Pediatrics, 2,* 81-106.

Horak, F. B. (1992). Motor control models underlying neurologic rehabilitation of posture in children. In H. Frossberg & H. Hirschfield (Eds.), *Movement disorders in children* (pp. 21-30). Basel, Switzerland: Karger.

Keith, R. A., Granger, C. V., Hamilton, B. B., & Sherwins, F. S. (1987). The Functional Independence Measure. *Advances in Clinical Rehabilitation, 1,* 6-18.

Lamb, R. (1985). Manual muscle testing. In J. Rothstein (Ed.), *Measurement in physical therapy* (pp. 47-55). New York: Churchill Livingstone.

Lawton, M. P., & Brody, M. (1969). Assessment of older people: Self-monitoring and instrumental activities of daily living. *Gerontologist, 9,* 179-186.

Miller, L. J., & Roid, G. H. (1993). Sequence comparison methodology with movement disorders. *American Journal of Occupational Therapy, 47,* 339-347.

Miller, P. J. (1985). Assessment of joint motion. In J. Rothstein (Ed.), *Measurement in physical therapy* (pp. 103-136). New York: Churchill Livingstone.

Pinkston, D. (1989). Evolution of the practice of physical therapy in the United States. In R. M. Scully & M. R. Barnes (Eds.), *Physical therapy* (pp. 2-30). Philadelphia: Lippincott.

Piper, M. C., & Darrah, J. (1994). *Motor assessment of the developing infant.* Philadelphia: Saunders.

Prechtl, H. F. R. (1977). The neurological examination of the full-term infant. (2nd ed.). *Clinics in Development Medicine, 63,* 10-33.

Public Health Service, U.S. Department of Health and Human Services. (1993). Categories of disability: Functional and physical limitations. In *Research plan for the National Center for Medical Rehabilitation Research* (pp. 60-68). (NIH Publication No. 93-3509). Rockville, MD: Author.

Rosenbaum, P. L., Russell, D. J., Cadman, D. T., Gowland, C., Jarvis, S., & Hardy, S. (1990). Issues in measuring change in motor function in children with cerebral palsy: A special communication. *Physical Therapy, 70,* 125-137.

Rothstein, J. (1985). *Measurement in physical therapy.* New York: Churchill Livingstone.

Rothstein, J. (Ed.). (1994). Physical disability. [Special issue]. *Physical Therapy, 74.*

Rothstein, J., & Echternach, J. (1993). *Primer on measurement: An introductory guide to measurement issues.* Alexandria, VA: American Physical Therapy Association.

Russell, D. J., Rosenbaum, P. L., Gowland, C., Hardy, S., Lane, M., Plews, N., McGavin, H., Cadmen, D., & Jarvis, S. (1993). *Gross motor function measure* (2nd ed.). Ontario, Canada: McMasters University, Children's Developmental Rehabilitation Programme, Chedoke-McMasters Hospitals and McMillan Rehabilitation Centre.

Sawner, K., & Lavigne, J. (1992). *Brunnstrom's movement therapy in hemiplegia.* Philadelphia: Lippincott.

Schnieder, J. W., Lee, W., & Chasnoff, I. J. (1988). Field testing the movement assessment of infants. *Physical Therapy, 66,* 321-327.

Shumway-Cook, A., & Horak, F. B. (1986). Assessing the influence of sensory interaction on balance. *Physical Therapy, 66,* 1548-1550.

Stolt, M. (1990). Reliability of auscultatory method of arterial blood pressure. *Hypertension, 3,* 697-703.

Tinetti, M. (1986). Performance-oriented assessment of mobility problems in elderly patients. *Journal of the American Geriatrics Society, 34,* 119-126.

Wade, D. T. (1992). *Measurement in neurological rehabilitation.* New York: Oxford University Press.

Weiner, D., Duncan, P. W., Chandler, J., & Studenski, S. (1990). Functional reach: A marker of physical frailty. *Journal of the American Geriatrics Society, 38,* 1120-1126.

Wilhelm, I. J. (1993). Physical therapy assessment in early infancy. In *Clinics in Physical Therapy.* New York: Churchill Livingstone.

Wolfson, L. I., Whipple, R, Amerman, P., & Kleinberg, A. (1986). Stressing the postural response: A quantitative method for testing balance. *Journal of the American Geriatrics Society, 34,* 845-850.

World Health Organization(WHO). (1980). *International classification of impairments, disabilities and handicaps.* Geneva, Switzerland: World Health Organization.

Wylie, C. M., & White, B. K. (1964). A measure of disability. *Archives of Physical Medicine and Rehabilitation, 8,* 834-839.

4

Rehabilitation Assessment and Planning for Neurogenic Communication Disorders

PATRICK J. DOYLE

INTRODUCTION

Communication disorders may result from neurological insult and disease, psychological and affective disturbances, structural-mass abnormalities including those following exogenous trauma, and iatrogenic causes (Hartman, 1988). Within these broad etiological categories numerous disorders have been described, and it is not possible to address all of them in the present context. Readers are referred to texts edited or written by Darby (1981a, 1981b, 1985) Metter (1985), and Costello (1985) for descriptions and clinical management principles of several communication disorders not discussed in the current chapter.

This chapter will focus on the clinical process of rehabilitation assessment and planning for three neuropathologies of communication: dysarthria, apraxia of speech, and aphasia. These disorders may coexist, and frequently share speech and language symptomatology. Similarly, the localization, etiology (e.g., metabolic, vascular, neoplastic), and nature (e.g., static versus progressive, focal versus diffuse) of the causative lesion often overlap diagnostic categories and have a profound influence on the course of the communication disturbance, focus of therapy, and expected long-term outcomes. Therefore, a thorough understanding of the patterns of behavioral deficits, neuroanatomical correlates, and disease processes associated with each disorder is necessary for their appropriate management.

PATRICK J. DOYLE   VA Pittsburgh Healthcare System, Highland Drive Division, 7180 Highland Drive, Pittsburgh, Pennsylvania 15206-1297.

Rehabilitation, edited by Goldstein and Beers. Plenum Press, New York, 1998.

Nomenclature

Historically, the nomenclature associated with neurogenic communication disorders has been plagued by controversy. No less than 28 different aphasia classification systems employing varying rationales and terminology have been proposed since Broca's 1865 description of "aphemia" (McNeil, 1988), and more than 20 labels have been applied to the current concept of apraxia of speech (Johns & LaPoint, 1976). Darley (1969b), in a paper entitled *Apraxia of Speech: 107 Years of Terminological Confusion,* differentiated apraxia of speech from aphasia and dysarthria and suggested that the use of imprecise and overlapping terminology had led to misunderstanding of the disorder and mistreatment of patients who were handicapped by it. Indeed, because the diagnosis or labeling of a communication disorder affects both the approach to patient management and the predicted outcomes, it is imperative that the nomenclature associated with neurogenic communication disorders be clearly delineated. This is especially important in view of the fact that the overall management of communicatively impaired, brain-injured adults is frequently the joint responsibility of a multidisciplinary team of professionals. It is from this perspective that the following clinically useful classification system is provided.

Prior to discussing the disorders individually, it is important to distinguish between neuropathologies of speech and neuropathologies of language. Apraxia of speech and dysarthria are considered motor speech disorders because the mechanisms of impairment are at the levels of programming motor (i.e., articulatory) patterns and executing basic motor processes that support speech, respectively. In contrast, aphasia is a language disorder, as the mechanism of impairment is at the level of symbolic operations. The behavioral characteristics and associated neuroanatomical lesion sites and disease processes of these pathologies are described by Darley (1969a), Wertz (1985), and others as follows:

Dysarthria

Dysarthria refers to a group of speech disorders resulting from disturbances in muscular control (i.e., weakness, slowness, or difficulties with coordination) of the speech mechanism. The term encompasses coexisting neurogenic disorders of several or all of the basic processes of speech, including respiration, phonation, resonance, articulation, and prosody. Dysarthria may result from damage to the central or peripheral nervous system. However, because the speech musculature is bilaterally innervated, most persistent and debilitating dysarthrias result from bilateral subcortical involvement. Seven different types of dysarthria (flaccid, spastic, ataxic, hypokinetic, hyperkinetic-dystonic, hyperkinetic-choreoathetosis, and mixed) have been identified based on clusters of prominent speech and voice deviations and their associated neuromuscular condition and disease processes (Darley, Aronson and Brown, 1969a; 1969b; 1975).

Apraxia of Speech

Apraxia of speech refers to an acquired neurogenic articulatory disorder in which the capacity to position speech muscles and sequence muscle movements for the volitional production of phonemes and phoneme sequences is impaired.

However, unlike the clinical presentation of dysarthria, significant weakness of oral musculature and slowness and/or incoordination of oral motor movements during reflexive or automatic acts is not observed. Prosodic alterations such as abnormal stress and intonation patterns are also frequently associated with the disorder either as a primary component or perhaps as a compensatory adaptation. Apraxia of speech is usually associated with focal, left hemisphere lesions of the premotor cortex (Canter, 1973; Rosenbek, Kent, & LaPoint, 1984) resulting from vascular, neoplastic, or infectious disease processes.

APHASIA

Aphasia refers to an acquired neurogenic language disorder in which the capacity to interpret and formulate language symbols and linguistic units, such as morphemes and sentential structure, is impaired across all input and output modalities. Further, the disruption of language processes is disproportionately severe relative to impairments of other intellectual functions. Aphasia is not attributable to psychosis, dementia, sensory loss, or motor dysfunction. It is manifested in reduced availability of vocabulary, reduced efficiency in applying phonologic, morphologic, and syntactic rules, and reduced auditory retention span. Aphasia is usually associated with focal, static lesions of the perisylvian and premotor cortex of the left cerebral hemisphere (Benson, 1979). However, aphasia following insular (H. Damasio & Damasio, 1980), thalamic (Cohen, Gelfer, & Sweet, 1980; Mohr, Watters, & Duncan, 1975) and other subcortical lesion sites (A. R. Damasio, Damasio, Rizzo, Vamey, & Gersh, 1982; Hier, Davis, Richardson, & Mohr, 1977; Naeser et al., 1982; Wallesch et al., 1983) has been reported. Similarly, cases of slowly progressive aphasia resulting from undefined or nonspecific atrophic processes have also been described (Chawluk et al., 1986; Gordon & Selnes, 1984; Heath, Kennedy, & Kapur, 1983; Mesulam, 1982).

THE APPRAISAL PROCESS

During different phases of patient management, the appraisal will serve different purposes. McNeil and Kennedy (1984) listed ten such purposes with respect to the assessment of dysarthria, but they apply to the other neurogenic disorders of communication equally well. For example, during the initial phases of patient management the appraisal may be aimed at detecting or confirming a suspected problem, establishing a differential diagnosis, or determining the suspected disease process. Once these tasks have been accomplished the appraisal may be directed toward establishing the severity of the communication disorder, candidacy for treatment, and prognosis for improved communication skills. Finally, the appraisal should help to determine the focus of treatment, document changes in patient performance that occur during the course of treatment, and establish criteria for discharge from speech and language therapy.

Accomplishment of these appraisal goals requires thorough and systematic collection of pertinent biographical, medical, and behavioral data regarding the patient (Wertz, 1985). Such information may be obtained from the patient's medical records, but ongoing consultation with the referring physician, other members of the rehabilitation team, the patient, and significant others is usually required to obtain a complete understanding of the medical, behavioral, and psychosocial contexts in which the communication disorder occurred, the current status of the patient, and the prognosis for successful speech and language rehabilitation.

Important biographical data includes information regarding the patient's age at onset of the communication disorder, handedness, level of education, occupation, premorbid intelligence, premorbid communicativeness, current social support network, living environment, and motivation to participate in therapy. Such information has both prognostic and treatment implications. For example, Rau and Schultz (1988) have discussed the importance of social networks and frequency of social contacts as important prognostic factors with respect to elderly stroke patients' participation in and response to speech and language therapy. Similarly, the patient's premorbid education, occupation, and intelligence may have considerable influence on the focus and goals of therapy.

Medical data are most important with respect to their diagnostic implications, and this is particularly true during the initial phases of patient management. Indeed, it is only when the history, physical, and laboratory findings are equivocal that the clinician focuses on the pattern of speech and language symptoms to establish the differential diagnosis (Kitselman, 1981). Nevertheless, the medical findings used to arrive at a differential diagnosis among neurogenic communication disorders also greatly influence the prognosis for recovery and the approach to patient management. Therefore, the speech-language pathologist seeks to obtain medical information regarding factors that have been shown to influence recovery and that is diagnostically specific. In this regard, the neurological data such as the etiology, site, extent, and dynamics of the causative lesion are most important. Other medical data having prognostic and treatment implications include the presence and nature of sensory and/or motor deficits, coexisting major medical problems such as diabetes, high blood pressure, or pulmonary disease, and the potential central nervous system side effects of the patient's pharmacotherapy. Finally, affective disturbances including anxiety and depression are frequently associated with brain injury (Wahrborg, 1991) and will ultimately affect both acute and long-term management of the communication disorder.

Speech, language, and related behavioral data comprise the final and most extensively investigated category of information collected by the speech-language pathologist during the appraisal process. In addition to their importance with respect to several of the aforementioned appraisal goals, speech, language, and related behavioral data serve as the primary source of information from which the nature and severity of the communication disorder is determined and the treatment plan developed. Further, measures of performance on specific speech and language tasks serve as a baseline from which to assess the effects of treatment. In the sections that follow, assessment procedures specific to dysarthria, apraxia of speech, and aphasia are reviewed.

Assessment of Dysarthria and Apraxia of Speech

Dysarthria and apraxia of speech constitute the category of neuropathologies of communication referred to as motor speech disorders. The procedures used most frequently to evaluate motor speech disorders in a clinical setting include a functional components evaluation (Netsell, 1984b), an analysis of perceptual characteristics of speech (Darley et al., 1975), and intelligibility testing (Yorkston & Beukelman, 1981). The extent to which the clinical presentation of the patient and the available medical data support a diagnosis of dysarthria versus apraxia of speech, and the extent to which speech production is compromised, will determine what combination of these procedures will be necessary to assess and manage the patient. Regardless of the battery employed, the results of the assessment procedures in conjunction with pertinent medical data should provide the clinician with the necessary information to (a) differentiate among the dysarthrias, (b) differentiate dysarthria from apraxia of speech, (c) determine the effect of the disorder on the intelligibility of speech production, and (d) focus treatment.

FUNCTIONAL COMPONENTS EVALUATION. The functional components evaluation is based upon a physiologic approach to the management of disordered speech motor control (Netsell, 1983, 1984a, 1984b). The model identifies three physiological systems (i.e., respiratory-phonotory, velopharyngeal, and articulatory) comprised of one or more functional components. The compromise of these components by central or peripheral nervous system lesions results in relatively systematic variations of normal speech processes. Functional components are defined by Netsell and Daniel (1979) as a structure or set of structures that work to generate or valve the speech airstream. The functional components of the respiratory-phonatory system are the diaphragm, abdomen, rib cage, and intrinsic and extrinsic muscles of the larynx. The velopharyngeal system includes the soft palate, velum, and posterior and lateral pharyngeal walls. The articulatory system is comprised of the posterior tongue, anterior tongue, lips, and jaw.

The purpose of the functional components evaluation is to determine the nature and extent of neuromuscular dysfunction involving each functional component and the effects of these disturbances on perceptual aspects of speech. The examiner tests each component in isolation and in controlled combinations using speech and nonspeech tasks to evaluate individual, alternating, and sequential oral-motor movements with respect to accuracy, symmetry, force, speed, range, timing, and coordination. Maximum performance measures such as the duration of vowel and fricative (e.g., [s] versus [z]) productions, number of syllables produced on one breath, vital capacity, and fundamental frequency range are employed to make inferences regarding the status of functional components comprising the respiratory-phonatory system. The observed patterns of motor performance and estimated severity of neuromotor dysfunction are recorded for each functional component and are considered along with the findings of the perceptual analysis for purposes of differential diagnosis and treatment planning.

ANALYSIS OF PERCEPTUAL CHARACTERISTICS OF SPEECH. Perceptual characteristics of speech must be described to assist in the differential diagnosis of the dysarthrias and apraxia of speech. Indeed, it is the identification of clusters of deviant speech characteristics that is used to distinguish among the dysarthrias and apraxia of speech, with patterns of motor performance observed during the functional components evaluation serving as confirmatory evidence. Perceptual ratings are conducted during (a) structured speech tasks in which the physiological subsystems of the speech mechanism are isolated, (b) articulatory programming tasks such as rapid production of syllables requiring different articulatory placement and repetition of complex multisyllabic words and sentences, and (c) contextual speech. The examiner notes parameters of voice and speech such as pitch, loudness, vocal quality, resonance, articulation, and prosody in order to identify clusters of deviant perceptual characteristics that have been shown to co-occur in certain neurological groups (Darley et al., 1969a, 1969b, 1975).

Differentiating dysarthria from apraxia of speech is a critical component of the assessment as it relates to rehabilitation planning, because each disorder has a different underlying mechanism of impairment and requires distinct approaches to patient management. Apraxia of speech is differentiated from the dysarthrias primarily on the basis of articulatory performance during sequential motion rate tasks and repetition of multisyllabic words and sentences. Apraxic patients typically produce articulatory substitutions, omissions, repetitions, or additions during these more difficult programming tasks, in the absence of muscle weakness or limited range of movement for individual oral motor postures. Patients for whom apraxia of speech is suspected on the basis of the functional components evaluation and perceptual analyses may be further evaluated using standardized measures such as the Apraxia Battery for Adults (Dabul, 1979) and the Comprehensive Apraxia Test (DiSimoni, 1989). These measures use tasks that require volitional and rapid manipulation of the articulators and are designed to confirm the diagnosis and quantify the severity of the disorder.

INTELLIGIBILITY TESTING. Measuring the effects of neuromuscular dysfunction and the resulting deviant perceptual characteristics of speech on overall speech intelligibility is another important consideration in the clinical assessment of motor speech disorders. Duffy (1995) has described a clinically useful 10-point rating scale for judging speech intelligibility and communicative efficiency. The scale allows clinicians to rate performance under "ideal" and "adverse" listening conditions and accounts for the complexity and/or predictability of the message being communicated. Although subjective rating scales are useful for screening purposes to determine the need for intervention, an objective approach to intelligibility assessment is necessary for establishing baseline levels of performance and for quantifying response to treatment. The tool that is used most frequently in clinical settings is the Assessment of Intelligibility of Dysarthric Speech (Yorkston & Beukelman, 1981). This measure quantifies speech intelligibility, speaking rate, and communicative efficiency using both single word and sentence level tasks. The measure requires an examiner who selects the word and sentence samples to be produced by the speaker and a listener who transcribes or responds in a multiple-choice format to the recorded samples. An intelligibility percentage score is derived by dividing the number of correctly transcribed words by the total number of words in the sample. The rate of intelligible speech is determined by dividing the number of intelligible words produced during the sentence task by the total duration of the sample. This score can then be used to calculate the communicative efficiency ratio, which is derived by dividing the rate of intelligible speech per minute by 190 (the mean rate of speech of normal adults who are 100% intelligible on the measure). The communicative efficiency ratio is especially useful as an outcome measure for quantifying communicative disability resulting from motor speech impairment as well as response to treatment.

Assessment of Aphasia

Approaches to the assessment of aphasia will differ depending upon one's conceptualization of the disorder and the specific purposes of the assessment. From a rehabilitation planning perspective it is important that the assessment address three issues. First, the presence of aphasia must be differentiated from language disturbances associated with dementia, delirium, and psychosis. The language disturbances associated with these conditions have different underlying mechanisms of disruption, different clinical courses, and require different approaches to patient management. In cases where the history, medical data, and radiographic findings are inconclusive with respect to differential diagnosis, assessment of orientation, memory, speech, language, and nonverbal reasoning abilities can reveal deficit profiles that serve to differentiate among these disorders (Halpern, Darley, & Brown, 1973; Kitselman, 1981). Second, the nature and severity of aphasic impairment must be described and quantified in sufficient detail to guide the selection of treatment strategies and to permit reliable measurement of the treatment response. For example, it is well documented that dysnomia may result from failure of lexical-semantic or phonological processes (Hadar, Jones & Mate-Kole, 1987; Howard, Patterson, Franklin, Orchard-Lisle & Morton, 1985; Lesser, 1989). Because the treatment strategies employed in the rehabilitation process will differ depending on the source of the naming deficit, it is imperative that the assessment procedures address the suspected underlying mechanism of the observed performance deficits in addition to quantifying their severity. Finally, the effect of the aphasic impairments on interpersonal communication and communication of basic needs should be assessed for purposes of estimating the patient's functional communicative abilities. In the paragraphs that follow, the clinical processes addressing each of these assessment goals are described in further detail.

Differentiating Aphasia from Other Acquired Neurogenic Language Impairments. Clinicians are frequently confronted with communicatively impaired individuals for whom history, medical, and radiographic data are unavailable or inconclusive. In such cases, differential diagnosis of the communication disorder must be made primarily based on behavioral data. While there is not one standard measure that reliably discriminates aphasia from the language impairments associated with dementia, delirium, and psychosis, there are several distinguishing behavioral features that may be used to differentiate aphasia from the language impairments associated with these other conditions. First, the patient with aphasia will manifest *language* deficits in all input and output modalities. Specifically, oral and written expression will reveal both phonologically and semantically based dysnomias and omission or improper use of morphological and syntactic structures. In contrast, morphological and grammatical errors are rarely seen in the language impairment associated with dementia, delirium, and psychosis (Irigaray, 1973; Schwartz, Marin, & Saffran, 1979; DiSimoni, Darley, & Aronson, 1977). Second, while the productive language of persons with aphasia may lack detailed information or contain paraphasic errors, the information that is communicated is relevant to the elicitation context. In contrast, the spontaneous speech of persons suffering from dementia, delirium, and psychosis is characterized by overpersonalization, irrelevancy, and preservation of ideas (Bayles, 1994; Halpern et al., 1973; Gerson, Benson, & Frazier, 1977; Reilly, Harrow, & Tucker, 1973; DiSimoni et al., 1977). Samples of spoken and written discourse elicited by picture description tasks provide the best source of data from which to evaluate these parameters of language because they permit analyses of the structural, lexical, and contextual parameters of expression. Most standardized tests for aphasia have spoken and written picture description tasks that may be used for this purpose. However, a set of pictures is usually needed to obtain a sample that permits reliable observations (Brookshire & Nicholas, 1994).

Individuals with aphasia may also be distinguished from persons with other neurogenic language impairments based on their performance on tasks requiring immediate and delayed recall of information in spoken discourse passages. Bayles, Boone, Tomoeda, Slauson, and Kaszniak (1989) reported that the performance of both fluent and nonfluent aphasic individuals did not differ significantly from immediate to delayed recall conditions. In contrast, persons with both mild and moderate degrees of dementia were found to perform significantly more poorly under the delayed recall condition as compared to their performance during the immediate recall condition, and significantly more poorly than normal elderly and aphasic adults on delayed recall performance. Finally, tests of mental status, delayed spatial recognition memory, and delayed verbal recognition memory differentiated aphasic adults from patients with mild and moderate forms of dementia (Bayles et al., 1989). These procedures and many other tasks have been standardized and comprise the Arizona Battery of Communication Disorders of Dementia (Bayles & Tomoeda, 1993). Performance profiles on this measure can assist the clinician in differentiating aphasia from language impairments associated with general intellectual, memory, and attentional deficits.

ASSESSING THE NATURE AND SEVERITY OF APHASIA. The purpose of this phase of the assessment process is to understand the mechanisms of impairment that underlie the behavioral manifestations of aphasia so that an appropriate treatment plan can be developed, and to quantify the severity of impairments so that the response to treatment can be reliably measured. Brookshire (1986) explains that to accomplish these goals most standardized aphasia tests such as the Western Aphasia Battery, (Kertesz, 1982), Boston Diagnostic Aphasia Exam (1983), and Minnesota Test for Differential Diagnosis of Aphasia (Scheull, 1972) employ a rather gross model of language that consists of an "input" or reception stage, a "central" or processing stage, and an "output" or expression stage. Individual language production and comprehension tasks are designed to systematically manipulate input parameters to observe their effect on performance. Input parameters that are typically manipulated include the modality of stimulus presentation (i.e., auditory, visual, tactile) and characteristics of the stimuli such as their frequency of occurrence in print or spoken language, level of abstraction, semantic relatedness, phonological complexity, rate of presentation, syllable length, linguistic form class, and so on. Similarly, input parameters may also be held constant and performance observed under conditions that manipulate output parameters or response requirements. It is through this systematic manipulation of input and output parameters and observing the quantitative and qualitative effects on patients' performance that inferences may be drawn regarding the mechanisms of impairment underlying the behavioral manifestations of the disorder. Treatment plans may then be developed that appropriately target the disrupted language processes. Treatment plans developed solely on the basis of aphasic behavior are doomed to failure because many aphasic behaviors result from different underlying mechanisms of impairment.

Quantifying severity of aphasia is also an important part of rehabilitation assessment, as it provides a baseline from which response to treatment may be measured. Most standardized aphasia tests provide an overall severity index. Unfortunately, these tests also lack the necessary validity and reliability data to be employed as a means of quantifying treatment response. One notable exception is the Porch Index of Communicative Abilities (Porch, 1981). This measure is highly quantified, has undergone extensive psychometric development, and is suitable for use as a dependent measure for both clinical and research purposes.

Summary

Rehabilitation assessment and planning for neurogenic communication disorders is a process that requires thorough and systematic collection of biographical, medical, and behavioral data known to influence the nature, severity, and course of the communication impairment. These data are used by the speech-language pathologist to describe the underlying mechanisms of impairment, establish a differential diagnosis of the communication disorder, quantify its severity, and plan an effective program of rehabilitation.

References

Bayles, K. A. (1994). Management of neurogenic communication disorders associated with dementia. In R. Chapey (Ed.), *Language intervention strategies in adult aphasia* (3rd ed., pp. 535–545). Baltimore: Williams & Wilkins.

Bayles, K. A., & Tomoeda, C. K. (1993). *The Arizona Battery for Communication Disorders of Dementia.* Tucson, AZ: Canyonlands.

Bayles, K. A., Boone, D. R., Tomoeda, C. K., Slauson, T. J., & Kaszniak, A. W. (1989). Differentiating Alzheimer's patients from the normal elderly and stroke patients with aphasia. *Journal of Speech and Hearing Disorders, 54,* 74–87.

Benson, D. F. (1979). *Aphasia, alexia and agraphia.* New York: Churchill Livingstone.

Brookshire, R. H. (1986). *An Introduction to Aphasia.* BRK Publishers, Minneapolis, MN.

Brookshire, R. H., & Nicholas, L. E. (1994). Speech sample size and test-retest stability of connected speech measures for adults with aphasia. *Journal of Speech and Hearing Research, 37,* 399–407.

Canter, C. J. (1973). *Dysarthria, apraxia of speech and literal paraphasia: Three distinct varieties of articulatory behavior in the adult with brain damage.* Paper presented to the meeting of the American Speech and Hearing Association, Detroit, MI.

Chawluk, J. B., Mesulam, M. M., Hurtig, H., Kushner, M., Weintraub, S., Saykin, A., Rubin, N., Alavi, A., & Reivich, M. (1986). Slowly progressive aphasia without generalized dementia: Studies with positron emission tomography. *Annals of Neurology, 19,* 68–74.

Cohen, J. A., Gelfer, C. E., & Sweet, R. D. (1980). Thalamic infarction producing aphasia. *Mount Sinai Journal of Medicine, 47,* 398–404.

Costello, J. M. (Ed.). (1985). *Speech disorders in adults: Recent advances.* San Diego, CA: College Hill.

Dabul, B. (1979). *Apraxia Battery for Adults.* Tigard, OR: CC Publications.

Damasio, A. R., Damasio, H., Rizzo, M., Varney, M., & Gersh, F. (1982). Aphasia with nonhemorrhagic lesions in the basal ganglia and internal capsules. *Archives of Neurology, 39,* 15-20.

Damasio, H., & Damasio, A. R. (1980). The anatomical basis of conduction aphasia. *Brain, 103,* 337-350.

Darby, J. K. (Ed.). (1981a). *Speech evaluation in medicine.* New York: Grune & Stratton.

Darby, J. K. (Ed.). (1981b). *Speech evaluation in psychiatry.* New York: Grune & Stratton.

Darby, J. K. (Ed.). (1985). *Speech and language evaluation in neurology: Childhood disorders.* New York: Grune & Stratton.

Darley, F. A. (1969a). *Aphasia: Input and output disturbances in speech and language processing.* Paper presented in dual session on aphasia to the American Speech and Hearing Association, Chicago.

Darley, F. A. (1969b). *Apraxia of speech: 107 years of terminological confusion.* Paper presented to the American Speech and Hearing Association, Chicago.

Darley, F. A., Aronson, A. E., & Brown, J. R. (1969a). Clusters of deviant speech dimensions in the dysarthrias. *Journal of Speech and Hearing Research, 12,* 462-496.

Darley, F. A., Aronson, A. E., & Brown, J. R. (1969b). Differential diagnostic patterns of dysarthria. *Journal of Speech and Hearing Research, 12,* 246-269.

Darley, F. A., Aronson, A. E., & Brown, J. R. (1975). *Motor speech disorders.* Philadelphia: Saunders.

DiSimoni, F. G. (1989). *Comprehensive apraxia test.* Dalton, PA: Praxis House.

DiSimoni, F. G., Darley, F. L., & Aronson, A. E. (1977). Patterns of dysfunction in schizophrenic patients on an aphasia test battery. *Journal of Speech and Hearing Disorders, 42,* 498-513.

Duffy, J. R. (1995). *Motor speech disorders: Substrates, differential diagnosis, and management.* St. Louis, MO: Mosby.

Gerson, S. N., Benson, D. F., & Frazier, S. H. (1977). Diagnosis: Schizophrenia versus posterior aphasia. *American Journal of Psychiatry, 134,* 966-969.

Goodglass, H., & Kaplan, E. (1983). *The Boston Diagnostic Aphasia Exam.* Lea and Febiger, Philadelphia.

Gordon, B., & Selnes, O. (1984). Progressive aphasia "without dementia": Evidence of more widespread involvement. *Neurology, 34,* 102.

Hadar, U., Jones, C., & Mate-Kole, C. (1987). The disconnection in anomic aphasia between semantic and phonological lexicons. *Cortex, 23,* 505-517.

Halpern, H., Darley, F. L., & Brown, J. R. (1973). Differential language and neurologic characteristic in cerebral involvement. *Journal of Speech and Hearing Disorders, 38,* 162-173.

Hartman, D. E. (1988). Referral for speech-language pathology. In D. E. Yoder & R. D. Kent (Eds.), *Decision making in speech-language pathology* (pp. 166-167). Philadelphia: Decker.

Heath, P. D., Kennedy, P., & Kapur, N. (1983). Slowly progressive aphasia without generalized dementia. *Annals of Neurology, 13,* 677-687.

Hier, D., Davis, K., Richardson, E., & Mohr, J. (1977). Hypertensive putaminal hemorrhage. *Annals of Neurology, 1,* 152-159.

Howard, D., Patterson, K., Franklin, S., Orchard-Lisle, V. & Morton, J. (1985). Treatment of word retrieval deficits in aphasia. *Brain, 108,* 817-829.

Irigaray, L. (1973). *Le language de dements.* The Hague, Netherlands: Mouton.

Johns, D. F., & LaPoint, L. L. (1976). Neurogenic disorders of output processing: Apraxia of speech. In H. Whitaker & H. A. Whitaker (Eds.), *Studies in neurolinguistics* (Vol. 7, pp. 161-200). New York: Academic.

Kitselman, K. (1981). Language impairment in aphasia, delirium, dementia, and schizophrenia. In J. K. Darby (Ed.), *Speech evaluation in medicine* (pp. 199-214). New York: Grune & Stratton.

Lesser, R. (1989). Some issues in the neuropsychological rehabilitation of anomia. In X. Seron & G. Deloche (Eds.), *Cognitive approaches in neuropsychological rehabilitation* (pp. 65-104). Hillsdale, NJ: Lawrence Erlbaum Associates.

McNeil, M. R. (1988). Aphasia in the adult. In N. J. Lass, L. V. McReynolds, J. L. Northern, & D. E. Yoder (Eds.), *The handbook of speech-language pathology and audiology* (pp. 738-786). Philadelphia: Decker.

McNeil, M. R., & Kennedy, J. G. (1984). Measuring the effects of treatment for dysarthria: Knowing when to change or terminate. *Seminars in Speech and Language, 5,* 337-358.

Mesulam, M. M. (1982). Slowly progressive aphasia without generalized dementia. *Annals of Neurology, 11,* 592-598.

Metter, E. J. (1985). *Speech disorders: Clinical evaluation and diagnosis.* New York: SP Medical and Scientific Books.

Mohr, J. P., Watters, W. C., & Duncan, G. W. (1975). Thalamic hemorrhage and aphasia. *Brain and Language, 2,* 2-17.

Naeser, M. A., Alexander, M. P., Helm-Estabrooks, N., Levine, H. L., Laughlin, S. A., & Geshwind, N. (1982). Aphasia with predominantly subcortical lesion sites: Description of three capsular/putaminal aphasia syndromes. *Archives of Neurology, 39,* 2-14.

Netsell, R. (1983). Speech motor control: Theoretical issues with clinical impact. In W. Berry (Ed), *Clinical dysarthria* (pp. 1-22). San Diego, CA: College Hill.

Netsell, R. (1984a). A neurobiological view of the dysarthrias. In M. McNeil, J. Rosenbek, & A. Aronson (Eds.), *The dysarthrias: Physiology, acoustics, perception, management* (pp. 1-36). San Diego, CA: College Hill.

Netsell, R. (1984b). Physiologic studies of dysarthria and their relevance to treatment. In J. C. Rosenbek (Ed.), *Seminars in speech and language: Current views of dysarthria* (pp. 279-291) New York: Theime-Stratton.

Netsell, R., & Daniel, B. (1979). Dysarthria in adults: Physiologic approach to rehabilitation. *Archives of Physical Medicine and Rehabilitation, 60,* 502-508.

Porch, B. E. (1981). *Porch Index of Communicative Ability.* Consulting Psychological Press, Palo Alto, California.

Rau, M. T. and Schulz, R. (1988). The psychosocial context of stroke-related communication disorders in the elderly: A social network perspective. *Seminars in Speech and Hearing, 9,* 167-176.

Reilly, F. E., Harrow, M., & Tucker, G. J. (1973). Language and thought content in acute psychosis. *American Journal of Psychiatry, 130,* 411-417.

Rosenbek, J. C., Kent, R. D., & LaPoint L. L. (1984). *Apraxia of speech: Physiology, acoustics, linguistics, management.* San Diego, CA: College Hill.

Scheull, H. M. (1972). *The Minnesota Test for Differential Diagnosis of Aphasia.* University of Minnesota Press, Minneapolis.

Schwartz, M., Marin, O., & Saffran, E. (1979). Dissociations of language functions in dementia: A case study. *Brain and Language, 7,* 277-306.

Wahrborg, P. (1991). *Assessment and management of emotional and psychosocial reactions to brain damage and aphasia.* San Diego, CA: Singular.

Wallesch, G., Kornhuber, H., Brunner, R., Kunz, T., Hollerbach, H., & Sugar, G. (1983). Lesions of the basal ganglia, thalamus, and deep white matter: Differential effects on language functions. *Brain and Language, 20,* 286-304.

Wertz, R. T., (1985). Neuropathologies of speech and language: An introduction to patient management. In D. F. Johns (Ed.), *Clinical management of neurogenic communicative disorders* (pp. 1-96). Boston: Little, Brown.

Yorkston, K. M., & Beukelman, D. R. (1981). *Assessment of intelligibility of dysarthric speech.* San Diego, CA: College Hill.

5

Dynamics of Rehabilitation Nursing

SHIRLEY P. HOEMAN

Many of the dynamic and unique roles nurses have enacted in rehabilitation in the past have prepared them for the roles which they will develop into the upcoming century. This chapter opens with an explanation of the nature and goals of rehabilitation nursing, followed by discussion of rehabilitation nursing as a professional specialty. The final portion of the chapter is devoted to two flourishing areas of practice for rehabilitation nurses, namely, case management and community based care.

HISTORICAL INFLUENCES ON REHABILITATION NURSING

Before the turn of the 20th century, nurses had been active across all levels of health-care delivery. Not only were nurses pioneers in case finding and care of children who had congenital or acquired disabilities, but they worked with a great deal of autonomy as key professionals in public health for surveillance and monitoring of persons with infectious, chronic diseases such as tuberculosis, and in occupational health.

Although the large medical hospitals and institutions had not been built yet, nurses worked in the handful of pediatric residential hospitals for children with "custodial" needs. Nurses were educated in their fundamental classes about ways to assist persons with chronic conditions, to function and manage daily activities in clients' homes, and to use community resources. However, rehabilitation was not recognized as a mainline medical specialty in health care.

Then, buoyed by events during the first part of the 20th century, the specialized practice of rehabilitation nursing flourished, as did other rehabilitation

---

SHIRLEY P. HOEMAN   Health Systems Consultation, 6 Camp Washington Road, Long Valley, New Jersey 07853; and School of Nursing, Fairfield University, Fairfield, Connecticut 06430.

Rehabilitation, edited by Goldstein and Beers. Plenum Press, New York, 1998.

disciplines. For example, nurses were actively involved both on the battlefields and in rehabilitation institutions as veterans returned home with disabling conditions following two world wars. Many nurses worked in Veterans Administration hospitals as that system developed methods and treatments which would promote veterans' reintegration into their family life and society. Rehabilitation nurses became experts in assisting persons with damage to the spinal cord or multiple injuries following traumatic events in war or daily life (Gritzer & Arluke, 1985; McCourt, 1993).

Rehabilitation nurses were foremost providers of the long hours of care for children in the "iron lung" who were affected during the poliomyelitis epidemic, an epidemic that captured the attention of the world. About this time, methods proposed by an Australian nurse and therapist, Sister Kenny, for treating poliomyelitis were receiving attention in the United States. Rehabilitation nurses teamed with therapists and physiatrists to provide therapeutics through exercises and hot packs, care where cure was not assured.

Rehabilitation nurses became the clinicians for persons who survived formerly fatal infectious diseases or trauma only to develop chronic, disabling conditions, and served as advocates for enabling persons to regain optimal levels of dignity, function, and independence. Rehabilitation nurses were the 24-hour companions and caregivers for clients in institutions such as Sister Kenny Institute in Minneapolis and Rusk Institute in New York City prior to broadened scope in physical and occupational therapy practices. Nurses were central in guiding individuals through activities of daily living, home assessments and trial living arrangements, wheelchair prescriptions and other equipment needs for home, education of families and individuals about managing bowel and bladder programs and medical treatments or procedures, and sexuality discussions (Hoeman, 1970). Nurse educators such as Stryker (1977) wrote the early rehabilitation nursing textbooks and training manuals for clients and families.

In mid-century, a number of seemingly unrelated events charted the direction of rehabilitation nursing and, indeed, influenced all of rehabilitation. For instance, legislation for vocational rehabilitation was drafted and major organizations and foundations were developed expressly to address rehabilitation goals. Examples were the Paraplegia Society and the National Foundation for Infantile Paralysis. Efforts in research and development produced a multitude of medical advances and technologies to improve the outlook for persons with disabilities; similar efforts resulted in discovery of the Salk vaccine for preventing poliomyelitis. Persons spoke about developing standards, increasing accountability, and measuring the quality of health care.

These events brought rehabilitation to a crossroads. The public directed questions at leaders of large, powerful national organizations, questions such as, "What will become of the March of Dimes now that the threat of polio has passed?" During organizational meetings held in free-standing rehabilitation institutes, staff asked, "Can rehabilitation survive as a specialty? Will third-party payment sources reimburse health teams for providing rehabilitative services to persons who have chronic, disabling conditions such as that following stroke, or to children with special needs (then called minimal brain dysfunction and school learning problems), or to persons with work-related injuries?" and so forth. These were questions people across the nation and in local communities asked and they opened a chapter of social inquiry about the value, commitment, and level of investment in rehabilitation in the United States long before the Americans with Disabilities Act (ADA) was conceived (Hoeman, 1970).

Rehabilitation as a Nursing Specialty

A mere quarter of a century ago, in 1974, rehabilitation nurses organized as a professional specialty. The Association of Rehabilitation Nurses (ARN) and the American Nurses Association (ANA) defined rehabilitation nursing as "the diagnosis and treatment of human responses of individuals and groups to actual or potential health problems stemming from altered functional ability and altered life style" (ANA & ARN, 1988, p. 4).

Today, there are more than 11,000 members of ARN, along with other nursing organizations with interests in rehabilitation, such as the Neuroscience Nurses' Association, the Rehabilitation Nurses Foundation (RNF), and the American Association of Spinal Cord Injury Nurses (AASCIN). ARN has developed national standards, scope of practice, core curriculum, and examination for nurses eligible for certification as either Rehabilitation Nurses (RN, CRRN) or as Advanced Practice Rehabilitation Nurses.

Rehabilitation nurses consistently collaborate as members of interdisciplinary teams, an approach that is a hallmark of the field of medical rehabilitation. Changes in the concepts for interdisciplinary team functions, such as innovations for product or program managers and forms of cross-training, are influenced by principles of nursing practice in rehabilitation settings. Uncontained by institutional boundaries, rehabilitation nursing practice extends well beyond hospitals and institutions in that nurses are present at nearly all levels and in models of health-care delivery as they seek ways to improve outcomes of care.

Goals of Rehabilitation Nursing

Goals of rehabilitation nursing include those to

1. assist persons with chronic, disabling conditions
2. improve their outcome by enabling them to
   —be active participants in reaching their maximum level of independent function
   —perform optimal amounts of self-care or self-directed care
   —maintain health, abilities, and well-being
   —develop effective coping strategies and adjust to an altered lifestyle
   —attain satisfaction with quality of life
   —make decisions or choices in concert with their lifestyle or preferences
   —manage access for optimal participation in society
   —prevent complications or further disability

Functional Health Patterns

Models have been suggested in the nursing literature as means for promoting optimal outcomes for persons in all levels of care. Functional health patterns are a useful model for rehabilitation nursing because they cover all aspects of a person's

TABLE 1. MODIFIED FUNCTIONAL HEALTH PATTERNS

Skin integrity	Sensory functions	Breathing
Circulation	Elimination	Eating
Regulation and control	Sleep	Cognition
Communication	Sensation and perception	Sexuality
Spirituality	Coping strategies	Self-care
Learning	Function	Activities of daily living

Note: Adapted from *The Elements of Nursing: A Core Curriculum* (3rd ed), by W. Roper, W. W. Logan, and A. Tierney, 1980, New York: Churchill Livingstone. Copyright 1980 by Churchill Livingstone. Also from *Manual of Nursing Diagnoses* (6th ed.), by M. Gordon, 1993, St. Louis, MO: Mosby-Year Book. Copyright 1993 by Mosby.

health in a holistic approach. The model contains provisions for each person's basic needs for maintaining life and for attaining high-level wellness and life satisfaction.

For instance, a rehabilitation nurse would assure that a person with dysphagia receives education about swallowing and special diets. However, the nurse would also work with the interdisciplinary team to monitor the person's nutritional intake, identify culturally acceptable foods, work with the client to balance fluid intake to promote elimination, evaluate dentures for fit and the ability to chew, monitor body weight, promote adjustment as an effective coping strategy, recommend relaxation and therapeutic exercises, or schedule rest periods to conserve energy as holistic interventions (Hoeman, 1996).

Because self-care or self-directed care is another concept reinforced by rehabilitation nurses, therapeutic interventions or activities are framed as short-term, long-term, or lifelong goals necessary for reaching a person's optimal level of independent functioning. Goals are set in collaboration with the person, family, community, and members of the interdisciplinary team. Mutually acceptable care and goals are planned with clients and families to assure sensitivity to cultural and ethnic differences, lifestyle preferences, and life satisfaction issues.

THE NURSING PROCESS

The nursing process is a scientific problem-solving process which is basic to nursing practice. A professional nurse applies the nursing process using a theoretical basis for practice, a problem-solving approach, critical thinking skills, and a holistic concept. The steps in the process are assessment, analysis, nursing diagnoses, planning, and evaluation.

ASSESSMENT

Assessment is not only a first step in the nursing process, it is an ongoing activity. Subjective assessments are built on observations, conversation, and intuitive understandings. Objective assessment includes data gathered from tests, examinations, procedures, laboratory results, and medically related information such as that concerning medications. Both objective and subjective assessments form the basis for one or more nursing diagnoses that relate closely with functional health patterns.

Assessment begins with a health history for physical and psychosocial areas and the person's (or caregiver's) current concerns. The nurse interviews a client and family while actively listening to their concerns and making professional observations. During subjective assessments cultural and life-span differences may be identified and spirituality and values may be recognized as factors in outcomes.

A family genogram and demographic data such as ethnic origins, family dynamics, composition, and lifestyle, and socioeconomic status are included. An extensive psychosocial history provides information about social situation and competencies, emotional or psychological responses, developmental status, cognitive development, and economic impact on health. In making an assessment with a client, a rehabilitation nurse uses the nursing process with specific attention to areas in which a person is at risk for health problems or has deficits, functional impairment, or altered status.

Time spent conducting a health history provides an opportunity for establishing mutual trust and rapport and for assessing how persons are adjusting to the situation. Individuals and family are encouraged to express their own goals and preferences, state their major concerns, and ask questions about their health or condition, and to receive answers to those questions.

Since nurses are responsible for around-the-clock care in institutions, and for ongoing supervision and monitoring of a client's health status in the community, they are accountable for assessments of a person's physical and medical status. In many instances, a rehabilitation nurse will be responsible for assessing how well a person is able to maintain an activity or level of function learned in therapy sessions when practiced at other times of day or night or when a person decides not to practice.

A person's physical assessment is directed in part by the history, but professional standards of care contain directives for nurses to continually assess a person for any changes, untoward developments, or improvements. Thus, health assessment is integrated with physical assessment of illness or dysfunction. Findings from the health history are supplemented by a physical examination, results from laboratory tests or procedures, results from specific questionnaires or tools (selected based on the person's problems), and direct observations or reports of functional abilities.

A wide range of instruments, measurements, tests or tools, and questionnaires are used in assessment. Rehabilitation nurses may use a client's responses to instruments or results from tests to validate or quantify psychosocial assessments. These comparisons are useful when data is gathered through subjective means or in areas such as cognition, coping effectiveness, or chronic pain, and with more subtle influences, such as social support. For instance, a person who demonstrates signs of ineffective coping may be asked to complete a depression scale, or an elderly client will be screened for cognitive impairments using nursing assessments and responses to questions on a short mental status test.

Rehabilitation nursing assessments include evaluation of the person's living situation, lifestyle choices for maximum independence, or the best ways for family members to learn how to perform procedures necessary in a person's care. Increasingly, assessments from one health professional are coordinated with evaluations made by other members of the team. In a truly collaborative approach, any member of the team is able to conduct a comprehensive intake evaluation, then refer complex or deviant situations to providers in the discipline in which the person will receive more expert attention (Hoeman, 1993).

On the other hand, persons with medical problems that require consultation with specialized providers may find their orders for treatment lead to complications, cause problems with other body systems, and contradict or ameliorate treatments ordered by other providers. Whenever specialized interventions are being provided without collaboration among providers, the provider may not be aware that the client has other physical or psychosocial problems that will be affected negatively by their treatment plan.

As part of a client's transition from an institution into the community, rehabilitation nurses assess the home environment or residence, the workplace, or school or day care center, as well as the arrangements for personal assistance. Data from the person's medical history, such as surgeries and illnesses, are integrated with information about the person's health maintenance and functional abilities. Key measures of functional health are based on self-care potential as evidenced by a person's ability to perform activities of daily living. A person's abilities to perform activities of daily living are indicators of the level, amount, type, and extent of assistance needed for optimal independence and lifestyle.

Many persons who have chronic conditions also have deficits or dual or multiple diagnoses that impact them in more than one of the functional health patterns. Multiple chronic conditions tend to increase in severity with age. A nursing assessment of the whole person, instead of only medical or body systems, benefits the client because of the continuity and consistency of the plan for care. Persons who have medical problems do not experience the effects of pathology apart from other body functions or other aspects of their lives.

Consumers and providers are becoming more aware of the effects of environmental and ecological factors on health outcomes. Researchers are asking questions about the relationships among health status or outcomes and a multitude of factors that influence a person's ability to reenter and become included in a community. For example, access, acceptability of services, community competence, available support systems, architectural and social-environmental barriers, toxic exposures, violent or threatening situations, transportation, emergency response systems, and opportunities to improve or utilize skills in education, vocation, and recreation pursuits are factors that are important to health.

A safe, accessible, and dignified environment is essential to well-being whether persons live at home, with family members, in group or boarding facilities, in nursing homes or assisted living arrangements, or otherwise. Rehabilitation nurses evaluate a client's lifestyle for maximum independence and inspect the situation for safety and to assure services are available for specific needs, such as supplying oxygen or electric generators for emergencies. Clients, family members, and caregivers who will use any assistive devices or adapted equipment are taught about safety, features, maintenance, and proper use.

Analysis

Following history and physical data collection, the nurse analyzes the data. Using a holistic approach, additional data may be collected before the nurse formulates an assessment. Content of the assessment logically leads to nursing diagnoses and/or an initial problem list that guide the plan.

Diagnosis

Nursing diagnoses logically flow from functional health patterns in the model. Once one or more diagnoses are formulated, the nurse is able to prepare an individualized plan and interventions and discuss it with the client and family. Examples of nursing diagnoses are Impaired Mobility, Altered Perceptions, and Impaired Urinary Continence.

Planning and Evaluation

In rehabilitation, the plan is comprised of goals mutually set by the person and the health provider according to short-term, long-term, or lifelong categories. The person and the family are participants. Only then does the nurse institute planned, purposeful interventions to improve outcomes. Evaluation and feedback are ongoing and systematic steps are built into the nursing process, with provision for systematic evaluation of outcomes.

A major role of rehabilitation nurses is to manage care by avoiding fragmentation, promoting continuity, and instituting preventive and health maintenance, even during tertiary levels of care. Nursing processes are designed to take maximum advantage of the person's strengths, resources, and opportunities in order to improve outcome for the whole person.

A Continuum of Care

Rehabilitation nursing has been from the onset a profession that extends practice and application across the continuum of care and promotes a person's maximum level of independent function and self-care. This professional discipline includes all levels of care, prevention, and intervention. Beginning with early detection and preventive behaviors, from the time a person encounters trauma or illness, perhaps enters the institutional setting, and finally transitions to home or community program, nurses are involved directly in the process. The practice of rehabilitation nursing extends from acute to long-term care settings, with persons in all developmental stages, ages, and diverse groups, and employs multiple levels of intervention with persons who have chronic or disabling conditions.

Nurses assess each person's status prior to admission (or decide whether they are to be admitted), during treatment at a facility, when scheduled to leave a program or institution, upon admission to services in the community or tertiary care center, and throughout the time of service. The older system of medical management referred to as discharge planning is held over from the times when leaving the hospital was a milestone and individuals had few options other than returning home or going to a nursing home facility. Today the system is known to extend from preadmission through the course of the person's course of health care.

Nurses have been active in preadmission programs whereby a nurse assesses a person for candidacy prior to admission to rehabilitation. The notion is that a person's plans for discharge begin even before admission. Using established criteria, a nurse meets with a client and family with a dual purpose. First, the nurse intends to predict a person's potential for achieving successful outcomes in a timely manner through early identification and rehabilitation services; second, the nurse conducts marketing of rehabilitation services so that eligible persons will be referred and will select the rehabilitation program. These are seeds of the case management role functions.

For example, with improving outcomes for clients in mind, nurses working in emergency, acute, and subacute care have come to rely on consultation from rehabilitation nurses about ways to prevent complications or long-term problems that if unattended would lead to further disability, for persons during the time they require acute treatment and procedures.

Preventing or reducing the severity and extent of problems with these types of primary interventions are distinctly client- or person-centered interventions that have enabled persons to focus energy and attention on their functional abilities rather than on treatment of additional problems or complications once the medical crisis has been ameliorated. When rehabilitation nurses educate families and other caregivers about ways to prevent problems for persons, such as ways to prevent damage to the skin, they include content about signs for early detection when problems do occur. The person who develops a pressure area of the skin, incontinence, or a contracture will have reduced, even lost, opportunity for functional independence regardless of how medically stable that person is.

This type of planning across the continuum of care is evident when persons receiving care in step-down units, subacute facilities, and long-term care settings are enabled to regain and maintain function with or without cure of their medical condition. In the subacute care example just mentioned, nursing expertise with medical procedures and treatments was combined with rehabilitative approaches so that when the person became medically stable, the least restrictive alternative was available. Because primary interventions were used in a tertiary setting, the person had opportunity to adjust to the situation and to reach optimal function without further disabilities.

One of the subtle understandings within rehabilitation nursing is that a person's gains often are small incremental steps which might pass unnoted in other situations. In fact, maintenance of function when losses are predictable actually may present a form of goal achievement. Clients, families, professionals, and community members who are able to reach this understanding have potential to respond in effective ways.

New Subspecialty Directions

One change within rehabilitation nursing itself has been the development of subspecialty areas. The relevance of subspecialties is evident in the number of thriving special-interest groups that have formed within the professional organization. These subspecialties include pediatrics, home health care including hospice, staff nursing, generalized nursing, gerontology, case managing, insurance nursing, education, administration, research, and pain management.

As care for clients becomes more complex, rehabilitation nurses have begun to move further into subspecialty or special-interest areas. Two of the fastest growing and most active subspecialties within rehabilitation nursing are home health care, including pediatric populations, and restorative care. Traditionally, nurses have been the provider of choice in the community and have made it part of their business to be knowledgeable about resources and services. Now rehabilitation and restorative nursing principles are becoming essential elements in home health care services, especially with the goal of maximizing functional outcomes for persons who live in a wide variety of community settings.

In fact, rehabilitation nursing is a growing professional presence in community and home health care; this presence is especially evident in services related to restorative care, functional abilities, and case management. Growth has been stimulated by increased numbers of persons streaming from acute care to the community "sooner and sicker." Rehabilitation nursing is also needed with same-day surgeries—persons who never enter the hospital facility, receive surgical interventions, and then return home. Initially, service providers in the community did not manage to meet the needs or prepare for the numbers of persons seeking care, the acute level of services, or need for continued assistance outside institutions. Subacute care units, assisted living situations, and similar programs now are options for persons who no longer require hospitalization, but who are not able to function independently in another setting.

A person's inability to manage outside the institution may be due to medical instability that ultimately may require skilled nursing care or it may be a matter of requiring assistance in a functional area. Many older persons could live in homes in the community if services for basic needs such as chores, food shopping or preparation, or assistance with bathing were provided. In many instances, however, barriers and needs in the person's home or residential environment, not factors associated with the person, prevent reentry into the community. Architectural, financial, social, safety, and other factors may become formidable barriers to independence.

As a group, providers who work in institutions or in primary care are unaware of the resources and services available in a community for assisting and supporting persons who have chronic, disabling conditions. As a result, utilization of community resources is less than optimal, clients do not meet their potential for independent function, and many are admitted unnecessarily to care facilities. Outcomes or goals for maximum independence seldom are achieved without clients and families becoming informed about the client's condition, the amount and kind of services needed, where to obtain care, what to expect, and how to manage and adjust to chronic, disabling conditions.

Restorative Care in the Community

Across the nation people are seeking a sense of belonging and community with others; they simply want to be home and interact. Isolation and lack of access to care are pervasive barriers to well-being. Rehabilitation nurses, advocates for their clients in the past and now, have played pivotal roles in assuring that clients acquire the access they need for care and services. These advocacy roles will have relevance in the next century.

For instance, in the community, nurses serve as rehabilitation specialists, case managers, home health nurses, hospice nurses, and advanced practitioners. Across the spectrum of a person's rehabilitation, nurses are involved in providing client-centered education, continuity, coordination, and care. As with other rehabilitation professionals, nurses are concerned with accountability and improving outcomes for persons with chronic, disabling, and developmental disorders.

Rehabilitation or restorative practice in home health care and community-based programs has been a nursing tradition and a bastion. Nurses have been instrumental in planning for a client to leave the hospital or rehabilitation institution for home, transitional setting, or long-term care facility. Around the world, the WHO commitment for community-based rehabilitation programs (CBR) and the national goals of Healthy People 2000 mesh with the Nursing Agenda for Health Care Reform. As national health services are reorganized, different labels and names may apply, but the need and desire for clients to make the transition to home and the community is consistent.

However, the advent of managed care has led nurses to reconsider their roles in providing restorative care, especially in community settings where therapists may be costly, too few in number, or interested in being assigned to complex cases. Ironically, it remains difficult for community agencies with professional nurses who provide restorative services such as range of motion exercises to receive reimbursement which often is awarded for therapists at additional cost and manpower or personnel loading.

There is a historical basis for this situation. As medical advances grew and acute-care hospitals were built, more nurses chose to work in the medical and technological environment provided in hospitals. An increased focus of content in basic nursing education was critical care and medical-surgical courses. Much of what had developed as rehabilitation nursing expertise in the war years and "polio days" was abandoned by nursing professionals, but it was not lost for those

TABLE 2. ESSENTIAL ELEMENTS OF PLANS FOR COMMUNITY-BASED CARE

—mutually agreed upon goals among person, family and caregivers, providers, and payers
—an appropriate referral considering the type, level, and amount of services needed
—a home and environmental assessment
—inclusion of the person and family with attention to their preferences and lifestyle
—allocation of resources and assistive devices or adapted equipment
—coordination with community services, vendors, or agencies
—support contingent on level of assistance and goals for functional independence

TABLE 3. COMMUNITY REHABILITATION COMPONENTS

Optimal level of independence, self-care, and function
Caregivers and attendants, family concerns and resources
Financial supports and barriers: life care plans
Home, environmental, and community assessment and plans
Resources in community and program options: all ages
        day care   early intervention
        respite   subacute, long-term care
        independent living   disability management programs
Transportation and access (rural and urban)
Diversity and cultural or ethnic groups
Life span issues, dual diagnoses and co-morbidity
Education, vocation, and recreation
Participant involvement of clients and family, as well as providers and purchasers

students who were becoming educated in occupational or physical therapy. As a result, nurses and therapists continue to sort out the various role functions in which each of the disciplines have mutual interests, especially in community settings.

Rehabilitation Nurses as Case Managers

Traditionally, rehabilitation nurses who held special interests in insurance and vocational training were active and highly visible within organizations such as ARN. The relationship of these special interests with rehabilitation nursing has historical roots which began following the institution of Medicare when the insurance industry recognized the unique role functions rehabilitation nurses performed and acknowledged their value by hiring nurses as specialists.

Today, this direction in rehabilitation nursing is realized in the nursing roles of case managers and insurance nurses. In these capacities rehabilitation nurses are able to continue to set an agenda for reform in health care, remain advocates for clients, and keep pace with changes in health care. Those rehabilitation nursing principles and processes which began with the conception of rehabilitation remain relevant and useful. Case management activities have been a key factor in promoting cost-effective care and services for persons as they move across the continuum of care toward their rehabilitation goals while retaining quality.

A nurse with rehabilitation expertise is an ideal case manager to coordinate resources with a person's needs and secure arrangements for access. In addition to working in collaboration with professionals and persons who comprise interdisciplinary teams, nurses who have expertise in specialty areas such as rehabilitation are learning how to serve as consultants, specialty experts, and referral resources for one another.

Case management as a concept and a nursing process cannot be separated from planning. The basis for case management is a systematic, comprehensive assessment of an individual's health status, resources, and limitations. The assessment includes interpersonal, environmental, and psychosocial elements which must then be projected over time.

For example, before a case management plan can be developed, the nurse assesses the person's level of function; the amount and type of assistance required is considered along with the family or other caregivers, the living conditions and environment, other support groups, and features of the family situation, among other factors.

The case management plan itself is a precise and detailed document which identifies client, family, and professional goals at various critical points in the person's projected needs. The plan contains information about who will provide or do what to meet the person's needs. Allocation of resources and services are scheduled so they will extend across the continuum of care. However, a well-thought-out plan also will present options and alternatives when the cost, availability, or utility of items become concerns.

No matter how well a plan is constructed, the implementation is crucial to success. Few professionals have as valuable an experience in coordinating and managing complex care models as do rehabilitation nurses. In fact, it has been said that case management is what rehabilitation nurses have done for decades, albeit informally. Armed with the medical systems knowledge, nursing process framework, payment system expertise, holistic approach, client advocacy, and client or family education experience, it is not surprising that rehabilitation nurses are sought as case managers.

The rehabilitation nurse may function as a case manager *internal* to the health system to promote more effective utilization and maximum benefits when a person enters a subacute or other residential setting. For instance, a nurse's attention to the person in the context of family and community during preadmission screening is integrated into the plan for care and discharge from subacute care. During the person's stay in the hospital or subacute setting, the nurse in the internal case manager role monitors the person for variances in terms of the subacute clinical or critical pathway.

External case managers are community-based nurses who have rehabilitation experience and work in conjunction with Medicare, insurance agencies, legal firms, or the like. Their charge is to manage resources to provide the standard of care for the most efficient cost-benefit. Fully informed and knowledgeable individuals and families are enabled to become active partners in the process and continuum of care. Individuals, families, and members of interdisciplinary teams in community programs or agencies will interface with external case managers.

Meeting goals for utilization and financial issues under Medicare and for various types of managed care is an activity in which success is dependent upon interdisciplinary team members' improving outcomes for persons during and beyond the subacute setting. To do otherwise invites a person's readmission, perhaps to a hospital, and complications or further impairment. Thus, the quality and appropriateness in interaction with and referral to an external case manager, coordinated by a rehabilitation nurse functioning as an internal case manager, is crucial.

One of the formats used by nurses who prepare external case management reports is the written *life care plan*. A life care plan is a complete recording of all professional services, caregivers, providers, equipment, medications, accessibility needs, treatments, supplies, and other things a person will need for a projected length of time, including lifelong care. The purpose or rationale is provided for each expense alongside the unit cost for the item and specifications or comments, such as preferred providers. It is important to emphasize that a life care plan does not exist independently from a more complete case management plan based on thorough assessment and that a plan will require reassessment when changes occur.

Another format for case management activities is a type of *critical or clinical pathway*, which began use as a cost and quality review for institutional settings. The idea of pathways is extending into community as providers seek ways to create a seamless system of care. Although many changes will occur as pathways are evaluated and refined, the general format of a pathway is arranged around lists of the individual's health-related problems correlated with expected outcomes and the resources anticipated to fulfill those outcomes. Each item on the list has been assigned an expected time in which the outcomes may be achieved and quality maintained; any over- or underexpenditure of the time is considered a deviation or a variance. Levels of acceptable tolerance are set for each item, for example, an 85% time variation.

Another portion of a clinical or critical pathway is prepared in the form of a table or chart that contains names of all activities, procedures, and treatments and the various professional disciplines the client can anticipate encountering based on their particular health problem list. Each potential encounter is entered on the pathway under the day of service on which it is anticipated to occur. This day is calculated according to the number of days which are allocated in the pathway as the appropriate and usual length of stay and the resources necessary for the problems on the client's list.

The notion is that by a given day, such as by Day 3 of services, a glance at a client's progress along a clinical pathway will indicate how their progress toward recuperation and discharge from the institution aligns with the preset estimates for the amount of time and resources needed, taking into account any standards for quality of care. The intent is that time and resources on the pathway are to be adjusted for individuals who are frail, have multiple problems or develop complications, or similar situations that may be considered variances. Often, a variation is used as an indicator of a need to review, or perhaps change, some area of practice. Clearly, clients who have one or more chronic, disabling conditions are at risk for having variations from the expected time and resources planned for a pathway. This potential for variances among certain populations is becoming both an economic and an ethical issue.

In one sense, the pathway documents are a form of business plans for a client and providers. The provider is responsible for recognizing how the individuality of a person may contribute to clinical variances and making adjustments within a plan or pathway. Not all situations are managed with the client as a priority. However, some providers have replaced the word "case" with the word "care" in an effort to acknowledge the person. The now worn phrase has caught attention: "No one wants to be a case and no one wants to be managed!"

RESOURCE MANAGEMENT

Services offered in the community through nursing agencies include skilled care, coordination of specialized professional care such as nursing or therapy or social work, supervision of home care aides or assistants or chore persons, monitoring of medications and procedures, coordination with vendors of equipment and supplies, management of types of services such as specialists in respiratory care or orthotics or hospice, education of caregivers and family members, and assessment of choices of services such as respite or adult day care or pain management. This system depends on accountability and feedback for effectiveness.

When multiple services and resources spread throughout the community are involved in a person's plan of care, the result is a complex network of separate and diverse providers and services that requires communication, management, and collaboration before it can be integrated into a useful system.

Perceptions of persons receiving services and family satisfaction are as important as utilization goals. One way to accomplish retaining personal preferences and meeting individual needs in case management is for the provider and client, with family as appropriate, to use rehabilitation principles to set mutually acceptable goals for short, long, and lifelong periods. Furthermore, when there are areas of dissonance or unresolved differences, parties involved may set contractual arrangements to manage the situation.

For example, a family may choose to use resources for a child in a manner that appears to the case manager to be too rapid a consumption and thus would not allow for resources to be available over the projected time for care. Contracting is a means of continuing the process while setting written responsibilities for participants and establishing limits. This type of negotiation also is a familiar function for nurses who have experience in community and rehabilitation practices.

EDUCATING AND ENABLING CLIENTS AND THEIR FAMILIES

In another role capacity, a rehabilitation nurse may educate other providers about types of impairments and medical conditions that are indicators that a client may be a candidate for rehabilitation. As a rule, health professionals have not been prepared to identify persons who may benefit from rehabilitation services. When a candidate is overlooked, the person has lost opportunity for attaining optimal functional levels and in some instances may be placed in a more restrictive environment or prevented from returning home.

Education is a primary function of a restorative nurse's role. Education is one of the more effective and efficient means of improving outcomes, but also tends to be underrated and overlooked on the list of things to be accomplished. When persons, family members, other caregivers, and members of the community are prepared through understanding about a person's condition, strengths, and needs for services or care they are able to respond in ways that preserve the person's dignity and promote the maximum level of functional abilities.

When family members are educated about performing a medical procedure for a person who cannot perform self-care, the person is educated about how to be more assertive in stating preferences for care. Similarly, the person is educated about ways to prevent problems, such as avoiding musculoskeletal complications following impaired mobility. Members of the community are involved; for example, members of a rescue squad team are educated about need for backup electricity for a person's mechanical respirator or a home care aide is educated about how to assist a person to transfer safely. Education is empowering and enabling.

Caregivers have a great deal of responsibility and tend to have heavy workloads and exposure to multiple stressors. Although the issue is too lengthy to dis-

TABLE 4. STEPS IN ESTABLISHING A HEALTH CONTRACT

Assess needs: clarify values and unmet needs
Establish mutually agreeable goals: Prioritize and set feasible goals
Explore strengths and resources: List assets, remove barriers, promote access, and resolve ineffective coping strategies
Implement the plan: Center on the person by being accountable, cost-effective, culturally relevant, and in line with goals
Evaluate the contract and decisions: Identify met and unmet goals with attention toward modifying, changing, or closing contractual agreement

Adapted from: Rehabilitation Nursing (2nd ed., p. 212), ed. by S. P. Hoeman, 1996, St. Louis, MO: Mosby-Year Book. Copyright 1996 by Mosby.

cuss in detail here, the nurse in the community includes evaluation of how effectively a caregiver and family members are coping with care of a person at home. At times, the nurse may be able to work with the family or may request other members of the interdisciplinary team to determine whether there are more efficient, less physically taxing ways of caring for the person.

For example, obtaining a mechanical lift or assistive devices may improve daily caregiving, or adding a wheelchair ramp may open access to the outdoors for both client and caregiver. Referrals for services such as respite care, counseling, support groups, and others may be needed. Community programs, churches, and other organizations are other resources.

ADVOCACY AND LEADERSHIP ROLES

Leadership roles for rehabilitation nurses have been activated in health education, health promotion, and early-detection functions. Nurses continue to educate clients and their families about prevention and maintenance, daily living, and procedures or techniques which are necessary for life satisfaction after a medical crisis or chronic condition is stabilized. They have gained notice as leaders who serve as health system advocates for quality of life for individuals and their families.

Families who are at home with persons who have impairments or altered function due to traumatic events, such as hemiplegia following a stroke, rely on nurses to teach them how to manage care and daily activities, as well as how to obtain access to services and resources. Rehabilitation nurses are advocates for preserving dignity for clients and proponents for persons' achieving the greatest degree of independence in self-care. In addition to providing content about how to perform techniques and procedures, the rehabilitation nurse is an educator of potentials and possibilities, along with assisting with adjustment to realities.

Unfortunately, the organized health system is driven by short-term profit goals rather than long-term person-centered goals. There is an old axiom that discharge planning begins on the first day of admission to an institution. This notion is carried through in, for example, the proliferation of critical or clinical pathways being developed for a large number of medical conditions. Clinical or critical pathways and similar predictor methods are useful in projecting the cost of meeting standards and providing care for persons depending on their age, overall condition, type of illness, and level of service. Although there is merit to the old axiom and to the model pathways, the process begins much earlier and continues throughout a health situation.

For instance, health researchers are beginning to link some categories or indicators of health risks with prevention, primary care, and early-detection behaviors which then can be extended all through the continuum of care, so as to correlate with rehabilitation outcomes. Basing discharge planning and/or clinical pathways on medical diagnoses alone creates serious problems for persons with multiple diagnoses and co-morbidity, dual diagnoses, and symptomatology or syndromes which have either no medical diagnosis or remain undiagnosed.

Consider that persons with chronic, disabling conditions have more need for services and resources that provide opportunity for maintaining function, life satisfaction, and preventive care than for medical treatments when cure is not an option. Consider further, how those persons who displayed symptoms prior to the development of diagnosis for Lyme disease or fibromyalgia, or those who experience symptoms for years before being diagnosed with multiple sclerosis, or those with recurrent symptoms, such as postpolio syndrome, may be disadvantaged by the tools of the system.

Other recipients of rehabilitation nursing practices are persons who comprise literally new populations of persons who are survivors of formerly fatal conditions, those who experience incurable chronic conditions, and those who choose to live to capacity with chronic or disabling conditions, be they school children with spina bifida or elders with arthritis, pulmonary disease, or the very old with co-morbidity.

In these instances, the functional health patterns mentioned previously offer more compassionate, appropriate, and efficient guides for management. The organized health system is remiss in not providing reimbursement for services when persons are in these situations or have these capacities. Rehabilitation nurses are qualified and poised to advocate for persons in this regard.

Future Directions

Global Trends

Global interest in rehabilitation nursing has only begun to emerge as more nations seek ways to manage with health-care advances and population changes reminiscent of the early days of rehabilitation in the United States. Chronic conditions, aging populations, and new aggregates of survivors will be challenges for many diverse nations. There is already a quest for ways to manage with less technology while promoting well-being and functional independence. Rehabilitation nurses will be called on to contribute globally in many directions over the next years.

Research Directions

In terms of research, a client's participation in rehabilitation is increasingly being regulated through preadmission assessments, discharge planning, and community reentry. As a result, researchers are gaining access to data which are increasingly controlled, valid, and reliable when used in evaluating outcomes for persons following rehabilitation. Increased quality and quantity of rehabilitation nursing research projects and studies in utilization of findings from research provide hope for clients and are a goal for the profession over the next decade.

Summary

Rehabilitation nurses collaborate with professionals from other disciplines as they work with individuals, families, and groups in the context of their culture and community; they educate and advocate for clients and their families while assisting them to manage care toward achieving optimal functional independence, dignity, and healthy outcomes.

REFERENCES

Gordon, M. (1993). *Manual of nursing diagnoses* (6th ed.). St. Louis, MO: Mosby-Year Book.

Gritzer, G., & Arluke, A. (1985). *The making of rehabilitation: A political economy of medical specialization, 1890-1980.* Berkeley, CA: University of California Press.

Hoeman, S. P. (1970). [Unpublished notes]. Memories of Sister Kenny Institute. Minneapolis, MN.

Hoeman, S. P. (1993). A research based transdisciplinary team model for infants with special needs and their families. *Holistic Nursing Practice, 7*(4), 63-72.

Hoeman, S. P. (Ed.). (1996). *Rehabilitation nursing* (2nd ed.). St. Louis, MO: Mosby-Year Book.

McCourt, A. E. (Ed.). (1993). *The specialty practice of rehabilitation nursing: A core curriculum* (3rd ed.). Skokie, IL: Rehabilitation Nursing Foundation.

Roper, N., Logan, W. W., & Tierney, A. (1980). *The elements of nursing.* New York: Churchill Livingstone.

Stryker, R. N. (1977). *Rehabilitative aspects of acute and chronic nursing care* (2nd ed.). Philadelphia: Saunders.

6

Geriatric Rehabilitation

JOAN C. ROGERS AND MARGO B. HOLM

INTRODUCTION

The role of geriatric rehabilitation in health care is to manage the residual effects of medical, psychiatric, and age-associated pathologies rather than the pathologies themselves. Rehabilitation begins with a functional assessment, that is, with measuring how pathology influences an older adult's ability to function. Historically, there has been substantial confusion about the meaning of the term *function*. It has been applied to body organs or systems, as in muscle function; body parts, as in the function of the upper extremity; task performance, as in eating functions; or role performance, as in homemaker functions. This lack of clarity about the meaning of function has obscured the role of rehabilitation and the focus of the functional assessment.

Attempts to clarify the construct of function, and to devise a conceptual model for rehabilitation, were initiated by Nagi (1969) and Wood (1975). Building on the work of these theorists, the World Health Organization (1980) endorsed a multidimensional model of chronic health conditions as exemplified in the *International Classification of Impairments, Disabilities, and Handicaps*. Accordingly, function is conceptualized in terms of impairment, disability, and handicap. Impairments occur at the level of organs or organ systems. An impairment is a loss or abnormality of an anatomic, physiologic, cognitive, or emotional structure or function. Examples are decreased muscle strength, loss of short-term memory, visual-perceptual dysfunction, and apathy. If an impairment is severe, it may lead to a disability, that is, an inability to carry out daily living tasks. Examples of disabilities are dysfunctions in feeding, dressing, preparing meals, or managing money. Social roles, such as homemaker or retiree, are comprised of tasks. When task performance is sufficiently dysfunctional,

---

JOAN C. ROGERS Western Psychiatric Institute and Clinic, 3811 O'Hara Street, Pittsburgh, Pennsylvania 15213.   MARGO B. HOLM Occupational Therapy Program, College Misericordia, 301 Lake Street, Dallas, Pennsylvania 18612.

Rehabilitation, edited by Goldstein and Beers. Plenum Press, New York, 1998.

handicap, that is social disadvantage, may result. Hence, the older adult with Parkinson's disease sees a neurologist for managing Parkinsonian symptoms and a rehabilitation specialist for managing the postural instability (impairment), the consequent inability to carry out household tasks (disability), and the resulting inability to fulfill the homemaker role and live independently in the community (handicap).

Measurement of function, which initiates rehabilitation, may occur at the level of impairment, disability, or handicap or may span several levels. The focus of the assessment depends on the dysfunction exhibited by patients and the specific goals of the rehabilitation program. One practical assessment approach begins with disability (Granger, Seltzer, & Fishbein, 1987). This yields a description of tasks that patients can and cannot perform. By analyzing the pattern of task abilities and disabilities, the rehabilitation specialist constructs a profile of the capabilities that enable task performance as well as the impairments that hinder it. For example, Mrs. Find, who has a traumatic brain injury secondary to a fall, completed all mobility, feeding, toileting, bathing, hygiene, and dressing tasks with ease. Bed making, dusting, sweeping, and preparing a light lunch also were readily completed. However, errors were made in organizing medications, interpreting a bill, and balancing a checkbook. This pattern of task performance suggests adequate motivation for task performance as well as the needed fine and gross motor capabilities. Deficits in medication and financial management, which involve tasks that have strong cognitive components, suggest cognitive impairment. Hence, disability assessment might be followed by a thorough neuropsychological evaluation to clarify the nature and extent of cognitive impairment.

Rehabilitation seeks to reverse the effects of impairment, disability, or handicap by correcting dysfunction, or to compensate for their effects by promoting function. For the initial assessment, instruments are needed that identify dysfunctions and provide data to guide intervention. Interim assessments are undertaken during rehabilitation to monitor the effects of the intervention and ascertain when function has stabilized and maximum benefit has been achieved. The discharge assessment captures the outcomes of rehabilitation. Because improvement is a primary goal of rehabilitation, the availability and use of functional assessment instruments that have the capability of detecting change is essential for demonstrating the effectiveness and efficacy of rehabilitation.

Assessment Methodologies and Their Limitations

Geriatric rehabilitation uses the full range of assessment methodologies available: testing (structured or unstructured), observation (naturalistic or performance testing), and questioning (interview or questionnaires). The primary mode of data collection tends to vary with the level of function that is being measured and this in turn is largely dependent on the nature of the phenomenon being measured. As a general rule, subject to many modifications, testing predominates for the evaluation of impairments, observation for disability, and questioning for handicap.

Observation of task performance yields data about the extent and nature of disability. Hypotheses about the nature of disability derived through observation are then investigated through testing to ascertain the extent and nature of the underlying impairment that is presumed to be causing the disability. Handicap is a more elusive concept than impairment or disability. It requires measurement of how patients interact with their environments to carry out their daily living activities. It is less practical to observe or test social role performance and questioning provides a feasible method for gathering data in this area.

As indicated, observation of daily living skills plays a pivotal role in the assessment strategy of rehabilitation. Admission, interim, and discharge measures may be taken by clinicians or trained evaluators. However, once patients leave rehabilitation, observation of task performance to obtain follow-up measures becomes difficult. Patients must either return to the rehabilitation site or clinicians must transport themselves and testing equipment to the discharge setting. The expense and inconvenience of these alternatives has fostered interest in using interviews and questionnaires to obtain postdischarge measures. If all data-gathering methods yielded comparable results, this plan would not be problematic. However, because a person's stated ability to perform tasks may differ substantively from observed performance of the same tasks, combining objective discharge measures with subjective postdischarge measures may obscure the progress made during rehabilitation (Collin, Wade, Davis, & Horne, 1988; Shinar et al., 1987). Disparity between objective and subjective test results increases as task complexity increases (Rogers, Holm, Goldstein, & Nussbaum, 1994). Further discrepancy may be introduced if proxies rather than patients themselves are used as respondents, as might occur when patients are language- or cognitively impaired or uncooperative. Patients tend to perceive their abilities more favorably than do proxies (Rubenstein, Schairer, Wieland, & Kane, 1984). There are no easy solutions to these problems, particularly in view of the evidence suggesting that performance tests conducted before discharge in the hospital yield results dissimilar to those conducted in the home shortly after discharge (Haworth & Hollings, 1979).

The rating dimensions on disability instruments generally use ordinal measurement. Accordingly, the assistance needed for a task may be graded as minimal, moderate, maximum, or total dependence. Use of such gross gradations makes it difficult to measure change because a substantial amount of change must occur before it is detected. In addition to lacking sensitivity to small changes, ordinal scales pose other problems. Ordinal measurement makes it possible to state that older adults who need maximum assistance at admission and moderate assistance at discharge have progressed during rehabilitation. However, it would be inappropriate to conclude that patients who progressed from needing maximum to minimal assistance improved twice as much as those who progressed from moderate to minimal assistance. The distance between points on ordinal scales is not equal as it is on interval scales. Unfortunately, there is a tendency to interpret ordinal scales as if they are interval, resulting in misleading conclusions (Merbitz, Morris, & Grip, 1989).

MEASUREMENT OF IMPAIRMENT

To perform tasks, persons must have the cognitive capacity to plan them, the affective ability to want to carry them out, the sensory capacity to obtain information relevant to tasks, and the motoric capability to actually carry them out. If cognitive, affective, sensory, or motoric capabilities are sufficiently impaired, and adaptive strategies have not been learned, task performance will be affected. The task performance deficits of older adults are typically caused by multiple impairments. An older adult may have difficulty walking due to reduced leg strength (motor impairment) which impairs balance, low vision (sensory impairment) which makes it hard to see where one is going, and forgetfulness (cognitive impairment) which impairs recall of cane use. Rehabilitation uses a multidisciplinary team of professionals to manage the extensive array of impairments that can impinge on task performance. The following discussion of impairments is not intended to be comprehensive. Rather, selected impairments are presented for the cognitive, affective, sensory, and motor dimensions of task performance and common instruments used within each of these dimensions are described.

COGNITIVE IMPAIRMENT

Evaluation of cognitive status may encompass consciousness level and mental status, with the latter including orientation, attention, memory, perceptual ability, psychomotor ability, judgment, problem solving, comprehension and use of language, complex reasoning skills, and mood and affect. Since dementia is strongly related to advanced age and is highly prevalent in the older population (Weissman, Myers, & Tischler, 1985), screening for cognitive impairment is generally included in the geriatric functional assessment.

The most widely used mental status instrument is the Mini-Mental State Examination (MMSE) (Folstein, Folstein, & McHugh, 1975). This 11-item test was devised to detect the presence and severity of cognitive impairment and monitor changes in cognitive status over time. It assesses orientation in time, immediate recall and short-term memory, and the ability to perform calculations, name objects, repeat a sentence, follow verbal and written commands, compose and write a sentence, and copy a polygon design. The test is administered by a trained examiner and provides for oral and performance responses. It is not timed and takes about 5 to 10 minutes to complete. The maximum score on the MMSE is 30 and the lower the score, the lower the patient's cognitive functioning. A score of 24 is indicative of cognitive impairment. The MMSE has been adapted for administration over the telephone (Roccaforte, Burke, Bayer, & Wengel, 1992).

In samples of psychiatric and neurologic patients, test–retest reliability with a 24-hour interval exceeded .89 and interrater reliability exceeded .82 (Folstein et al., 1975). Adequate correlations were found between MMSE scores and verbal IQ (.78) and performance IQ (.66) scores on the Wechsler Adult Intelligence Scale. MMSE scores also correlate with structural changes in the brain (Tsai & Tsang, 1979). Using a clinical diagnosis of dementia or delirium as the criterion, the sensitivity of the MMSE was found to be 87% and specificity 82% in detecting cognitive impairment (Anthony, Le Resche, Niaz, Van Koff, & Folstein, 1982). Tombaugh and McIntyre (1992) provide a comprehensive review of the psychometric properties of the MMSE.

Another commonly used mental status instrument is the Short Portable Mental Status Questionnaire (SPMSQ) (Pfeiffer, 1975). The SPMSQ is a part of the OARS Multidimensional Functional Assessment Questionnaire (G. G. Fillenbaum, 1988), a comprehensive geriatric interview tool for community-based assessment. It consists of 10 items that measure orientation to time and place, long-term memory of personal and historical information, and calculations. It is given by a trained interviewer and the person responds orally. Test administration requires about 10 minutes. The maximum score is 10 errors; 3 errors are indicative of impairment and 4 errors of dementia. Error scores are modified based on educational level and ethnic heritage.

The test–retest reliability coefficient with a 4-week interval was .82 in one elderly sample and .83 in another (Pfeiffer, 1975). In a sample of community-living older adults, test sensitivity was found to be as low as 50% and specificity better than 90% (Fillenbaum, 1980; Erkinjuntti, Sulkava, Wikstrom, & Autio, 1987). In another study, sensitivity was found to be 82%, but the SPMSQ could not adequately distinguish those who were mildly impaired from those who were intact or moderately impaired (Smyer, Hofland, & Jonas, 1979).

Affective Impairment

Assessment of affect may span negative feelings, such as depression and suicidality, which impair task performance, and positive feelings, such as personal control and self-efficacy, which enable task performance. Although depression is the most common psychiatric disorder in the elderly, it often goes unrecognized and untreated. The loss of interest in activities that accompanies depression severely hinders rehabilitation, which requires the active participation of patients.

Measurement of depression in the elderly poses some unique problems. First, depression may be confused with dementia, because it is often accompanied by subjective experiences of memory loss and cognitive impairment (Kahn, Zarit, Hilbert, & Niederehe, 1975). Second, somatic symptoms that are diagnostic for depression in younger persons, such as sleep disturbances, are common in nondepressed elderly persons. Third, questions that are appropriate for younger persons, such as those dealing with sexuality, may elicit defensiveness in the elderly. Although the Center for Epidemiologic Studies Depression Scale (Radloff, 1977), Zung Self-Rating Scale for Depression (Zung, 1965), and Hamilton Rating Scale for Depression (Hamilton, 1967) have been used in geriatric practice and research, attempts to remediate these problems have led to the development of geriatric-specific instruments.

The Geriatric Depression Scale (GDS) (Yesavage et al., 1983) was specifically designed for older adults. With this population in mind, the GDS is short, uses a simple yes/no response format, and is devoid of somatic items. The presence of depressive symptoms is indicated by a positive response on 20 of the 30 items and by a negative response on the remaining 10 items. The GDS takes about 5 minutes to complete. It may be self- or interviewer administered. Questions symptomatic of depression are allocated a score of one point and the scores are summed to obtain a total score. A score of 11 or greater is indicative of depression (Brink et al., 1982). A score of 11 had a sensitivity of 84% and specificity of 95% (Brink et al. 1982).

A high degree of internal consistency of the GDS is suggested by Cronbach's coefficient alpha of .94. Test–retest reliability with a 1-week interval was found to be .85 (Brink et al., 1985). Concurrent validity of the GDS was established through acceptable correlations with established instruments, namely, the Hamilton Rating Scale for Depression and the Zung Self-Rating Scale for Depression (Yesavage et al., 1983). Using the Research Diagnostic Criteria for Affective Disorders as classification variables, nondepressed subjects scored significantly lower on the

GDS than mildly and severely depressed subjects and severely depressed subjects scored higher than nondepressed and mildly depressed subjects. A correlation of .82 was obtained between the classification variable and the GDS (Yesavage et al., 1983). Subsequent studies have indicated that the GDS can be validly applied to the physically ill and those with mild to moderate dementia (Sheikh & Yesavage, 1986). Although the utility of the GDS for screening is well established, its capability for detecting changes in depression over time and following treatment has yet to be ascertained. A shorter, 15-item version of the GDS is available and uses a cutoff score of 5 (Sheikh & Yesavage, 1986).

Sensory-Motor Impairment

Adequate input from the visual, auditory, olfactory tactile, proprioceptive, and kinesthetic senses is required to initiate, monitor, and conclude task performance. For the most part sensory testing for older adults is comparable to that for younger adults, although greater attention is given to sensory functioning in real-life situations as opposed to laboratory or clinical conditions. For example, tests of visual and auditory acuity may be supplemented by questionnaires such as the VF-14, an index of visual impairment (Steinberg et al., 1994) and the Sent-Ident, a measure of hearing impairment (Erber, 1992) to examine application of the senses to daily living challenges.

Range of motion, manual muscle and reflex testing, and gait analysis are basic to the examination of motor function. As with sensory testing, testing procedures are similar for young and older adults. However, with the increased incidence of falls and task dependencies in later maturity, greater attention is given to the analysis of balance, gait, and dexterity.

The Performance-Oriented Assessment of Balance (POAB) and Performance-Oriented Assessment of Gait (POAG) (Tinetti, 1986) were designed to assess balance and gait characteristics in older adults. Items were selected that required no equipment, were easy to administer, reliable, and sensitive to change, and imitated the movements used in normal daily activities. The POAB assesses sitting balance, rising from a chair, immediate and delayed standing balance, balance with eyes closed, turning balance, balance when nudged on the sternum, neck turning, one-leg standing, back extension, reaching up, bending down, and sitting down. Movements are graded as normal, adaptive, or abnormal. The POAG rates initiation of gait; step height, length, symmetry, and continuity; path deviation, trunk stability, walk stance, and turning while walking. Gait parameters are graded as normal or abnormal.

Content validity was based on the expert opinion of bioengineers, orthopedists, neurologists, rheumatologists, and physical therapists. Interrater reliability calculated on 15 subjects yielded agreement of 85% on individual items (Tinetti, 1986). The instruments are useful for identifying persons at risk for falls (Tinetti, Speechley, & Ginter, 1988). Clinically, in addition to identifying maneuvers that patients are likely to have difficulty with when performing daily living tasks, the instruments suggest potential medical, rehabilitation, or environmental interventions to improve mobility.

A newer and simpler clinical measure of balance frailty is that of functional reach (FR) (Duncan, Weiner, Chandler, & Studenski, 1990). FR is defined as the "maximal distance one can reach forward beyond arm's length while maintaining a fixed base of support in the standing position" (Duncan, Studenski, Chandler, & Prescott, 1992, p. M93). FR is measured by having subjects reach as far forward as they can, without moving their feet, in a plane parallel to a yardstick fixed to a wall at shoulder height. Conceptually, reaching tasks simulate the age-sensitive leaning tasks typically used to assess postural control and measured in the laboratory with center of pressure excursion.

The intraclass correlation coefficient for test–retest reliability was .92 and for interobserver reliability .98 (Duncan et al., 1990). FR correlated acceptably with center of pressure excursion (.71). Additional evidence of concurrent validity was furnished by correlations with other measures of physical capacity, including walking speed (.71), life space (.71), tandem walk (.67), mobility skills (.65), one-footed standing (.64), and IADL (.66) (Weiner, Duncan, Chandler, & Studenski, 1992). In a sample of elderly community-dwelling male veterans, FR predicted recurrent falls independent of age, depression, or cognitive status (Duncan et al., 1992). FR has also been shown to correlate with balance changes in hospitalized male veterans ($N = 28$) undergoing physical rehabilitation (Weiner, Bongiorni, Kochersberger, Duncan, & Studenski, 1991).

The Timed Manual Performance Test (TMP) (Williams, Hadler, & Earp, 1982) consists of 27 manual tasks that simulate those encountered in daily living. Part one (TMP Doors) involves opening and closing nine types of door fasteners (e.g., padlock). These are fixed to small doors on a plywood board which is positioned vertically before the patient. Part two (TMP Table) consists of five items borrowed or adapted from the Jebsen–Taylor Test of Hand Function (Jebsen, Taylor, Trieshmann, Trotter, & Howard, 1969): copying a sentence, turning over index cards, picking up paper clips, pennies, and bottle caps and placing them in a container, stacking four checkers, and transporting kidney beans from one bowl to another using a spoon. For the TMP Doors, 9 opening and 8 closing actions are timed to the nearest second. For the TMP Table, 5 actions are timed to the nearest second for each hand for a total of 10 actions. The 27 item scores are summed to yield a total score. A short version of the TMP is available consisting of 5 door and 6 table items (Gerrity, Gaylord, & Williams, 1993).

High correlations have been found between TMP scores and dependency, with residential living site (e.g., community living, nursing home) as the dependency measure (Ostwald, Snowdown, Rysavy, Keenan, & Kane, 1989). In a cross-sectional study of elderly women, differences in TMP performance explained most of the variation in health care requirements (Williams et al., 1982). In a 12-month longitudinal study, TMP estimated changes in health-care intensity of residents of an intermediate care nursing home better than professional judgment (Williams & Hornberger, 1984). The TMP has also been shown to predict the need for additional long-term care services (Scholer, Potter, & Burke, 1990; Williams, 1987), hospitalization, and mortality (Williams, Gaylord, & Gerrity, 1994)

MEASUREMENT OF DISABILITY

Disability refers to dysfunction in performing daily living tasks. Disability measurement is a vital element in the medical management and rehabilitation of older adults. An older adult's disability status is a marker of health, a determinant of the ability to live independently, and an indicator of the need for and use of health-care services. Theoretically, the measurement of daily living activities should encompass all tasks that older adults routinely engage in. From a practical standpoint, however, disability measurement focuses on three categories of activities of daily living—functional mobility, personal self-care, and instrumental activities. Functional mobility involves the ability to move the body from one place to another. It includes moving in various directions (e.g., forward, backward) on different walking surfaces (e g., sidewalk, gravel) by walking or maneuvering a wheelchair, ascending and descending stairs, and transfers (e.g., commode, car). Personal self-care includes tasks associated with basic survival—feeding, toileting, bathing, hygiene, dressing, and communicating essential needs. Instrumental activities of daily living (IADL), involves tasks associated with managing a household, living independently in the community, working (e.g., part-time employment, volunteer jobs) and pursuing leisure interests (e.g., cleaning house, socializing). IADL tasks are more complex than mobility and personal self-care and require more highly integrated neuropsychologic and cognitive skills. About 23% of community-dwelling older adults report difficulty in mobility or personal self-care and about 27% report difficulty in IADL (Fitti & Kovar, 1987).

Disability Scaling Approaches

The most common scale used to measure disability is the level of independence displayed by the person during task performance. Ratings of independent, able, or no help from another person mean that a person can perform a task independently. Independence usually connotes that a task is completed safely and within reasonable time. Ratings of dependent, unable, or needs help from another person mean that a person cannot perform a task independently. Independence-dependence may be rated on a continuum, extending from no to total dependence, rather than dichotomously.

If help is needed to perform a task, the kind of assistance may also enter into the rating. By ranking different kinds of help in a hierarchy, ranging from help that is least assistive to help that is most assistive, finer gradations of task dependence can be obtained. Supervision, for example, is regarded as less assistive than verbal prompting and verbal prompting is less assistive than physical assistance. Using an assistive device to perform a task may reduce the level of independence depending on the particular rating scale; however, the need for nonhuman help implies greater independence than the need for human help. Instruments using a hierarchy of assistance scale provide useful information for planning interventions, since systematic reductions in the level of assistance reflect improvements in performance.

Even when tasks are performed independently, they may be executed with more or less efficiency. The time needed to complete tasks generally provides an overall index of efficiency. As the time needed to complete tasks increases, the efficiency of task performance decreases. However, in the presence of behaviors like impulsivity or manic symptoms, time becomes a poor parameter of efficiency, as speed is generally achieved at the expense of quality. The level of difficulty (e.g., no, some, a lot) with which a task is executed provides another index of task efficiency.

In the task-descriptive approach to measurement, each point on a scale is defined in terms of specific task behaviors. Dressing performance might be scaled as follows: locates, selects and dons appropriate clothing; locates and dons but cannot select appropriate clothing; dons but cannot locate or select appropriate clothing; cannot locate, select, or don appropriate clothing. A variant of task-descriptive scaling uses task analysis. With this method, tasks are segmented into their critical behaviors and each behavior is rated. A task analysis for dressing might consist of locating, selecting, and donning clothing. Performance on each behavior might be rated as dependent or independent. Each of these scaling approaches is illustrated in the disability instruments selected for presentation.

DISABILITY INSTRUMENTS

Until recently, the Barthel Index and Index of ADL were the predominant disability instruments used in rehabilitation practice and research. The Barthel Index (BI), also known as the Maryland Disability Index (Mahoney & Barthel, 1965), was designed to assess the gains made during rehabilitation by patients with neuromuscular and neuroskeletal disorders. The BI consists of three mobility (bed transfers, walking on a level surface or moving a wheelchair, ascending and descending stairs) and seven personal self-care items (feeding, personal grooming, toileting, bathing, dressing, bowel control, bladder control). Each item is rated as independent or with human help; the use of equipment is noted but does not reduce the score. For scoring, items are weighted based on assumptions about the time and effort required by the caregiver. The lowest score is 0 and indicates complete dependence; the highest score is 100 and indicates independence. A score of 60 is suggested as the pivotal point between independence and more marked dependence (Granger & Greer, 1976). Adaptations of the BI include expanding the content, extending the rating scale, and changing the weight given to various items (Collin et al., 1988; Granger & Greer, 1976; Granger, Albrecht, & Hamilton, 1979). The BI was designed to be rated by health-care personnel based on observation. It has been adapted for self-report and proxy report (McGinnis, Marymae, Seward, DeJong, & Osberg, 1986; Roy, Togneri, Hay, & Pentland, 1988), telephone interview (Shinar et al., 1987), and medical record review (Granger et al., 1979).

Interobserver reliability was reported as .99 (Collin et al., 1988) and test-retest reliability as .89 (Granger et al., 1979). Validity and reliability studies have generally used modified versions of the instrument. Collin and colleagues (1988) obtained a significant Kendall's coefficient of concordance ($W = .93$) between ratings obtained by four different data collection methods. High correlations were obtained between the BI and other measures of self-care status (Granger et al., 1979; Gresham, Phillips, & Labi, 1980) and clinical judgments of functional level (Wylie & White, 1964). The BI has also been shown to predict discharge status and death (Granger & Greer, 1976; Granger, Sherwood, & Greer, 1977; Wylie & White, 1964).

The Index of ADL (Katz, Ford, Moskowitz, Jackson, & Jaffe, 1963) was developed to measure disability in chronically ill and aging populations. Performance in bathing, dressing, toileting, transferring, continence, and feeding is rated. Each task is rated on a 3-point scale of independence. Ratings are determined by the most dependent performance observed over the prior two weeks Performance in the six tasks is summarized on an 8-point independence-dependence scale, which takes into account the number of dependencies and their relative complexity.

Katz and colleagues (1963) reported that interobserver differences in ratings occurred less than once in 20 evaluations. The hierarchy of task complexity used in the scoring system has been demonstrated empirically, with feeding and continence being less complex than transferring and going to the toilet and these tasks in turn being less complex than dressing and bathing (Katz & Akpom, 1976). Tests of concurrent validity yielded correlations of .50 between the Index of ADL and a mobility scale and .39 with a house confinement scale (Katz, Downs, Cash, & Grotz, 1970). Katz concluded that the Index of ADL predicted outcomes over 2 years as well as or better than measures of physical or mental function (Katz et al., 1970). Brorsson and Asberg (1984) reported a reasonable prediction of mortality from Index of ADL scores.

In 1983, the American Congress of Rehabilitation Medicine and the American Academy of Physical Medicine and Rehabilitation joined forces to devise a universal functional assessment tool. The goal was to develop a more comprehensive measure of disability with the expansion of functional mobility and personal self-care and the inclusion of communication and cognitive function. The resulting instrument, the Functional Independence Measure (FIM) (Granger, 1987), has become the standard for medical rehabilitation. It is a part of the Uniform Data System for Medical Rehabilitation and there is precedence for using it to establish standards for functional independence following rehabilitation (Long et al., 1994). The FIM was not designed for use with persons who have mental impairment only.

The FIM samples 18 critical tasks. Of these, 13 are motor tasks related to self-care (feeding, grooming, bathing, dressing upper body, dressing lower body, toileting), sphincter control (bladder and bowel management), mobility (bed, chair, wheelchair; toilet; and tub, shower transfers) and locomotion (walk/wheelchair, stairs). The remaining 5 are cognitive tasks related to communication (comprehension, expression) and social cognition (social interaction, problem solving, memory) (Linacre, Heinemann, Wright, Granger, & Hamilton, 1994). The FIM uses the type and amount of assistance required to perform tasks to construct a measure of the severity of disability and burden of care. Each task is scored on a 7-point scale. A score of 7 reflects complete independence; 6, modified independence which implies some delay, safety risk, or device usage; 5, supervision; 4, minimal assistance (subject 75%+ effort); 3, moderate assistance (subject 50%+ effort); 2, complete dependence requiring maximal assistance (subject 25%+ effort); and 1, total (subject 0%+ effort) assistance. The potential for detecting change in disability is increased by the 7-level rating scheme. However, the 7-level hierarchy readily converts to a 4-level one: independence (7), modified independence (6), modified dependence (5, 4, 3), and dependence (2, 1). The FIM requires direct observation of task performance and completion by a trained observer—clinicians, patients, or family members. It takes about 30 minutes to complete. The FIM has been used successfully with various neurorehabilitation populations including stroke (Czyrny, Hamilton, & Gresham, 1990), spinal cord injury (Davidoff, Roth, Haughton, & Ardner, 1990), and head injury (Byrnes & Powers, 1989). A telephone version has been developed (Jaworski, Kult, & Boynton, 1994).

Using the 4-level rating scheme, the intraclass correlation for total scores was found to be .86 at admission and .88 at discharge. The average Kappa statistic for each of 18 items was .54 (Hamilton, Granger, Sherwin, Zielezny, & Tashman,

1987). Later reliability testing of a revised instrument reported a total FIM score intraclass correlation of .95 (Hamilton & Granger, 1991). The FIM has been transformed from an ordinal to an interval scale through Rasch analysis and the structure of the motor and cognitive scales has been shown to be stable at admission and discharge (Linacre et al., 1994). Compared to other disability measures, this psychometric advantage allows the FIM to be used more accurately to analyze change. FIM admission scores are useful for predicting functional performance at discharge and length of stay in rehabilitation, although predictive capacity varies with type of impairment. Function in motor tasks was a more important predictor of length of stay than function in cognitive tasks. However, the unique contribution of cognitive function was evident in selected patient groups. Shorter stays were associated with greater cognitive function for patients with traumatic brain injury and lesser cognitive function for patients with stroke and neurologic impairments (Heinemann, Linacre, Wright, Hamilton, & Granger, 1994).

The prototype instrument for IADL measures is the Instrumental Activities of Daily Living (IADL) Scale constructed by Lawton and Brody (1969) to assess the instrumental competence of older adults. The following eight tasks are included on the scale: using the telephone, shopping, preparing food, performing housekeeping tasks, doing the laundry, using transportation, taking medications, and handling finances. The rating scale for each task employs a task descriptive approach with three to five levels of description depending on the task. For example, the competence levels for the laundry task are launders own items completely, launders small items, laundry must be done by others. Data for the ratings are obtained by a clinician from the best available informant. The items form an 8-point Guttman scale for women and a 5-point scale for men. For men, the food preparation, laundry, and housekeeping items are excluded.

Extensive reliability testing of the IADL Scale has not been undertaken. An interrater correlation of .85 was obtained by two raters using interview data. In terms of validity, moderate correlations were obtained between the IADL Scale and measures of pathology (.40), personal self-care (.61), mental status (.48), and behavior (.36).

The Functional Status Index (FSI) (Jette, 1987) was devised for the subpopulation of older adults with rheumatic disease. Hence, unlike the other instruments discussed, the FSI is a disease-specific measure. It focuses on self-reported functional performance in 18 tasks in 5 categories. The categories are mobility (walking, climbing stairs, rising from a chair), personal care (donning pants, donning and buttoning an upper extremity garment, bathing), home chores (vacuuming, reaching low, laundering clothes, doing yardwork), hand activities (writing, dialing a telephone), and social-role activities (performing a job, driving a car, attending meetings or appointments, visiting). Functional performance is conceptualized along three dimensions: dependence, difficulty, and pain. Dependence refers to the degree of help needed to perform a task. Difficulty implies a continuum of exertion ranging from easy to hard. Pain refers to the degree of discomfort or sensation of hurting experienced when performing the activity. Dependence is rated on a 5-point scale; difficulty and pain on 4-point scales. Dependence, difficulty, and pain increase as the score increases. To minimize daily fluctuations in performance, patients rate their average performance over the prior week. Because decreases in dependence, difficulty, and pain are frequently the goals of arthritis rehabilitation, the FSI seeks to increase sensitivity to change by expanding the dimensions of task performance rated.

Using data obtained from adults with rheumatoid arthritis, internal consistency coefficients ranged from .66 to .69 across all three dimensions, except for dependence in hand activities, which was .23. Using a 1- to 3-day retest interval, test–retest reliability coefficients of .72 to .88 were obtained, except for social-role activities which was .40. The range of interrater reliability coefficients was .64 to .89. Convergent validity of the FSI was ascertained by correlating the FSI dependence, difficulty and pain scores with measures of ARA stage of disease, ARA functional class, and physician estimates of disease activity and overall functional ability. As expected, correlations between the FSI and other functional measures were higher (.40–.49) than those between the FSI and the disease measures (.25–.32). Concurrent validity was also established in a sample of elderly patients recovering from hip fracture. Percent agreement between FSI self-report, dependence scale scores and performance test scores ranged from a low of 71% for stair use to a high of 95% for donning socks (Jette, 1987).

Because task performance requires adequate cognitive function, analysis of the pattern of deficits in daily living skills provides data useful for staging cognitive impairment similar to that of neuropsychological tests. The Functional Assessment Staging of Dementia (FAST) (Reisberg et al., 1984) is an interview measure developed to evaluate the deterioration in functional status that occurs with Alzheimer's-type dementia. It is an expansion of the functioning and self-care axis of the Brief Cognitive Rating Scale (BCRS) (Reisberg et al., 1983) and consists of seven major functional levels that correspond to the stages of the BCRS and the Global Deterioration Scale (Reisberg, Ferris, de Leon, & Crook, 1982). Levels 6 and 7 are divided into five and six substages respectively, thus yielding a 16-item ordinal FAST scale. The levels span instrumental and basic activities of daily living, making the FAST applicable to the continuum of normal to severely deteriorated adult functioning. A patient is scored at the highest level of deficit. Deficits that are unrelated to cognitive impairment are not scored. The FAST is useful for staging uncomplicated Alzheimer's-type dementia, describing the progression of dementia, distinguishing Alzheimer's-type dementia from other dementias, and identifying complications that deviate from the natural progression of the disease.

Reliability studies on the 16-level FAST yielded intraclass correlation coefficients of .87 for interrater agreement and .86 for intrarater consistency (Sclan & Reisberg, 1992). With samples of normal elderly and those with Alzheimer's-type dementia (GDS stages 2 through 6), a correlation coefficient of .87 was obtained between the FAST and the MMSE. Correlations ranging from .59 to .73 were obtained with various psychometric measures and .83 to .94 with various clinical assessments (Reisberg et al., 1984). The advent of newer techniques for assessing cognitive impairment in severely impaired persons who were previously untestable with conventional mental status and psychometric instruments has enabled the extension of validity to the final FAST substages. Correlations ranging between –.60 and –.79 were obtained between the FAST and subtests of a modified version of the Ordinal Scales for Psychological Development (OSPD) (Uzgiris & Hunt, 1975), whereas the correlation with the total OSPD score was –0.79. Thus, there is strong evidence of concurrent validity of the FAST throughout the entire range of cognitive deterioration (Reisberg et al., 1984; Sclan & Reisberg,

1992). Evidence of the validity of the ordinal pattern of functional decline was derived from a sample of 56 patients, 50 of whom demonstrated the predicted order of decline. Coefficients of reproducibility and scalability were .99 and .98 respectively. Although minor deviations from the pattern of decline may occur in uncomplicated dementia, major deviations in ordinality suggest alternative pathologic causes (Sclan & Reisberg, 1992).

Measurement of task performance has been hampered by the lack of performance measures for the more complex IADL. Recently, two instruments have been developed to fill this void. The Direct Assessment of Functional Status (DAFS) (Loewenstein et al., 1989) was designed to enable direct assessment of a broad spectrum of ADL and IADL in persons with dementia. Seven activity domains are evaluated. Content selection was based on precedence regarding functional assessment of older adults, expert judgment, feasibility of testing in an outpatient setting, and behaviors of clinical significance. It incorporates samples of time orientation, communication, transportation, financial management, shopping, grooming, and eating. Within each domain, basic and more advanced behaviors are included. Each behavior (e.g., telling time) is operationalized in subbehaviors (e.g., 3:00 P.M.) and each subbehavior is rated dichotomously as correct or incorrect. Summary functional scale scores are generated for the domains and individual subscale scores for the behaviors comprising a domain. The DAFS takes about 30 minutes to administer and yields a maximum score of 93 points; the transportation scale is optional and is not included in the total score.

Summary scale interrater reliabilities for elderly memory-impaired subjects ranged from .93 to 1.00 and test–retest reliabilities, with a 3- to 7-week interval between testings, ranged from .72 to .91. For the normal controls, interrater reliabilities were 1.00 except for road sign identification which was .99, and test–retest reliabilities ranged from .92 to 1.00. Interrater reliabilities for the individual subscales were comparable to those obtained for the summary scales. Significant correlations for memory-impaired subjects between the DAFS and the total Blessed Dementia Rating Scale (−.59) and a subscale limited to the functional behaviors on this scale (−.67) were used to establish convergent validity. Evidence of convergent validity was also obtained from comparisons of impaired and nonimpaired subjects on specific behaviors represented on individual subscales. Persons with dementia of the Alzheimer's type scored significantly lower than controls on the subscales, except for telling time, identifying coins, eating, and dressing and grooming.

Like the DAFS, the Structured Assessment of Independent Living Skills (SAILS) (Mahurin, DeBettignies, & Pirozzolo, 1991) was devised to assess functional status in older adults with dementia. It consists of standardized, quantifiable behavioral tasks that are frequently impaired in dementing disorders. Item selection was based on a review of existing instruments, relevance to functional competence, practicality of testing in a laboratory setting, and provision of a hierarchy of tasks within each domain. The 50 items are evenly divided in 10 subscales: fine motor skills, gross motor skills, dressing, eating, expressive language, receptive language, time and orientation, money-related skills, instrumental activities, and social interaction. The items are criterion-referenced and each is scored on a rating scale ranging from 0 (impaired) to 3 (normal), with a maximum total score of 150 points. In addition to the total score, scores are calculated for the 10 subdomains, a combined motor score (fine motor, gross motor, dressing, and eating skills), a motor time score, and a combined cognitive score (expressive language, receptive language, time and orientation, money-related). Scoring is based on accuracy and where appropriate time also contributes to the score. The SAILS takes about 1 hour to administer.

Interrater reliability on 10 subjects with Alzheimer's dementia was .99 for the total and motor time scores. test–retest reliability for 10 control subjects with a week interval between tests was .81 for the total score and .97 for motor time. Within the dementia group, all subscales except social interaction correlated significantly with the total score with correlations ranging from .55 to .94. Subjects with Alzheimer's dementia scored more poorly than controls for total, motor (including time), and cognitive scores. Scores were lower on all 10 subscales with the most impressive deficits manifested on expressive language, receptive language, time and orientation, money, and interaction. Significant correlations were revealed between the SAILS and the MMSE (.60), the Global Deterioration Scale (.69), the WAIS-R Verbal IQ (.73), WAIS-R Performance IQ (.76), and Full Scale IQ (.79). A battery of neuropsychological tests conducted on the dementia subjects yielded significant correlations between the total SAILS score and test of visuospatial construction, visual discrimination, and visual memory. The correlation between the SAILS and a measure of depression, the GDS, was not significant.

For older adults residing in long-term care, the Minimum Data Set for Nursing Home Resident Assessment and Care Screening (MDS) (Morris et al., 1991) is the primary instrument used for functional assessment and care planning. Completion of the MDS is mandated by the Health Care Financing Administration (HCFA) for skilled nursing facilities that provide nursing, medical, and rehabilitative services to Medicare and/or Medicaid recipients. Section E, Physical Functioning and Structural Problems, includes items on bed mobility, transfers, locomotion, dressing, eating, toilet use, personal hygiene, and bathing. Each item is rated on two scales for the 7 days immediately preceding the assessment. The ADL self-performance scale uses five independence-dependence gradations; interrater reliability coefficients of at least .75 were achieved. The ADL support scale employs four gradations to rate the extent of help in terms of no setup or physical help, setup help only, one-person physical assistance, and two-or-more-person physical assistance. For residents requiring cognitive support, there is a separate item concerning task segmentation requirements. The MDS provides the opportunity to record problems in body control (e.g., loss of balance, contractures) and the use of mobility appliances or devices. Rehabilitation potential and change in ADL function are also recorded. Continence is covered in Section F. When MDS data suggest the presence of clinical problems, such as delirium, pressure ulcers, or falls, care planning is triggered by reference to the Resident Assessment Protocols (RAPs), which provide guidelines for more fully examining the problem. The MDS is to be completed at admission, annually thereafter, and when a significant change in health status occurs.

Measurement of Handicap

Less attention has been directed toward the standardized assessment of handicap than impairment and disability. Ratings of work status become less significant in the postretirement years as measures of social integration. Living situation in terms of variables such as own home, assisted living, personal-care home, and long-term-care facility, provide global measures of community living status.

Characteristics of the physical environment have been prominent in environmental assessment, particularly as these relate to causing disability or handicap. Attention has been directed toward evaluating architectural barriers (e.g., stairs, door thresholds), nonfunctional product features (e.g., low commode seats, glare), and safety (e.g., clutter on walkways, frayed electrical cords). For those living alone or in restricted living situations, who are at risk for sensory deprivation, qualities indicative of adequate sensory and social stimulation are also assessed (Rogers, 1989). Although some environmental characteristics may be assessed in isolation, most take on meaning only in the context of how older adults use the environment. The presence of a grab bar on the bathtub is a safety feature, but if it is not used when entering and exiting the bathtub it does not serve this purpose. The complexity of the environmental assessment is underscored by the need to match personal qualities with environmental features to determine the "goodness of fit."

Although the influence of the social environment on task and social role performance has long been recognized, even less attention has been directed toward systematically assessing these features than has been given to the physical environment. The presence of an oversolicitous spouse may be noted, but the implications of this for function may not be appraised.

Future Prospects

Although the science of functional assessment has progressed considerably over the past few decades, much remains to be accomplished. Paramount directions for functional assessment research are conceptualizing an integrated view of function, developing new and refining existing instruments, and clarifying the utility of different data collection methods.

Integrated View of Function

The fundamental constructs of rehabilitation science have been delineated as impairment, disability, and handicap ensuing from medical, psychiatric, or age-associated pathology. To further advance the field, research is needed that examines these constructs in an integrated manner in various patient populations, and follows their trajectory as people move through adulthood and later maturity and accumulate age- and disease-associated biopsychosocial changes. This research would yield insight into the complex matrix of factors that contribute to well-being and disablement in older adults with disabilities.

Instrumentation

Disability instruments are commonly restricted to fundamental, essential, tasks. Unfortunately, there is no universally accepted list of these tasks, although consensus is greater for functional mobility and personal self-care tasks than it is for IADL tasks. An instrument that rates 6 tasks provides fewer opportunities to demonstrate disability than one that includes 12 tasks. Similarly, an instrument that concentrates on simple tasks as opposed to complex ones also provides less opportunity to detect disability. Even when the same tasks are included on instruments, they may not be operationalized in the same way. On the BI, for example, feeding independence requires the ability to cut foods, whereas on the Index of ADL dependence in cutting food is allowed. Hence, older adults who are unable to cut food would be rated as more functional on the Index of ADL than on the BI. These issues in instrument content and item operationalization have critical consequences when disability indices are used to determine eligibility for health or social services (Kane, Saslow, & Brundage, 1991; Travis & McAuley, 1990). Greater consensus is needed regarding the number and measurement of fundamental and essential daily living tasks and the role of personal preference in defining fundamental and essential tasks.

Instruments such as the DAFS and the SAILS reflect advances in applying performance testing techniques, previously restricted to mobility and personal self-care tasks, to the more complex IADL. Further, being devised for the dementia population, they reflect a new direction in disability assessment which has been heavily rooted in physical impairment to the neglect of mental impairment. These gains are somewhat offset, however, by the unnatural ambiance of the testing situation, including the provision of step-by-step commands and simulations of task performance. Further innovation is needed to approximate greater ecological validity. Once a feasible approach has been devised for measuring task complexity, the gender bias of the present generation of IADL disability instruments can be readily corrected through the addition of more masculine-oriented tasks (e.g., fixing a leaking faucet).

Measures of clinical progress are critical for allowing rehabilitation clinicians to gauge the effectiveness of their interventions as well as overall program effectiveness. Hence, improving the sensitivity of functional instruments to change is essential. Use of Rasch analysis to transform ordinal to interval scales, as has been accomplished with the FIM, warrants replication.

Task performance depends as much on the characteristics of the social and physical environment as it does on the abilities of the person. Standardized testing procedures require similar testing conditions at each time of testing so that any effects detected can be attributed to changes in task performance rather than changes in the environment. Controlling the effects of the environment is difficult. Some caregivers are more facilitative of independence than others and a change in caregiver from one testing time to the next may influence the amount of help received. Disability instruments are needed that capture environmental characteristics, so that any changes in task performance can be related to environmental conditions. New computer systems, such as the Observe software program (Bakeman & Gottman, 1986) hold promise in this regard, since they enable the simultaneous recording of patient–environment events.

Methodology

The current mandate to examine the effectiveness and efficacy of rehabilitation interventions, and the shrinking funds for health care services and research, combine to reinforce the use of less costly functional assessment procedures. Hence, it is likely that although testing and observation will continue to be the methods of choice during active rehabilitation, they will likely be replaced by interviews and questionnaires after discharge. To understand the risks to clinical care and research conclusions when data obtained through objective and subjective methods are combined, research is needed that examines the relative validities of these methods, the direction of bias from baseline measures, and the factors explaining this bias.

Summary

Functional assessment plays a pivotal role in the health and rehabilitative care of older adults. Functional status is predictive of health, independent living capacity, resource utilization, and death. Functional assessment of older adults may involve measurement of impairment, disability, handicap, or a combination of these dimensions of disablement. Cognitive, affective, sensory, and motor impairments are assessed to ascertain whether the substrates of task performance are present, impaired, or absent. Disability in occupations associated with functional mobility, personal self-care, home management, work, and leisure are evaluated to ascertain the influence of medical, psychiatric, and age-associated pathologies on the ability to carry out day-to-day tasks and routines. The effects of contextual factors, such as architectural barriers and societal attitudes, on task performance are assessed under handicap. Although the full range of assessment methodologies is used in geriatric rehabilitation, testing predominates for impairment, observation for disability, and questioning through interview or questionnaire for handicap. Selected instruments or procedures for appraisal of impairment, disability, and handicap were presented in this chapter. Cole, Finch, Gowland, and Mayo (1994), Ernst and Ernst (1984), Kane and Kane (1981), Lewis (1994), and McDowell and Newell (1987) provided compendia of instruments. Rehabilitation clinicians use the profile of impairments, disabilities, and handicaps to formulate clinical hypotheses about the nature and causes of disablement and to devise interventions to improve the functional performance of older adults. Salient directions for functional assessment research were delineated as conceptualizing a cohesive framework of disablement by articulating the complex transaction between impairment, disability, and handicap; developing and refining instrumentation by rethinking content, devising strategies for measuring complex tasks, inventing more sensitive scales, and incorporating contextual variables; and exploring the differential yield and care plans resulting from data obtained when the same functional constructs are measured by different methods.

References

Anthony, J. C., Le Resche, L., Niaz, U., Van Koff, M. R., & Folstein, M. (1982). Limits of the "Mini-Mental State" as a screening test for dementia and delirium among hospital patients. *Psychological Medicine, 12,* 397–408.

Bakeman, R., & Gottman, J. M. (1986). *Observing interaction: An introduction to sequential analysis.* Cambridge, England: Cambridge University Press.

Brink, T. L., Yesavage, J. A., Owen, L., Heersema, P. H., Adey, M., & Rose, T. L. (1982). Screening tests for geriatric depression. *Clinical Gerontologist, 1,* 37–43.

Brink, T. L., Curran, P., Dorr, M. L., Janson, E., McNulty, U., & Messina, M. (1985). Geriatric Depression Scale reliability: Order, examiner, and reminiscence effects: A critical review. *Clinical Gerontologist, 3,* 57–59.

Brorsson, B., & Asberg, K. H. (1984). Katz Index of Independence in ADL: Reliability and validity in short-term care. *Scandinavian Journal of Rehabilitation Medicine, 16,* 125-132.

Byrnes, M. B., & Powers, F. F. (1989). FIM: Its use in identifying rehabilitation needs in the head-injured patient. *Journal of Neuroscience Nursing, 21,* 61-63.

Cole, B., Finch, E., Gowland, C., & Mayo, N. (1994). *Physical rehabilitation outcome measures.* Toronto, Ontario, Canada: Canadian Physiotherapy Association.

Collin, C., Wade, D. T., Davis, S., & Horne, V. (1988). The Barthel Index: A reliability study. *International Disability Studies, 10,* 61-63.

Czyrny, J., Hamilton, B. B., & Gresham, G. E. (1990). Rehabilitation of the stroke patient. In M. G. Eisenberg & R. C. Grzesiak (Eds.) *Advances in Clinical Rehabilitation* (pp. 64-96). New York: Springer.

Davidoff, G. N., Roth, E. J., Haughton, J. S., & Ardner, M. S. (1990). Cognitive dysfunction in spinal cord injury patients: Sensitivity of the functional independence subscales vs. neuropsychological assessment. *Archives of Physical Medicine and Rehabilitation, 71,* 326-329.

Duncan, P. W., Weiner, D. K., Chandler, J., & Studenski, S. (1990). Functional reach: A new clinical measure of balance. *Journal of Gerontology: Medical Sciences, 45,* M192-M197.

Duncan, P. W., Studenski, S., Chandler, J., & Prescott, B. (1992). Functional reach: Predictive validity in a sample of elderly male veterans. *Journal of Gerontology: Medical Sciences, 47,* M93-M98.

Erber, N. A. (1992). An adaptive screening of sentence perception in older adults. *Ear and Hearing, 13,* 58-60.

Erkinjuntti, T., Sulkava, R., Wikstrom, J., & Autio, L. (1987). Short Portable Mental Status Questionnaire as a screening test for dementia and delirium among the elderly. *Journal of the American Geriatric Society, 35,* 412-416.

Ernst, M., & Ernst, N. S. (1984). Functional capacity. In D. J. Mangen & W. A. Peterson (Eds.), *Research instruments in gerontology* (Vol. 3, pp. 9-84). Minneapolis: University of Minnesota Press.

Fillenbaum, G. (1980). Comparison of two brief tests of organic brain impairment, the MSQ and the Short Portable MSQ. *Journal of the American Geriatrics Society, 28,* 381-384.

Fillenbaum, G. G. (1988). *Multidimensional functional assessment of older adults: The Duke older Americans resources and services procedures.* Hillsdale, NJ: Erlbaum.

Fitti, J. E, & Kovar, M. G. (1987). The supplement on aging to the 1984 national health interview survey. *Vital and health statistics, Series 1, No. 21.* DHHS Publication No. PHS 87-1323. Washington, DC: Public Health Services.

Folstein, M. F., Folstein, S. E., & McHugh, P. R. (1975). "Mini-Mental State," A practical method for grading the cognitive state of patients for the clinician. *Journal of Psychiatric Research, 12,* 189-198.

Gerrity, M. S., Gaylord, S., & Williams, M. E. (1993). Short versions of the Timed Manual Performance Test: Development, reliability, and validity. *Medical Care, 31,* 617-628.

Granger, C. (1987). *The Functional Independence Measure.* Buffalo: Research Foundation of the State University of New York.

Granger, C. V., & Greer, D. A. (1976). Functional status measurement and medical rehabilitation outcomes. *Archives of Physical Medicine and Rehabilitation, 57,* 103-108.

Granger, C. V., Sherwood, C. C., & Greer, D. S. (1977). Functional status measures in a comprehensive care program. *Archives of Physical Medicine and Rehabilitation, 58,* 555-561.

Granger, C. V., Albrecht, G. L., & Hamilton, B. B. (1979). Outcome of comprehensive and medical rehabilitation: Measurement by the PULSES Profile and the Barthel Index. *Archives of Physical Medicine and Rehabilitation, 60,* 145-154.

Granger, C. V., Seltzer, G. B., & Fishbein, C. F. (1987). *Primary care of the functionally disabled: Assessment and management.* Philadelphia: Lippincott.

Gresham, G. E., Phillips, T. F., & Labi, M. L. C. (1980). ADL status in stroke: Relative merits of three standard indexes. *Archives of Physical Medicine and Rehabilitation, 61,* 355-358.

Hamilton, B. B., & Granger, C. V. (1991). *A rehabilitation uniform data system.* (Research Grant Rep. No. G008435062). Buffalo: State University of New York, UDS Data Management Service.

Hamilton, B. B., Granger, C. V., Sherwin, F. S., Zielezny, M., & Tashman, J. S. (1987). A uniform national data system for medical rehabilitation. In M. J. Fuhrer (Ed.), *Rehabilitation outcomes: Analysis and measurement* (pp. 137-147). Baltimore: Brooks.

Hamilton, M. (1967). Development of a rating scale for primary depressive illness. *British Journal of Social and Clinical Psychology, 6,* 278-296.

Haworth, R. J., & Hollings, E. M. (1979). Are hospital assessments of daily living activities valid? *International Rehabilitation Medicine, 1,* 59-62.

Heinemann, A. W., Linacre, J. M., Wright, B. D., Hamilton, B. B., & Granger, C. (1994). Prediction of rehabilitation outcomes with disability measures. *Archives of Physical Medicine and Rehabilitation, 75,* 133-143.

Jaworski, D. M., Kult, T., & Boynton, P. R. (1994, Winter). The Functional Independence Measure: A pilot study comparison of observed and reported ratings. *Rehabilitation Nursing Research,* 141-147.

Jebsen, R. H., Taylor, N., Trieschmann, R. B., Trotter, M. J., & Howard, L. A. (1969). An objective and standardized test of hand function. *Archives of Physical Medicine and Rehabilitation, 50,* 311-319.

Jette, A. M. (1987). The Functional Status Index: Reliability and validity of a self-report functional disability measure. *Journal of Rheumatology, 14*(Suppl. 15), 15-19.

Kahn, R. L., Zarit, S. H., Hilbert, N. M., & Niederehe, G. (1975). Memory complaint and impairment in the aged. *Archives of General Psychiatry, 32,* 1569-1573.

Kane, R. A., & Kane, R. L. (1981). *Assessing the elderly.* Lexington, MA: Lexington Books.

Kane, R. L., Saslow, M. G., & Brundage, T. (1991). Using ADLs to establish eligibility for long-term care among cognitively impaired. *Gerontologist, 31,* 60-66.

Katz, S., & Akpom, C. A. (1976). Index of ADL. *Medical Care, 14,* 116-118.

Katz, S., Ford, A. B., Moskowitz, R. W., Jackson, B. A., & Jaffe, M. W. (1963). Studies of illness in the aged. The Index of ADL: A standardized measure of biological and psychosocial function. *Journal of the American Medical Association, 185,* 914-919.

Katz, S., Downs, T. D., Cash, H. R., & Grotz, R. C. (1970). Progress in development of the Index of ADL. *Gerontologist, 10,* 20-30.

Lawton, M. P., & Brody, E. M. (1969). Assessment of older people: Self-maintaining and instrumental activities of daily living. *Gerontologist, 9,* 179-186.

Lewis, C. B. (1994). *Clinical measures of functional outcomes "The functional toolbox."* Washington, DC: Learn.

Linacre, J. M., Heinemann, A. W., Wright, B. D., Granger, C. V., & Hamilton, B. B. (1994). The structure and stability of the Functional Independence Measure. *Archives of Physical Medicine and Rehabilitation, 75,* 127-132.

Loewenstein, D. A., Amigo, E., Duara, R., Guterman, A., Hurwitz, D., Berkowitz, N., Wilkie, F., Weinberg, G., Black, B., Gittelman, B., & Eisdorfer, C. (1989). A new scale for the assessment of functional status in Alzheimer's disease and related disorders. *Journal of Gerontology: Psychological Sciences, 44,* P114-P121.

Long, W. B., Sacco, W. J., Coombes, S. S., Copes, W. S., Bullock, A., & Melville, J. K. (1994). Determining normative standards for the Functional Independence Measure transitions in rehabilitation. *Archives of Physical Medicine and Rehabilitation, 75,* 144-148.

Mahoney, F. I., & Barthel, D. W. (1965). Functional evaluation—The Barthel Index. *Maryland State Medical Journal, 14,* 61-65.

Mahurin, R. K., DeBettignies, B. H., & Pirozzolo, F. J. (1991). Structured assessment of independent living skills: Preliminary functional abilities in dementia. *Journal of Gerontology: Psychological Sciences, 46,* P58-P66.

McDowell, I., & Newell, C. (1987). *Measuring health: A guide to rating scales and questionnaires.* New York: Oxford University Press.

McGinnis, G. E., Marymae, B. A., Seward, B. A., DeJong, G., & Osberg, J. S. (1986). Program evaluation of physical medicine and rehabilitation departments using self-report Barthel. *Archives of Physical Medicine and Rehabilitation, 67,* 123-125.

Merbitz, C., Morris, J., & Grip, J. C. (1989). Ordinal scales and foundations of misinference. *Archives of Physical Medicine and Rehabilitation, 70,* 308-312.

Morris, J. N., Hawes, C., Murphy, K., Nonemaker, S., Phillips, C., Fries, B. E., & Mor, V. (1991). *Resident Assessment Instrument training manual and resource guide.* Natick, MA: Eliot.

Nagi, S. Z. (1969). *Disability and rehabilitation.* Columbus, OH: Ohio State University Press.

Ostwald, S. K., Snowdown, D. A., Rysavy, S. D. M., Keenan, N. L., & Kane, R. L. (1989). Manual dexterity as a correlate of dependency in the elderly. *Journal of the American Geriatrics Society, 37,* 963-969.

Pfeiffer, E. (1975). A short portable mental status questionnaire for the assessment of organic brain deficit in elderly patients. *Journal of the American Geriatrics Society, 23,* 433-441.

Radloff, L. S. (1977). The CES-D Scale: A new self-report depression scale for research in the general population. *Applied Psychological Measurement, 1,* 385-401.

Reisberg, B., Ferris, S. H., de Leon, M. J. & Crook, T. (1982). The Global Deterioration Scale for assessment of primary degenerative dementia. *American Journal of Psychiatry, 139,* 1136-1139.

Reisberg, B., London, E., Ferris, S. H., Borenstein, J., Scheier, L., & de Leon, M. J. (1983). The Brief Cognitive Rating Scale: Language, motoric, and mood concomitants in primary degenerative dementia. *Psychopharmacology Bulletin, 19,* 702–708.

Reisberg, B., Ferris, S. H., Anand, R., de Leon, M. J., Schneck, M. K., Buttinger, C., & Borenstein, J. (1984). Functional staging of dementia of the Alzheimer's type, *Annals of the New York Academy of Science, 435,* 481–483.

Roccaforte, W. H., Burke, W. J., Bayer, B. L., & Wengel, S. P. (1992). Validation of a telephone version of the Mini-Mental State Examination. *Journal of the American Geriatrics Society, 40,* 697–702.

Rogers, J. C. (1989). The occupational therapy home assessment: The home as a therapeutic environment. *Journal of Home Health Care Practice, 1,* 73–81.

Rogers, J. C., Holm, M. B., Goldstein, G., & Nussbaum, P. D. (1994). Stability and change in functional assessment of patients with geropsychiatric disorders. *American Journal of Occupational Therapy, 48,* 914–918.

Roy, C. W., Togneri, J., Hay, E., & Pentland, B. (1988). An inter-rater reliability study of the Barthel Index. *International Journal of Rehabilitation Research, 11,* 67–70.

Rubenstein, L., Schairer, C., Wieland, G. D., & Kane, R. (1984). Systematic biases in functional status assessment of elderly adults: Effects of different data sources. *Journal of Gerontology, 39,* 686–691.

Scholer, S. G., Potter, J. F., & Burke, W. J. (1990). Does the Williams Manual Test predict service use among subjects undergoing geriatric assessment? *Journal of the American Geriatrics Society, 38,* 767–772.

Sclan, S., & Reisberg, B. (1992). Functional Assessment Staging (FAST) in Alzheimer's disease: Reliability, validity, and ordinality. *International Psychogeriatrics, 4*(Suppl. 1), 55–69.

Sheikh, J. I., & Yesavage, J. A. (1986). Geriatric Depression Scale (GDS) recent evidence and development of a shorter version. In T. L. Brink (Ed.), *Clinical gerontology: A guide to assessment and intervention* (pp. 165–173). New York: Haworth.

Shinar, D., Gross, C. R., Bronstein, K. S., Licata-Gehr, E. E., Eden, D. T., Cabera, A. R., Fishman, I. G., Roth, A. A., Barwick, J. A., & Kunitz, S. C. (1987). Reliability of the activities of daily living scale and its use in telephone interview. *Archives of Physical Medicine and Rehabilitation, 68,* 723–728.

Smyer, M. A., Hofland, B. F., & Jonas, E. A. (1979). Validity study of the Short Portable Mental Questionnaire for the elderly. *Journal of the American Geriatrics Society, 27,* 263–269.

Steinberg, E. P., Tielsch, J. M., Schein, O. D., Javitt, J. C., Sharkey, P., Cassard, S. D., Legro, M. W., Diener-West, M., Bass, E. B., Damiano, A. M., Steinwachs, D. M., & Sommer, A. (1994). The VF-14: An index of functional impairment in patients with cataract. *Archives of Ophthalmology, 112,* 630–638.

Tinetti, M. E. (1986). Performance oriented assessment of mobility problems in elderly patients. *Journal of the American Geriatrics Society, 34,* 119–126.

Tinetti, M. E., Speechley, M., & Ginter, S. F. (1988). Risk factors for falls among elderly persons living in the community. *New England Journal of Medicine, 319,* 1701–1707.

Tombaugh, T. N., & McIntyre, N. J. (1992). The Mini-Mental State Examination: A comprehensive review. *Journal of the American Geriatrics Society, 40,* 922–935.

Travis, S. S., & McAuley, W. J. (1990). Simple counts of the number of basic ADL dependencies for long-term care research and practice. *Health Services Research, 25,* 349–360.

Tsai, L., & Tsang, M. T. (1979). The Mini-Mental State Test and computerized tomography. *American Journal of Psychiatry, 136,* 436–439.

Uzgiris, I., & Hunt, J. McV. (1975). *Assessment in infancy: Ordinal Scales of Psychological Development.* Urbana: University of Illinois.

Weiner, D., Bongiorni, D., Kochersberger, G., Duncan, P., & Studenski, S. (1991). Does functional reach improve with rehabilitation? (Abstract). *Journal of the American Geriatrics Society, 39,* A10.

Weiner, D. K., Duncan, P. W., Chandler, J., & Studenski, S. A. (1992). Functional reach: A marker of physical frailty. *Journal of the American Geriatrics Society, 40,* 203–207.

Weissmann, M., Myers, J. K., & Tischler, G. L. (1985). Psychiatric disorders (DSM-III) and cognitive impairment among the elderly in a U.S. urban community. *Acta Psychiatrica Scandinavica, 71,* 366–369.

Williams, M. E. (1987). Identifying the older person likely to require long-term care services. *Journal of the American Geriatrics Society, 35,* 761–766.

Williams, M. E., & Hornberger, J. C. (1984). A quantitative method of identifying older persons at risk for increasing long-term care services. *Journal of Chronic Disease, 37,* 705–711.

Williams, M. E., Hadler, N. M., and Earp, J. A. L. (1982). Manual ability as a marker of dependency in geriatric women. *Journal of Chronic Diseases, 35,* 115–122.

Williams, M. E., Gaylord, S. A., & Gerrity, M. S. (1994). The Timed Manual Performance Test as a predictor of hospitalization and death in a community-based elderly population. *Journal of the American Geriatric Society, 42*, 21-27.

Wood, P. H. N. (1975). *Classification of impairments and handicaps.* Document WHO/ICDO/REVCONF/75.15. Geneva, Switzerland: World Health Organization.

World Health Organization. (1980). *The international classification of impairments, disabilities, and handicaps.* Geneva, Switzerland: World Health Organization.

Wylie, C. M., & White, B. K. (1964). A measure of disability. *Archives of Environmental Health, 8*, 834-939.

Yesavage, J. A., Brink, T. L., Rose, T. L., Lum, O., Huang, V., Adey, M., & Leirer, O. (1983). Development and validation of a geriatric depression screening scale: A preliminary report. *Journal of Psychiatric Research, 17*, 37-49.

Zung, W. W. K. (1965). A self-rated depression scale. *Archives of General Psychiatry, 12*, 63-70.

III

Assessment for Rehabilitation

7

Functional Assessment

MICHAEL MCCUE AND MICHAEL PRAMUKA

OVERVIEW

Functional assessment is a term which has various meanings dependent on the setting or context in which it is used. In the behavior modification/behavior therapy literature, functional assessment refers to a functional analysis of behavior including identification of antecedent conditions and consequences. For example, the term functional assessment has been used frequently in reference to identification of challenging behaviors in the classroom setting for special needs students (Demchak, 1993; Foster-Johnson & Dunlap, 1993; Kern & Lee, 1994; Tobin, 1994). This approach has also been used to assist in conceptualization and treatment of Attention-Deficit/Hyperactivity Disorder (Maag & Reid, 1994). Functional assessment is used synonymously with assessment of activities of daily living (ADL) in the literature on rehabilitation with the elderly. The term functional assessment has also been used to reflect functional outcome measures in various populations. For example, scales such as the Glasgow Coma and Outcome scales (Teasdale & Jennett, 1974) and the Functional Independence Measure (Forer, 1982; Granger, Hamilton, & Keith, 1986) have been identified as functional assessment scales that are used primarily to conduct program evaluation activities in rehabilitation facilities and to document patient change.

For the purposes of this chapter, functional assessment is seen as a clinical process conducted to identify the functional impact of disabling conditions. The purpose is to identify appropriate rehabilitation strategies directed at the impact of disability (rather than identifying neuropsychological "deficits"). An additional purpose is to increase the self-awareness of individuals with disabilities and their involvement in the rehabilitation process.

MICHAEL MCCUE   Center for Applied Neuropsychology, First and Market Building, 100 First Avenue, Suite 900A, Pittsburgh, Pennsylvania 15222.   MICHAEL PRAMUKA   James A. Haley Veterans Administration Hospital, 13000 Bruce B. Downs Boulevard, Tampa, Florida 33612.

Rehabilitation, edited by Goldstein and Beers. Plenum Press, New York, 1998.

In this chapter, we propose a definition of functional assessment, identify clinical approaches to functional assessment of cognitive disability, and address the utility of neuropsychological assessment in the functional assessment process, including a discussion of the ecological validity of neuropsychological evaluation. Finally, we present an example of a functional assessment of executive abilities using a simulation of everyday problem-solving ability.

Functional Assessment Defined

Functional assessment may be defined as "the analysis and measurement of specific behaviors that occur in real environments and are relevant to life or vocational goals" (Halpern & Fuhrer, 1984). Functional assessment is a clinical process which involves understanding the interaction between purposeful behavior and conditions imposed by the environment (e.g., people, rules, physical barriers, or schedules). Because demands placed on a person differ from one environment to another and from one task to another, functional assessment is always a highly individualized process.

Functional assessment is undertaken to determine the *impact* of disease or disability on behavior. Neuropsychological assessment, neurological evaluations, and psychiatric evaluations are primarily diagnostic and prognostic measures focusing on disability. In contrast, functional assessment, which relates to purposeful behavior such as maintaining a checkbook, is a measure of the degree of impediment experienced by the individual as a result of the disability. The *environmental specificity* and *goal directedness* of functional assessment separate it from other types of assessment. Put simply, functional assessment is the measurement of what persons can or cannot do (their strengths and weaknesses) in particular situations, under certain conditions, and in response to specific demands.

The goal of functional assessment is to identify obstacles to effective functioning. Obstacles occur as deficits associated with disease or disability (e.g., problems in prospective memory) interfere with the individual's ability to meet demands and conditions imposed by the environment in which the individual must function (e.g., function effectively in the face of employer expectations), the goal the person aspires to attain (e.g., return to employment after injury), or the specific task requirements of a particular situation the individual must master (e.g., manage multiple sales accounts). The challenge of functional assessment is not only to identify the individual's strengths and weaknesses, but to fully understand the demands and conditions of the environment in which the individual expects to function. The purpose is to delineate functional obstacles that are likely to occur so that rehabilitation intervention can be systematically applied.

Functional assessment should reveal information about assets and limitations or potential problem areas. When communicating findings, functional limitations should be conceptualized in behavioral rather than in diagnostic terms. For example, "an inability to dress independently" is preferable to "apraxia." When possible, active involvement of the individual in the assessment process should be encouraged to improve validity and reliability of information gathered, to enhance the individual's understanding of and investment in rehabilitation and to assist the person in increasing accuracy in self-appraisal and insight into strengths and limitations and consequent need for intervention or accommodation.

Approaches to Functional Assessment for Cognitive Disability

Functional assessment is accomplished through a variety of approaches with individuals who experience cognitive disability. Each of these assessment procedures can be used to evaluate how an individual might respond in the face of real-life problems and demands. These activities, unlike psychological tests, may require the individual to deal with multiple priorities, unforeseen circumstances, and interpersonal interactions. The following discussion will address direct observation in the natural environment, situational and simulated assessments, interviewing, and the use of rating scales and other mechanisms for eliciting observations on functional performance. The psychologist's expertise in the functional assessment process is uniquely valuable in interpreting and understanding complex performance in the face of cognitive demands.

Observational Procedures

Undoubtedly the most effective way of understanding an individual's functional capacity to perform a particular task or to respond to a specific environmental demand is to directly observe them in the process of actually performing the task in their natural environment over a period of time. The individual's ability to perform in his or her own environment may be markedly different from performance in other settings and should be assessed through direct observation. Furthermore, observations may provide more valid results than self-report measures in persons with impaired cognition that may interfere with their ability to accurately assess their own functioning.

Observations can be completed in the individual's home or community setting (e.g., school, workplace) depending on the questions to be addressed. In order to fully understand complex cognitive behaviors, observations should be made in the context of cognitive domains (e.g., memory, language), and should consider both areas of strength and those that pose difficulty. To make the best use of data from observations in the natural environment (and to allow for inferences about other functional abilities not directly observed), the *process* an individual engages in to meet environmental and task demands should be observed. Observation of process variables such as organizational abilities, communication style, and endurance may provide more relevant data than traditional measurement of productivity or task outcome. Observation in the natural environment also provides an opportunity to observe the individual's spontaneous use of compensatory strategies or accommodations, as well as identify possible problem situations that might easily necessitate such strategies.

Observing a person in his or her natural environment (e.g., home or workplace) risks confounding the individual's typical response or performance style. Effort must be made to minimize this by making the individual comfortable, and observing nonobtrusively. An evaluator might participate in activities around the individual, for example, converse with a spouse, rather than merely observe.

In addition to observing people in their natural environments, incidental observations of individuals in the context of therapy appointments or structured testing visits can yield valuable functional information. Behavior on breaks and in the waiting room may be more relevant to understanding functional capacities than behavior during standardized testing. The way the individual performs in structured versus unstructured time conditions, responds to feedback, seeks assistance, and changes behavior with increasing comfort in the setting have important functional implications and should be noted. Every opportunity should be taken to observe the individual's cognitive style in both formal and informal settings. Examples of situations in which relevant observations can be made include:

- calling to set up or cancel an appointment, or failing to call
- the way in which the individual gets to and from the appointment, including who accompanies them and how transportation was handled
- behavior in the waiting room, hallways, or break areas, including interactions with others
- activities during meal breaks
- behavior during structured activity, including work, classroom, or testing activities

During each of these activities, the evaluator has an opportunity to test hypotheses concerning the individual's cognitive strengths and limitations. The objective is to identify patterns of performance which suggest how problems or strengths in cognition may be contributing to handling of obstacles in the individual's everyday activities.

Form 1 lists some questions which can be used to guide observations during formal and informal activities.

SITUATIONAL ASSESSMENT

Situational assessment of functional abilities in simulated environments is also frequently used to assess prevocational capacities, specific vocational abilities, educational competencies, or ADL skills. This approach can be conducted in a rehabilitation workshop or hospital setting, in the patient's room, or in specially designed simulated environments. Difficulties such as cost and staffing requirements associated with direct assessment in the patient's home or workplace are minimized through the use of simulations. However, the individual's performance in a simulated setting may not reflect his or her actual capacity (or incapacity) to perform the task. For example, when assessing ADLs in a simulated kitchen environment, one must consider that in the person's kitchen, location of utensils and ingredients is very familiar. Finding necessary materials may be an overlearned, automatic process at home, whereas if the individual is required to perform in a simulated kitchen the cognitive demands of learning and remembering the location of materials become elements of the task. Failure to meet these cognitive demands (memory and learning), rather than the requirements of the task itself (cooking), may interfere with adequacy of performance and consequently affect the validity of the functional assessment. In this case, it may be falsely determined that the individual cannot prepare a meal and is therefore not independent in this aspect. Care must be taken to understand possible threats to validity of simulations in functional assessment. Despite potential confounds, simulations offer the opportunity to directly observe functional skills.

FORM 1. GUIDE FOR ELICITING FUNCTIONAL INFORMATION THROUGH INCIDENTAL BEHAVIORAL OBSERVATION

1. Did the individual attend to directions?
   Did he or she attend after cuing?
   Did he or she attend initially, then fade?
2. Were the directions understood?
   Could he or she explain the task back clearly?
   Was he or she able to demonstrate the task?
3. If the directions were not understood initially, what if anything, made the directions more clear?
   Repetition?
   Simplifying the language?
   Breaking directions down into smaller pieces?
   Providing direction in writing?
   Using pictures, graphics or diagrams?
   Demonstrating the directions?
4. Was the task begun independently, or was a cue needed to initiate?
5. Was the task begun prematurely or impulsively before directions were completed and understood?
6. After beginning, how long was he or she able to stay on task before requiring redirection?
7. Why was redirection necessary?
   Distraction by another activity?
   Self-distraction?
   Loss of interest or motivation?
   Confusion about how to proceed?
   Forgetting the instructions?
8. Was he or she able to generalize the instructions to a new situation?
9. Was he or she able to seek additional information that was needed?
10. Was he or she able to problem solve when confronted with uncertainty?
11. Was he or she able to persist with the task until completion?
12. Was the task completed in a timely fashion?
13. Was the task completed accurately?
14. Did he or she spontaneously use any accommodations or strategies to help himself or herself?
15. Was he or she able to detect his or her own errors?
16. Was he or she able to correct his or her own errors
17. Could he or she correct errors with feedback?
18. Was his or her overall rating of own performance accurate?
19. Was he or she able to identify what contributed to his or her difficulty if any?
20. Could he or she identify any strategies or approaches that would help?
21. How was the task completed on repetition?

FUNCTIONAL INTERVIEWING

Interviewing individuals and informants is another way of obtaining information about functional abilities and competencies in the natural environment without the sense of direct observation. Spouses, housemates, employers, coworkers, caregivers, and adult children can provide rich information concerning an individual's ability to perform a certain task, how an individual goes about performing a task, how the individual's environment is structured, what obstacles he or she encounters, and the supports the individual needs and has access to.

A functional interview is an extension of the information-gathering process that starts with collecting background information. The functional interview differs from the traditional clinical interview in its emphasis on the individual's current environment and the relationships between that environment and current problems. It also differs in its emphasis on exploring the interface between cognitive strengths and limitations and everyday life; its purpose is not to provide a historical view of social, vocational, educational, medical, or familial background. In order to conduct an interview that elicits truly functional information, it is critical to keep these two agendas separate. The goal of a functional interview is to provide specific information on what individuals can or cannot do. Examples of what might be obtained include a specific picture of daily life schedule and activities, statements detailing the impact of problems on everyday life experiences, and a list of previously attempted remedies.

When conducting a functional interview, information is usually available from many sources: the individuals themselves, various records, current or previous testing, and reports from family and friends. The type of information is often presented either as a deficit-diagnostic issue (memory problems, math learning disability), or as a description of everyday behavior or experience (class failure, slowness on the job). Both types of data provide valuable material for an interview. Presentation of deficits, impairments, or diagnoses should be followed with questions that explore their daily impact. In this strategy, no assumptions are made about the understanding or meaning a diagnosis or stated deficit has for an individual. The roles of deficits are discussed in terms of the individual's environment and personal goals. In some cases, this results in recognition that there is little daily impact of a deficit and therefore little need for further analysis or intervention. A lack of recognition of the impact, however, may indicate instead that the individual has little insight into the daily ramifications of the deficit, and may therefore suggest that future efforts explore the impact of the deficit in greater detail. Alternatively, discussion of diagnoses and deficits may elucidate significant connections to performance in everyday life and provide direction for intervention.

The second strategy is to use presenting complaints about life competence or everyday problems and move in reverse to the impairments behind the problems. Although the general issue of cognition is paramount in this discussion, such an approach makes no assumptions about the etiology of everyday competence. Issues in mental health, motivation, living circumstances, significant interpersonal relationships, financial status, and personal values may be sources of everyday problems resulting from cognitive problems.

Using both strategies in functional interviews, the examiner walks individuals through a careful investigation of what they know and believe about impairment and everyday competence. Daily life demands and possible obstacles to meeting these demands are explored from several directions, so that the interactions between impairment and environment become clearer for both the individual and the clinician.

The manner of questioning is important in that it may influence the response. For example, asking an informant whether an individual "can" perform a task may produce a different (and perhaps less accurate) response than asking if the person regularly "does" a task. Asking about *how* an individual goes about a task (the approach, strategies and accommodations used) can provide the neuropsychologist with information to make generalizations about other functional skills and future behaviors.

Rating Scales and Questionnaires

Rating scales and questionnaires represent a very common approach to quantifying functional behaviors and obtaining information from informants in a structured fashion. Rating scales may be completed by clinical staff or informants (e.g., teacher, job coach, employer) who have firsthand knowledge of the individual's functional skill level. A variety of rating scales for rehabilitation are available, such as the Functional Assessment Inventory (Crewe & Athelstan, 1984), Patient Assessment of Own Functioning Inventory (PAF) (R. K. Heaton & Pendleton, 1981), Patient Competency Rating Scale (PCRS) (Fordyce, 1983), Social Problem-Solving Inventory (SPSI) (D'Zurilla and Nezu, 1990), and the Functional Assessment Measure (FAM) (Hall, 1992). A review of additional scales for addressing the impact of cognitive disabilities can be found in Hall (1992).

With respect to the elderly, instruments such as the OARS-JADL scale (Duke University Center for the Study on Aging, 1978) have been commonly used in an attempt to quantify functional capacities through clinician ratings. A great variety of format and item content is seen in rating scales used with the elderly. Many of the scales focus primarily on basic ADL items including mobility, self-care, feeding, and hygiene; these scales may be less useful to the psychologist. Scales that focus on higher-level skills, such as instrumental activities of daily living (e.g., balancing a checkbook, planning a meal, responding to mail) are more relevant to questions typically posed to the neuropsychologist, and the use of such instruments may facilitate the functional assessment process.

Ecological Validity Studies

While not developed for this purpose, psychological and neuropsychological test results provide valuable preliminary information regarding functional abilities and limitations. The *ecological validity* of tests refers to the capacity to provide accurate and reliable information regarding an individual's ability to function in the natural environment. Mixed results are reported in the literature regarding the ecological validity of neuropsychological measures. At best, test data provide only very general information regarding an individual's functional abilities outside of the testing situation. Following is a brief review of that literature.

There is evidence that a significant relationship exists between results of neuropsychological assessment and vocational functioning in neuropsychiatrically disabled individuals. S. Heaton, Chelune, and Lehman (1978) found that in a sample of 381 subjects with mixed neuropsychiatric diagnoses, a comprehensive neuropsychological evaluation composed of measures of adaptive ability and personality which is commonly used in neurological and psychiatric diagnosis (Halstead–Reitan Battery) was able to discriminate between vocationally successful and unsuccessful (employed versus unemployed) subjects. Newman, Heaton, and Lehman (1978) followed up 78 patients with mixed neurological disorders (including head injuries, stroke, epilepsy, anoxia, slow-growing tumors, and poisoning) to determine vocational status six months after neuropsychological evaluation. Significant relationships between performance on the neuropsychological tests and employment/unemployment patterns, wage income, and indices of cognitive and perceptual work requirements were reported. A number of other studies have also demonstrated a strong relationship between neuropsychological tests and vocational outcome in persons with epilepsy (Dennerll, Rodin, Gonzalez, Schwartz, & Lin, 1966; Dikmen & Morgan, 1980; Schwartz, Dennerll, & Lin, 1968). In a study of 30 male medical and psychiatric patients, Morris, Ryan, and Peterson (1982) reported that performance on selected neuropsychological tests (Halstead Category

Test, Trailmaking Test, and the Average Impairment Rating of Halstead–Reitan battery) was able to identify patient success or failure on work samples with hit rates ranging from 67% to 77% correct. Studies that involve neuropsychological methods in relation to vocational outcome provide reasonable justification for the clinical use of these tests as a component of the assessment process in the vocational rehabilitation of individuals with cognitive disability.

There is also a modest literature on the prediction of functional outcome in patients with various neurological disorders. M. B. Acker (1986) reviewed ten studies that support a predictive relationship between neuropsychological measures and rehabilitation outcome. Of these, Mackworth, Mackworth, and Cope (1982) found that neuropsychological tests of complex verbal and motor coordination functions were most predictive of impairment in vocational and independent living outcome in head-injured persons. M. Acker (1982) reported significant correlations between scores on a battery of neuropsychological tests and outcome status as well as level of functioning in the community. DeTurk (1975) found that a summary score made up from a battery of neuropsychological measures was significantly correlated with outcome in self-care, social interaction, community, and vocational responsibilities ($r = .76$; $p > .001$). Several other studies found significant relationships between neuropsychological test scores and specific outcomes such as fiscal management (Wang & Ennis, 1985) and functional praxis (Baum & Hall, 1981; Lorenze & Concro, 1962; Willlams, 1967).

Ben-Yishay, Gerstman, Diller, and Haas (1970), utilizing multiple regression methods, obtained a .9-level multiple R for the relationship between their predictors and outcome measures: self-care, ambulation, and length of in-hospital stay. The authors concluded that the value of multiple psychometric tests far outweighed the contribution of clinical and demographic variables. Weintraub, Baratz, and Mesulam (1982) attempted to form associations between cognitive tasks (subtests of the WAIS, the Mattis Dementia Rating Scale and clock drawing) and a Record of Independent Living devised by the authors. They categorized ADL problems according to initiation, memory, and visuopractic skill components, finding that patterns of cognitive impairment paralleled patterns of ADL difficulties.

In a study of elderly hospitalized patients with either dementia, depression, or mixed neurological conditions, neuropsychological assessment was found to be a valid predictor of those activities of daily living that have a strong cognitive component (McCue, Rogers, & Goldstein, 1990). The significant predictive relationship between specific neuropsychological skill measures and the ability to perform higher-level, cognition-based daily living functions further supports the potential utility of neuropsychological tests for prediction of specific vocational outcomes.

Use of Neuropsychological Assessment to Generate Inferences about Functional Capacities

The nature of functional activities is such that any content area generally involves several cognitive, perceptual, and motor requirements. For example, as Weintraub and colleagues (1982) have pointed out, such activities as dressing have initiation, memory, and visuopractic skill components. Thus, the individual may have difficulty initiating dressing behavior, may forget where clothes are kept, or may don clothes in a way that is not spatially correct (e.g., putting on trousers backward). This multiability nature of most everyday activities makes it necessary to develop some model for relating those activities to level and pattern of cognitive skills. Ideally, the neuropsychological tests should predict the content deficit as well as the specific pattern of underlying neuropsychological deficits. For example, for a particular patient, tests should be predictive of difficulty with dressing because of visuopractic difficulties and not because of a memory problem regarding location of clothing or initiation of dressing behavior. Different patterns may be predicted for other patients.

When it is stated that neuropsychological test performance should be predictive of functional status, this does not mean that prediction should be solely to specific content areas. Obviously, the best predictor of how an individual should do at a particular content area such as dressing would be the level of specific task performance at some previous time. Thus, the best predictor of how well a patient dresses at occasion B would ordinarily be how well dressing was accomplished at occasion A. Rather, test performance should predict specific cognitive deficits that may have implications for performance in certain content areas (e.g., communication skills). Therefore, during treatment or disposition planning, it becomes possible to identify appropriate placements and develop particular accommodation and management strategies based on identification of particular problem areas and the cognitive deficits associated with them. Inferences that particular patterns of neuropsychological tests performance are associated with performance of functional activities are commonly made on theoretical grounds rather than on the basis of empirical demonstrations. A systematic understanding of the individual's cognitive strengths and weaknesses is likely to contribute significantly to clinical predictions regarding functional skills. It is understood that the underlying cognitive structures needed to perform any practical task may be extremely variable, depending on the specificity of the task demands.

One could argue that prediction of skill level of some specific activity can be best assessed by observing the individual perform the activity itself. There obviously is merit to that view, but neuropsychological tests may have the capacity to predict a variety of behaviors involving, for example, short-term memory and perceptual-motor coordination rather than predicting behavior in single specific content areas. Therefore, neuropsychological testing may provide a more efficient prognostic tool relative to assessing each specific skill. Furthermore, within the context of a specific skill, even slight variations in task requirements might make the difference between success and failure. Thus, while functional assessment is able to assess how an individual will perform a specific task in a specific situation, it is uncertain whether assessment of this skill can be generalized to other situations or to slightly different tasks. Neuropsychological tests, in consideration of their generic nature, may assist in understanding patients' abilities over a variety of functional tasks and situations.

Neuropsychological assessment for functional purposes then becomes a process of generating clinical inferences about what an individual is likely to be able to do and not do and about the specific support and accommodations needs the individual may require for task completion. These inferences or clinical hypotheses about real-world competencies proposed on the basis of psychological and neuropsychological tests require knowledge and expertise in three areas. The first is clear clinical knowledge of the skills and abilities that are being measured. For example, in order to make predictions about how difficulties in memory would interfere with fiscal management of personal affairs, one would need a thorough appreciation for the types, degree, and intensity of disorders of memory. Second, predictions are contingent on a clear understanding of what the tests measure, including the limitations of the instruments used. Such expertise includes an appreciation for the range of behaviors required for adaptive or intact performance, and strong interpretive skills for addressing difficulties or performance failures. Finally, an understanding of and appreciation for the demands of the situation or environment one is attempting to make predictions about is important. Performance on standardized tests must be understood in light of the demands of environment predicted concerned for the test to be functionally relevant. For example, the clinician must know that adherence to a medically advised diet requires a patient to be skilled in organization, planning, judgment, and initiation, as well as basic cooking and shopping.

Psychologists and neuropsychologists are typically well trained and experienced in the tests utilized and in understanding cognition. However, in order to generate sound hypotheses about functional abilities from psychological and neuropsychological tests, the psychologist or neuropsychologist must either develop knowledge and expertise in evaluating the demands of the environment, or obtain this information in the referral and dialogue process. This may be done by conferring with other professionals (e.g., occupational therapists) or by talking with informants familiar with the environmental and task demands faced by the patient. Without a clear sense of the outcome environment, test scores are of limited use in contributing to our understanding of how an individual will function in the work or everyday living environments.

Cognitive Task Analysis

In order to understand the impact of cognitive disability and the obstacles facing the individual, a functional assessment must identify the demands that will be placed on the individual in the environment. Although environmental demands or "task analysis" has long been used to describe the physical characteristics and demands of jobs, the functional assessment of persons with cognitive disabilities requires a parallel assessment of the task demands from a cognitive perspective.

A central idea in this activity is that the task or environment itself requires some level of ability in a specific cognitive area or areas. It is not the individual that is initially assessed, but the cognitive demands of the task for any person performing it. Once these demands have been determined, the specific strengths and weaknesses of the individual in question con be compared with the task demands. The question for the examiner is not "What can this individual do?" but, rather, "What does this task require of anyone to complete it successfully?" Because the activity here is specifically oriented to cognition, it is critical for the evaluator to have a good grasp of cognitive domains and how each is recognized in everyday environments.

In conducting such a cognitive task analysis, the examiner considers many factors that contribute to the demands of the situation. For example, the specific cognitive domains that are involved in the activity may be critical. The work of a physician may initially be rated as "high" on executive functions. In many clinic settings, however, the physician's work requires a very high level of judgment and problem solving, but is very low in organizing and planning; most of the daily activities are actually prescribed for the physician by the environment. The level of task complexity should also be considered. For any specified domain, the level of complexity may range dramatically. For example, attention to auditory input is critical to both receptionists and simultaneous translators, but is probably much more demanding and complex in the latter position.

Language processing demands should also be explored. For receptive language, the complexity of a communication (telephone numbers, familiar instructions, contract negotiation in a second language), and the conditions under which the communication will be received (calm, noisy, hurried, spoken versus written versus typed, etc.) should be understood. The degree of task structure is important. At issue here is determining how much structure is provided by the task or the environment itself and the level of structure that must originate from the individual performing the task. This may be different for the same job in different environments (a nurse conducting telephone interviews has been provided much more structure than a nurse doing home visits). One should consider what supports and assistance are available for training, crisis assistance, or guidance in the natural environment.

Employment has multiple reinforcers available to encourage ongoing work efforts, but across environments, they differ significantly in type, match to individual values, and recognizability. Environments should be evaluated for the frequency of reinforcement (daily or frequent to never), types of reinforcement (wages, perquisites, social contact, verbal praise, altruistic satisfaction, feedback from task, intrinsic satisfaction), and availability to individuals (provided on a regular basis, only upon request, contingent upon some behavior, etc.). Reinforcers may also differ in temporal value (short-term payoff versus long-term rewards) and in equity of distribution in the environment.

Applying structure to performing a cognitive task analysis may be beneficial. Form 2 presents a format for assessing cognitive demands. Although not comprehensive in detailing cognitive domains, the format outlines important areas to assess, particularly in the area of executive functioning. When using such a form, recall that the task and environment should be rated for anyone performing the task. Only after such an analysis has been completed should an individual's assets and weaknesses be contrasted with what the environment demands. Existing information sources such as the *Dictionary of Occupational Titles* (*DOT;* Employment and Training Administration, 1991; reference) and the *Occupational Outlook Handbook* (*OOH;* Bureau of Labor Statistics, 1996) can be utilized to provide many descriptions of tasks, jobs, and career fields. Although descriptions are oriented to physical and educational requirements, these are also excellent resources for beginning a cognitive task analysis for a job. Words used in the job descriptions to describe jobs at a nonphysical level (judges, chooses, selects, interviews, evaluates, communicates, etc.) can be used to understand what cognitive domains are taxed by the task, and at what level. For greater specificity, job descriptions or job analyses can provide important data for a cognitive task analysis.

FORM 2. COGNITIVE DEMAND ANALYSIS

Task/Job Title: _____   Date:

Location: _____   Evaluator: _____

Domains involved   *Level*

Organization/planning	NA	L	H
Problem solving/judgment	NA	L	H
Memory and learning	NA	L	H
Attention/concentration/vigilance	NA	L	H
Perceptual/motor demands			
Visual	NA	L	H
Auditory	NA	L	H
Tactual	NA	L	H
Language processing demands			
Oral comprehension	NA	L	H
Written comprehension	NA	L	H
Vocabulary	NA	L	H

Functional Assessment of Executive Abilities: A Simulation of Everyday Problem-Solving Ability

The naturalistic problem-solving simulation, the American Multiple Errands Test (AMET) (Aitken, Chase, McCue, & Ratcliff, 1993) is an illustration of how functional assessment may be accomplished using simulations of everyday cognitive demands.. The AMET was adapted from a test developed in the United Kingdom by Shallice and Burgess (1991a) which was designed to assess the ability of patients with brain injury to act effectively on their own initiative and to organize nonroutine activities in a real-life setting. It proved to be very sensitive to the executive dysfunction that caused serious difficulties in everyday life for their subjects who were of above-average intelligence and performed relatively well on a wide range of conventional neuropsychological tests. The AMET requires individuals to shop for a standard set of 6 items as quickly and cheaply as possible in an urban retail environment, while obeying 6 rules. They must also mail a postcard to the examiner containing information that can reasonably be obtained during the shopping trip and rendezvous with the examiner at a specified place and time.

Individuals are not told how to acquire the information and some task requirements are only implied (e.g., the need to buy a pen to write a postcard). A total of 16 subtasks can be identified (Form 3), each of which receives an error score of 0, 1, or 2 for successful completion, partial completion, or failure. In addition, subjects are scored for rule breaks (e.g., leaving the test site), inefficiencies (e.g., going into the same shop twice), and misinterpretations (e.g., thinking that all tasks must be completed before meeting the examiner at the designated time).

The task is explained to individuals before they enter the shopping area and each is given a list of the tasks to be performed and the rules to be followed. All of

Form 3. AMET Tasks

1. Buy a birthday card
2. Buy a head of lettuce
3. Buy a loaf of Italian bread
4. Buy a package of bar soap
5. Buy a postcard (with or without a stamp)[a]
6. Buy a pen or pencil[a]
7. Buy a stamp (if a stamped postcard is purchased credit is given here)[a]
8. Buy a newspaper[a]
9. Buy a pound of apples
10. Buy a pack of gum
11. Be at the bakery 15 minutes after the test begins
12. Mail the postcard[a]
13. Name of the store in the shopping area likely to have the most expensive item[b]
14. Price of a pound of tomatoes[b]
15. The name of the U.S. city predicted to reach the highest temperature tomorrow[b]
16. Yesterday's winning daily number in the PA lottery[b]

[a] Items are implied and not included on the task list.
[b] Items are "Relevant Information" to be written on the postcard.

the individual's money and other belongings are temporarily removed. They are then given a small sum of money. It is explained that only items bought during the test and the subject's own watch may be used to complete the assignments. Instructions are repeated upon arrival at the shopping area. Individuals are asked to paraphrase the instructions to insure understanding and are permitted to carry the list of tasks and rules with them. An examiner remains with them, observing and recording the individual's activity and debriefing them upon completion.

The AMET was modeled very closely after the original test and also appeared to be sensitive to cognitive dysfunction in a group of individuals with neurological and learning disorders who were known to have difficulty successfully completing everyday activities. An additional aim was to supplement the information provided by standard neuropsychological tests. The patient groups in this study had standard neuropsychological test scores that were indicative of deficits in higher-level cognitive skills. However, these tests did not significantly correspond with patient complaints of disability and clinically did not suggest the specific ways in which these deficits would be manifested in everyday activities. This information was considered to be essential for optimal rehabilitation planning, especially as it relates to independent living and vocational placement. The AMET was in fact very useful for these purposes, and provided useful clinical information regarding the individual's naturalistic problem-solving abilities; this information is essential for developing an accurate awareness of functional capacities and limitations, engaging in realistic goal setting, and for planning rehabilitation interventions.

In preliminary studies using the AMET (Aitken et al., 1993; McCue, 1995), the simulation was found to discriminate between groups of individuals with brain injuries and individuals with learning disabilities, both of which had documented limitations in everyday living and vocational functions, and a group of control subjects with no history of neurological, learning, or psychiatric disability. Individuals in the patient groups underwent components of the Halstead–Reitan Neuropsychological Battery, the WAIS-R, and the Paced Auditory Serial Addition test

(PASAT) following standard administration procedures. Individuals and informants also completed the Patient Competency Rating Scale (Fordyce, 1983), a self-report questionnaire designed to identify the functional impact of cognitive disability on everyday performance. Correlations between errors on the AMET and standard neuropsychological tests revealed no strong relationships between problem-solving tests such as the Category test ($r = -.03$); the WAIS-R (FSIQ, $r = -.23$; VIQ, $r = -.22$; PIQ, $r = -.14$); and the Trail Making test (Trails B, $r = .00$).

Regarding discriminating between normals and patient groups, performance on the AMET was significantly better for the normals than the two patient groups in all areas (Figure 1). On the 16 tasks that comprise the AMET, the performance of the patient groups differed significantly on ambiguous and implied tasks and on tasks that required integrating information from the environment into problem solving. Performance on concrete tasks involving specific purchases did not discriminate the groups. The total number of task failures, however, and the number of errors made were highly significant in discriminating the controls from the patient groups. The total time to complete the test and the number of misinterpretations made also differed between the groups. With the exception of total time, the normals performed best, the LD group next and the neurologically impaired group least effectively.

In order to explore the validity of the AMET for identifying functional problems encountered by patients, the relationship of AMET scores with patient (PCRS) and informant (RCRS) ratings was also examined. Total errors and task failure scores were compared with ratings of executive difficulties in scheduling

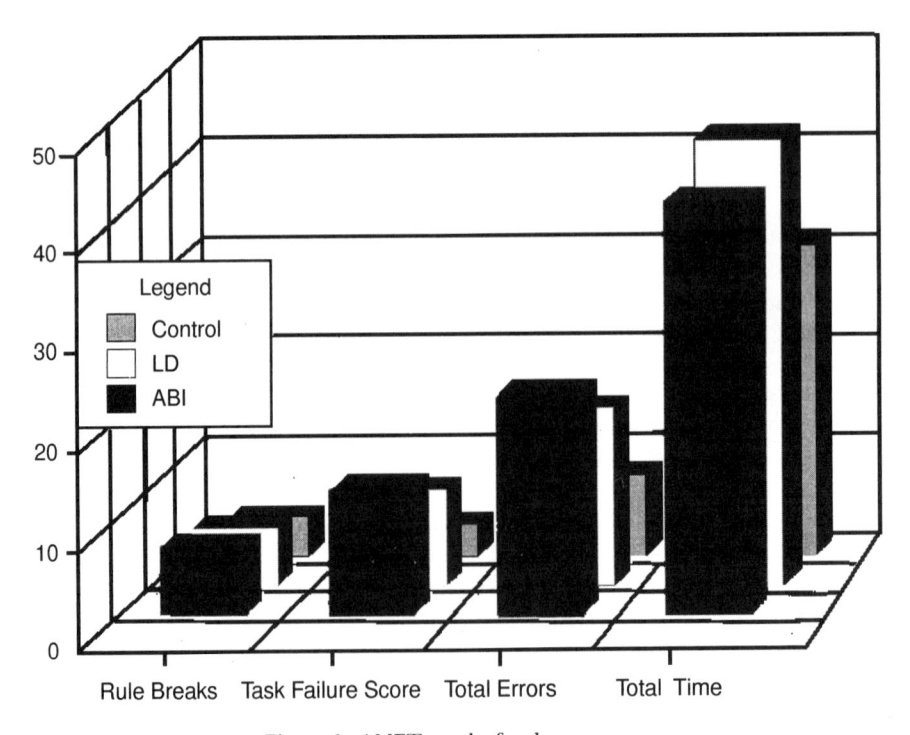

Figure 1. AMET results for three groups.

daily activities, consistently meeting daily responsibilities, keeping appointments on time, adjusting to change, and remembering daily schedules. Results indicated that the number of AMET task failures was highly correlated with informant ratings of functional abilities in keeping appointments ($r = -.70$), remembering a schedule ($r = -.53$), scheduling activities ($r = -.62$) and meeting responsibilities ($r = -.63$). Thus the AMET appeared to be sensitive to areas of executive functioning difficulty that were not identified in psychometric assessment.

SUMMARY

The problems encountered by individuals with cognitive disorders frequently relate to functional capacity. Functional assessment is required to make predictions regarding functional skills and limitations and to assist in identifying rehabilitation and independent living needs. Direct and simulated observation procedures can be used to address functional questions, but these approaches may be costly and typically fall within the expertise of other professionals (e.g., occupational therapists). Furthermore, findings may fail to generalize to situations and demands not directly observed, particularly when observations are not made by individuals who have an appreciation for how brain–behavior relationships may influence the conduct of complex cognitive behaviors in the environment. Rating scales and informant interviews may also provide an economical and valid source of functional information. The neuropsychologist, using standard procedures, also can make significant contributions to the determination of functional capacity.

A critical component of the assessment process is establishing not only the appropriate diagnosis, or elucidation of the cognitive deficits that might be associated with diseases, but determining how the disease or disability impairs or impedes functioning in the natural environment in the face of real-life demands. Cognitive manifestations of disability often are not easily quantified, and the impact of specific cognitive deficits is difficult to ascertain. Assessment must provide not only diagnostic formulations, but also detail as to how deficits might interact with task and environmental demands to impact the individual's functioning in real-life situations.

Most psychological and neuropsychological tests were not developed to predict behavior in the natural environment, and extensive validation does not exist for this purpose, but there is increasing evidence to suggest that standardized measurements of cognitive and behavioral skills on psychological and neuropsychological tests are well correlated with performance in the daily living and work environments. Furthermore, the ability to make such predictions about real-world behavior is enhanced when test data are combined with specific knowledge about the environment and its demands.

REFERENCES

Acker, M. (1982). Discipline report: Clinical neuropsychology. In *Head injury project final report*. San Jose, CA: Santa Clara Valley Medical Center.

Acker, M. B. (1986). Relationships between test scores and everyday life functioning. In B. P. Uzell & Y. Gross (eds.), *Clinical Neuropsychology Intervention* (pp. 85-117). Boston: Nijhoff.

Aitken, S., Chase, S., McCue, M., & Ratcliff, G. (1993). An American adaptation of the Multiple Errands test: Assessment of executive abilities in everyday life (Abstract). *Archives of Clinical Neuropsychology, 8*(3), 212.

Baum, B., & Hall, K. (1981). Relationship between constructional praxis and dressing in the head-injured adult. *American Journal of Occupational Therapy, 35*, 438-442.

Ben-Yishay, Y., Gerstman, L., Diller, L., & Haas, A. (1970). Prediction of rehabilitation outcomes from psychometric parameters in left hemiplegics. *Journal of Consulting and Clinical Psychology, 34*, 436-441.

Bureau of Labor Statistics. (1996). *Occupational Outlook Handbook.* (Bulletin 2470, 1996-97 ed.). Washington, DC: Department of Labor.

Crewe, N. M., & Althelstan, G. T. (1984). *Functional Assessment Inventory manual.* Menomonie: University of Wisconsin-Stout.

Demchak, M. A. (1993). Functional assessment of problem behaviors in applied settings. *Intervention in School and Clinic, 29*(2), 289-95.

Dennerll, R. D., Rodin, E. A., Gonzalez, S., Schwartz, M. S., & Lin, Y. (1966). Neuropsychological and psychological factors related to employability of persons with epilepsy. *Epilepsy, 7*, 318-329.

DeTurk, J. (1975). Neuropsychological measures in predicting rehabilitation outcome. *Dissertation Abstracts International, 36*(2), 437B. (University Microfilms No. 7514653).

Dikmen, S., & Morgan, S. F. (1980). Neuropsychological factors related to employability and occupational status in persons with epilepsy. *Journal of Nervous and Mental Disease, 168*, 236-240.

Duke University Center for the Study on Aging. (1978). *Multidimensional functional assessment: The OARS methodology* (2nd ed.). Durham, NC: Duke University Press.

D'Zurilla, T. J., & Nezu, A. M. (1990). Development and preliminary evaluation of the Social Problem-Solving inventory. *Psychological Assessment: A Journal of Consultants and Clinical Psychology, 2*, 156-163.

Employment and Training Administration. (1991). Dictionary of Occupational Titles (4th ed., rev., Vol. 1). Washington, DC: U.S. Employment Service.

Fordyce, D. J. (1983). *Psychometric assessment of denial of illness in brain injured patients.* Paper presented at the 91st annual convention of the American Psychological Association, Anaheim, CA.

Forer, S. (1982). Functional assessment instruments in medical rehabilitation. *Journal of the Organization of Rehabilitation Evaluators, 2*, 29-41.

Foster-Johnson, L. & Dunlap, G. (1993). Using functional assessment to develop effective, individualized interventions for challenging behaviors. *Teaching Exceptional Children, 25*(3), 44-50.

Granger, C. V., Hamilton, B. B., Keith, R. A., Zielezny, M., & Sherwin, F. S. (1986). Advances in functional assessment for medical rehabilitation. In A. M. Jette (ed.), *Topics in geriatric rehabilitation* (pp. 59-74). Rockville, MD: Aspen.

Hall, K. M. (1992). Overview of functional assessment scales in brain injury rehabilitation. *Neuro-Rehabilitation, 2*(4), 98-113.

Halpern, A. S., & Fuhrer, M. J. (1984). *Functional assessment in rehabilitation.* Baltimore: Brooks.

Heaton, R. K., & Pendleton, M. G. (1981). Use of neuropsychological tests to predict adult patients' everyday functioning. *Journal of Consulting and Clinical Psychology, 49*(6), 807-821.

Heaton, S., Chelune, G., & Lehman, R. (1978). Using neuropsychological and personality tests to assess the likelihood of patient employment. *Journal of Nervous and Mental Disorders, 166*, 408-416.

Kern and Lee and others. (1994). Student Assisted Functional Assessment Interview. *Diagnostique, 19*(2-3), 29-39.

Lorenze, E., & Concro, R. (1962). Dysfunction in visual perception with hemiplegia: Its relation to activities of daily living. *Archives of Physical Medicine and Rehabilitation, 43*, 512-518.

Maag, J. W., & Reid, R. (1994). Attention-Deficit/Hyperactivity Disorders: A functional approach to assessment and treatment. *Behavioral Disorders, 20*, 5-22.

Mackworth, N., Mackworth, J., & Cope, N. (1982). Cognitive-visual assessment of head injury recovery to predict social outcome by measuring verbal speeds and sequencing skills. In *Head injury project final report.* San Jose, CA: Santa Clara Valley Medical Center.

McCue, M. (1995). Ecologically valid assessment of problem-solving abilities: The American Multiple Errands Test (Abstract). *Journal of the International Neuropsychology Society, 1*(2), 149.

McCue, M., Rogers, J. C., & Goldstein, G. (1990). Relationships between neuropsychological and functional assessment in elderly neuropsychiatric patients. *Rehabilitation Psychology, 35*(2), 91-99.

Morris, J., Ryan, J., & Peterson, R. (1982). *Neuropsychological predictors of vocational behavior.* Paper presented at meetings of the American Psychological Association, Washington, DC.

Newnan, O. S., Keaton, R. K., & Lehman, R. A. (1978). Neuropsychological and MMPI correlates of patients' future employment characteristics. *Perceptual and Motor Skills, 46,* 635-642.

Schwartz, M., Dennerll, R., & Lin, Y. (1968). Neuropsychological and psychological predictors of employability in epilepsy. *Journal of Clinical Psychology, 24,* 174-177.

Shallice, T., & Burgess, P. W. (1991a). Deficits in strategy application following frontal lobe damage in man. *Brain, 114,* 727-741.

Teasdale, G., & Jennett, B. (1974). Assessment of coma and impaired consciousness: A practical scale. *Lancet, 2,* 81-84.

Tobin, T. (1994). Recent developments in functional assessment: Implications for school counselors and psychologists. *Diagnostique, 19*(2-3), 5-28.

Wang, P., & Ennis, K. E. (1985). Competency assessment in clinical populations. In B. P. Uzell & Y. Gross (Eds.), *Clinical neuropsychology of intervention* (pp. 119-133). Boston: Nijhoff.

Weintraub, S., Baratz, R., & Mesulam, M. (1982). Daily living activities in the assessment of dementia. In S. Corkin et al. (eds.), *Aging: Vol. 19. Alzheimer's disease: A report of progress.* New York: Raven.

Williams, N. (1967). Correlation between copying ability and dressing activities in hemiplegia. *American Journal of Physical Medicine, 46,* 4.

8

Assessment of Sensory and Motor Function

PAUL D. HANSEN AND LYNETTE S. CHANDLER

INTRODUCTION AND OVERVIEW

The assessment of individuals with neurological deficits requires focused evaluation of the extent to which pathophysiology of cells or tissues is expressed as neurological impairment, as well as an understanding of the disabilities and limitations imposed by the sensory-motor losses (see Table 1) (National Institutes of Health, 1993; World Health Organization, 1980). This chapter provides an overview of the clinical evaluation of sensory motor system impairment and the pathology that points to the impairment underlying the disability. Furthermore, we discuss the combined effects of sensory-motor impairment on function as evaluated by gait. Limitations imposed by the environment are not the focus of this chapter.

PRINCIPLES OF A SENSORY-MOTOR EVALUATION

Sensory-motor impairments present clinically with a variety of symptoms from disorders of movement through depression. The evaluation of these presentations requires an understanding of the pathophysiology that may cause these diverse symptoms. Although the evaluation of the sensory-motor system is the domain of several disciplines, for example, general practice physicians, neurologists, physiatrists, psychiatrists, neuropsychologists, physical and occupational therapists, each professional performs and interprets the results of the evaluation from his or her own frame of reference in a manner that best meets the treatment goals of the patient. Our focus as physical therapists is on the evaluation of impairments that lead to movement disabilities and the evalua-

---

PAUL D. HANSEN AND LYNETTE S. CHANDLER   School of Occupational and Physical Therapy, University of Puget Sound, 1500 North Warner Street, Tacoma, Washington 98416.
Rehabilitation, edited by Goldstein and Beers. Plenum Press, New York, 1998.

tion of residual functions that will allow the patient to overcome those movement disabilities.

Assessment of a neurologically impaired patient requires a thorough physical and neurological evaluation of which the sensory-motor components are isolated segments. The application of a neurological examination requires knowledge of evaluation techniques, normal neuroanatomy, and neuropathology. The complexity of and techniques for completing a neurological examination are reviewed by most medical neurology texts and several texts, such as DeMyer (1994), are devoted entirely to the topic. Each text presents slightly different approaches to the examination. We have provided an outline of the neurological evaluation for reference (Table 2).

This chapter's emphasis is on those evaluation areas that physical therapists frequently include in a sensory-motor examination, understanding that their exam must be complemented by the work of other disciplines. The chapter begins with general comments on the examination, then proceeds into the formal evaluation. The exam begins with a history and observation, and then progresses through the assessment of range of motion/muscle tone, sensation, strength, coordination, reflexes, and gait analysis.

In addition to the clinical examination, laboratory, neuroimaging, or other specialized tests are often called for to confirm the clinical diagnosis. As these tests are rapidly evolving, and are, for the most part, beyond the diagnostic domain of physical therapy, the reader is referred to current medical neurology texts such as Adams and Victor (1993) for discussions and indications for these tests.

The function of the neurological evaluation in the rehabilitation setting extends beyond localization and etiology determination. The rehabilitation clinician also utilizes the neurological examination to determine residual normal function. Thus, the rehabilitation clinician must have, in addition to a knowledge of the neurological evaluation, neuroanatomy, and neuropathology, a thorough understanding of the requirements for normal function and rehabilitative strategies.

Based on the neurological evaluation, rehabilitation treatment is directed at those items in the diagnosis, be it the pathophysiology, impairment, disability, or environmental limitations that are amenable to treatment and will enhance or maintain patient function. The medical treatment of a bacterial infection of the coverings of the brain (bacterial meningitis), for example, is directed at pharmacologically killing the causative organism, that is, a treatment of the pathophysiology. In some cases the pathophysiology is not amenable to any known treatment.

TABLE 1. DISABILITY MODEL

Hierarchical designation	Level
Patho-physiology or disease	Cells and tissue level
Impairment	Organ system level: sensory-motor
Disability	Person level
Limitation	Environmental levels either physical, social, political, and/or cultural

Note: The impact of sensory motor system impairment can be evaluated at several levels.

A person with lower-extremity paralysis that resulted from a spinal cord injury can strengthen the upper extremities and trunk to substitute for the lower-extremity impairment, can use a wheelchair to minimize the disability, and can install ramps at home to decrease environmental limitations. Each of these treatment strategies addresses increasing the patient's mobility.

VALIDITY AND ITS EFFECT ON DIAGNOSTIC ACCURACY

Most of the sensory-motor aspects of the neurological evaluation are significantly impacted by a lack of construct and content validity. If a test has validity it measures what is intended to be measured. While a strength test is designed to test muscle force capabilities, it also tests the motivation of the patient to perform the task (Rothstein, 1985). Sensory tests likewise not only test the integrity of the sensory pathways, but also the cooperation of the patient (DeMyer, 1994). The lack of validity presents a problem if the clinician tries to use only one evaluative test to determine the presence or absence of pathology. To overcome this problem there continue to be attempts at making evaluative tools pathognomonic (possess construct validity) for a given disease or syndrome. Hence, the neural domain circumscribed by the given test would be limited (Figure 1A).

It at first seems that diagnostic accuracy would be enhanced if there were an exclusive test for each pathology. Actually the opposite is true. Using broad evaluative tools, also called coarse coding (Hinton, McClelland, & Rumelhart,

TABLE 2. COMPLETE NEUROLOGICAL EXAMINATION

History
Mental status
Speech
Cranial nerves
Motor systems
  Inspection
    Gait
    Posture
    Build (body Gestalt)
    Skin
  Palpation
  Strength
  Range of motion/muscle tone
  Muscle stretch reflexes
  Cerebellar
  Nerve root stretching
Sensory system
  Superficial: light touch, temperature, pain
  Deep sensory: vibration, position sense fingers & toes, stereognosis, Romberg
  Distribution: dermatomal, peripheral, central, nonorganic
Functional assessment
Special assessments:
  Evoked potential, MRI, PET

Note: A sensory motor examination cannot be taken out of context of a complete neurological exam.

1986), provides substantial overlap of the domain (multiple constructs) that each test evaluates (Figure 1B). A more precise diagnosis can then be specified by the pattern of overlap between the positive and negative tests. For example, three evaluative tools with high sensitivity could be used to describe three pathologies (+A, +B, or +C) or three course tools could be used to describe seven pathologies (+A-B-C, +A+B-C, +A+B+C, -A+B-C, -A+B+C, -A-B+C, +A-B+C). Thus, for a singular neural lesion, the less focused tools may provide a more specific diagnosis.

The issue of validity is critical in the clinical evaluation of the neurological patient. Most of the tools used in a clinical evaluation lack well-defined diagnostic sensitivity. While the tools are less selective, the skilled clinician overlaps the domains of a multitude of tests so that a high degree of diagnostic precision is obtained at the conclusion of the evaluation.

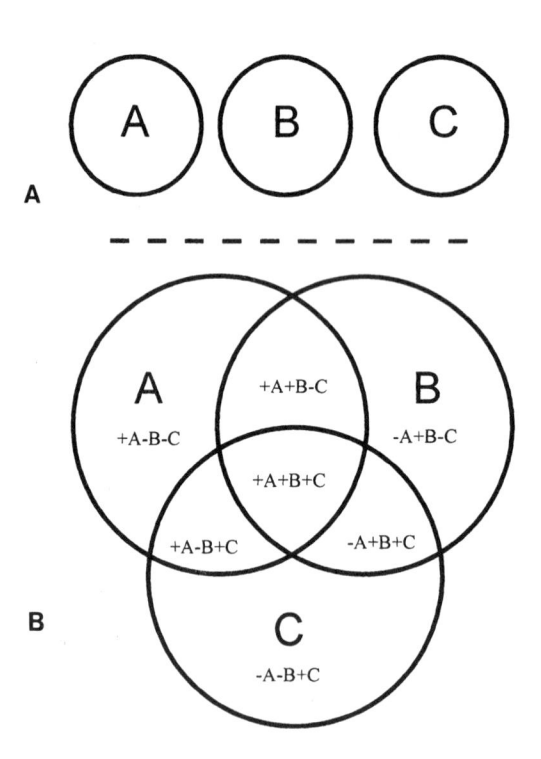

Figure 1. A. With three pathognomic tests (A, B, C) three conditions can be described (+A, +B, +C). B. With three coarse tests with overlapping construct validity (ABC) seven conditions can be described (+A-B-C, +A+B-C, +A+B+C, -A+B-C, -A+B+C, -A-B+C, +A-B+C).

Evaluative tools with coarse coding also act to increase the breadth of the evaluation. With these tools the clinician is obligated to evaluate domains that will not have pathological findings. This forces the clinician to view alternatives to the primary diagnosis. Thus, in using coarse evaluative tools, conditions that are not considered at the outset often are uncovered during the evaluation.

A final advantage of using more inclusive evaluative tools is that the negative tests, those that fail to find dysfunction, indicate where the patient's residual strengths lie. These strengths are critical for the rehabilitation clinician to utilize in maximizing a patient's residual function so as to minimize the patient's disability.

By relying on general evaluative tools, a clinical neurological evaluation depends on the ability of the examiner to look at the totality of the examination in arriving at a diagnosis. Test results that are normal become as important as pathological test results in determining the clinical impression.

SUGGESTIONS FOR PERFORMING THE NEUROLOGICAL EVALUATION

As with most clinical evaluative tools, sensory-motor clinical testing requires cooperation between the clinician and the patient. Besides this trusting relationship, there are several admonitions that aid testing.

1. Perform a neurological evaluation with as many normal patients as possible. This will give you the knowledge to differentiate between normal function, normal variance, and dysfunction when it is seen.
2. Perform the evaluation methodologically. Develop an order in which you perform the evaluation, for example, from rostral to caudal or sensory to motor. Execute the evaluation each time in this order. A structure to the evaluation helps insure that components are not left out with any patient.
3. In the nonemergent patient always begin the evaluation with the patient's least affected component. For example, if a patient appears to have motor involvement of the right leg, begin your motor evaluation by looking at the upper extremities and then the left lower extremity before assessing the right leg. By performing the evaluation in this manner the clinician assesses the patient's ability to follow commands and determines the patient's general state and "normal" function in that individual.

After working with many patients, the clinician should be able to recognize a gestalt, of the overall impression of each patient (DeMyer, 1994). Because a clinical evaluation may not discern an identifiable pathology or impairment when in fact one exists, it is important that the clinician recognize this gestalt. If the clinician suspects that "something is amiss" with the patient, that patient needs further assessment. Thus, by being aware of an overall impression of the patient, the clinician can better interpret the results of the clinical evaluation.

THE SENSORY-MOTOR EXAM

This chapter addresses two general domains of the neurological evaluation, the motor system and the sensory system, which are primarily assessed through clinical tests. Although the motor and sensory aspects of the evaluation are presented separately, they should not be viewed in isolation, for they must be integrated into the totality of the neurological examination (see prior discussion regarding coarse, nonspecific clinical tests). Sensory-motor evaluation attempts to assess the integrity of the neural pathways and structures that carry and process the sensory-motor information by testing the function carried by the pathway or performed by the struc-

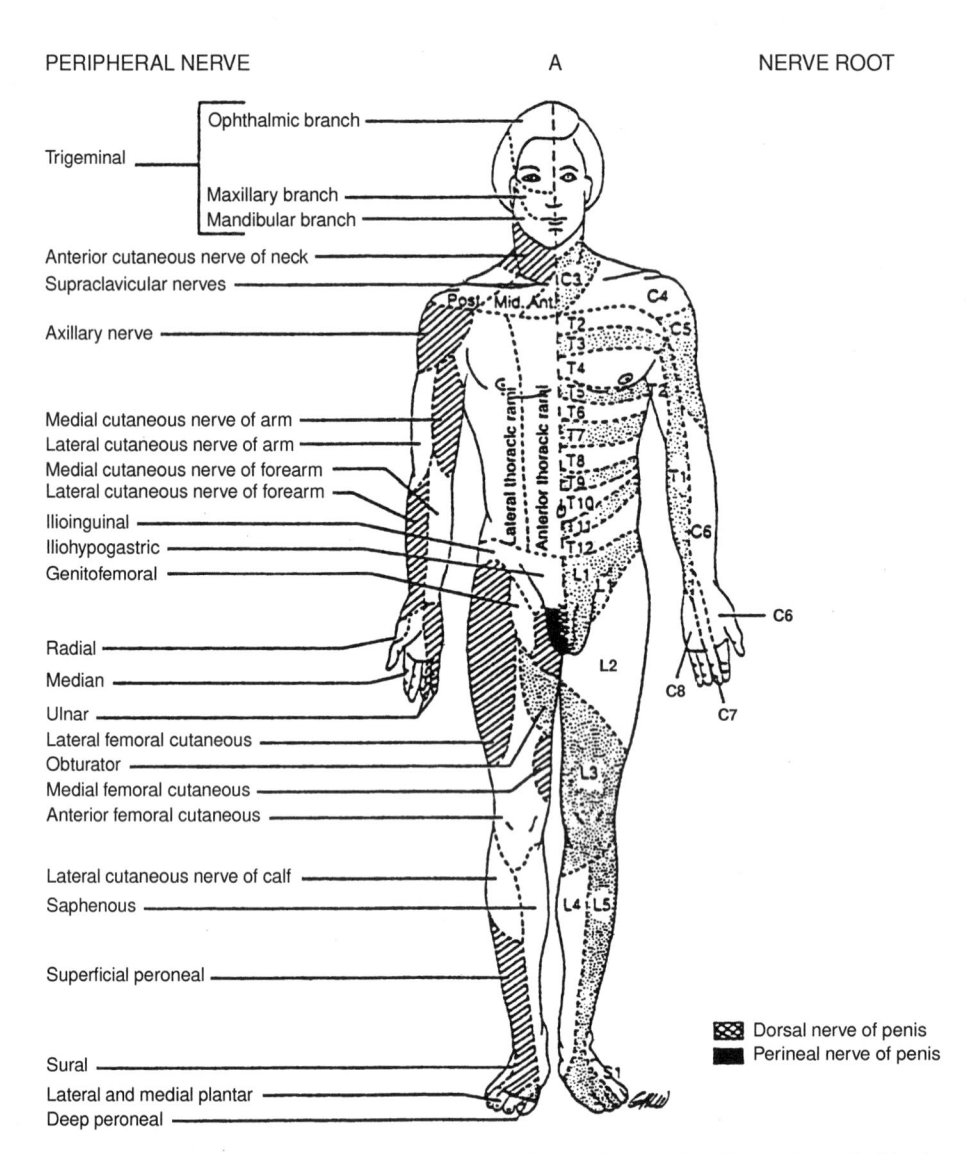

Figure 2. A representative dermatome and peripheral nerve innervation chart. Although this chart has clear lines of demarcation between the different dermatomes, there is actually extensive overlap; for example, the C5 dermatome may extend from the lateral upper arm to the base of the thumb (see Cyriax, 1978 for discussion). The peripheral nerve innervation also exhibits overlap between adjacent nerves. Note that the peripheral nerve innervation is not equivalent to spinal nerve root from which the dermatome is defined. (From *Clinical Neurology* [2nd ed., pp. 186–187], by D. A. Greenberg, M. J. Aminoff, and R. P. Simon, 1993, Norwalk, CT: Appleton & Lange. Copyright 1993 by Appleton & Lange. Reprinted with permission.)

ture. In general, a lack of function (a negative Jacksonian sign) suggests that the domain or the pathway leading from that domain has been damaged. Exaggerated or excessive function (a positive Jacksonian sign) suggests that other structures that typically modulate the domain being tested have been damaged and therefore have released their control over the region of interest. In acute neural injuries that result in neural shock, only negative signs are evident until the neural shock recedes (De-Myer, 1994).

The sensory-motor evaluation assesses both the peripheral nervous system and the central nervous system. The peripheral nervous system arises from the spinal nerve roots, which become the peripheral nerves upon exiting through the intervertebral foramen. In a cylindrical animal such as the snake, the spinal nerve roots are directly continuous with the peripheral nerves that innervate consecutive

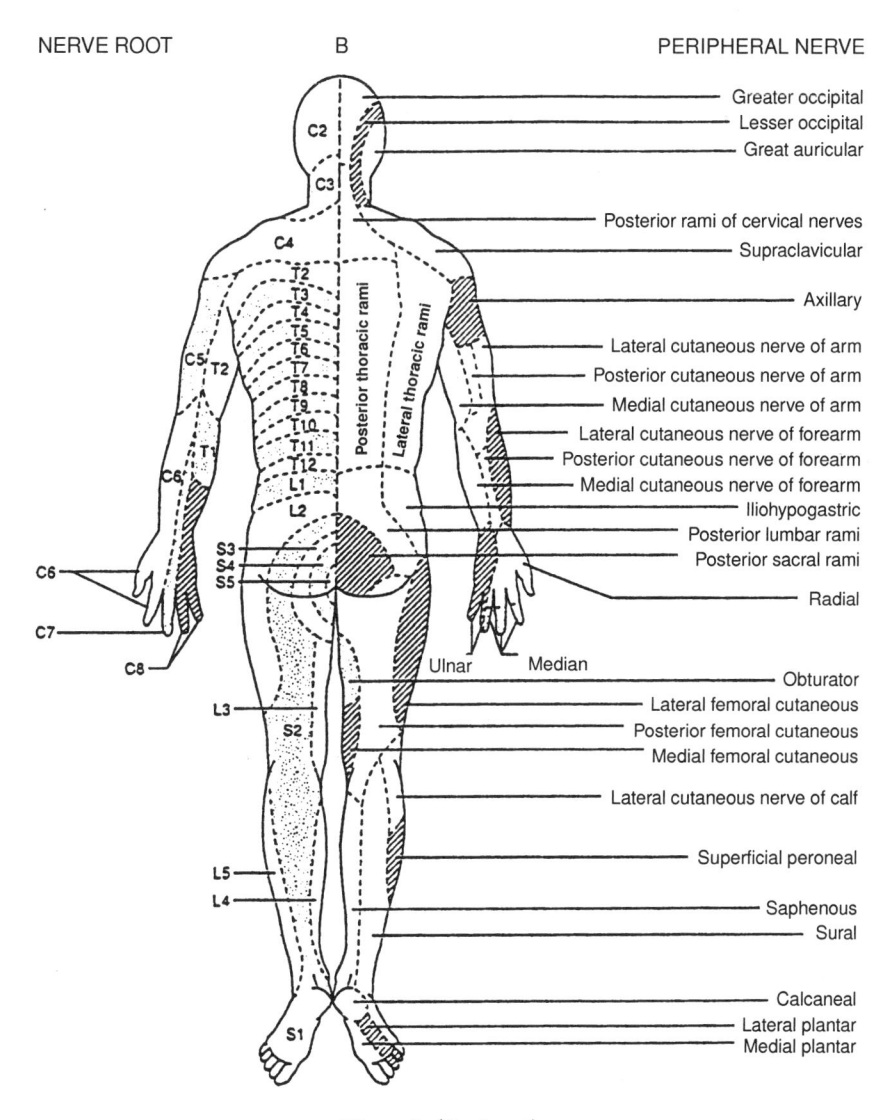

Figure 2. *(Continued).*

bands of tissue along the body's longitudinal axis. The skin that is innervated by each spinal nerve root is called a dermatome. Each dermatome is named according to the spinal nerve root from which it arose. In the human the outgrowth of the extremities distorts the bandlike appearance of the spinal nerve roots, although the rostral to caudal ordering is preserved. In the cervical and lumbar regions, the spinal nerve roots pass through the brachial and lumbar plexuses respectively, effectively mixing the representation of the spinal nerve roots in each peripheral nerve. The result of this mixing is that the innervation patterns of the peripheral nerves and the spinal nerve roots are not equivalent (see Figure 2).

A working knowledge of dermatomal and peripheral nerve innervation patterns provides the clinician with a means of differentiating potential sites (root versus peripheral nerve) of involvement in the peripheral nervous system. Furthermore, deeper tissues tend to refer pain into specific dermatomes according to the tissue from which they embryologically developed. This pattern of pain referral information can direct the clinician to look at specific organs or tissues (Cyriax, 1978).

The central nervous system (CNS) structure can be viewed as a loosely organized hierarchy. In the spinal cord the motoneurons and primary sensory axons carry discrete, well-localized information such as the motor neurons for the biceps muscle or pressure sensation from the tip of the index finger. At higher processing levels the CNS concerns itself with more abstract concepts, such as a task's movement goal or the sensation produced by a coin. Typically, there are multiple, topographically mapped pathways between the multiple higher centers (each of which abstracts information in slightly different ways) and lower centers, each of which carry relatively specific information, although there is considerable overlap. By assessing the function of each neural region and the pathways that pass through those regions, the clinician can localize the site of dysfunction in the nervous system.

To assess the motor system, range of motion/muscle tone, strength, and coordination are assessed. To assess the sensory system, separate assessment of each sensory pathway and of higher-level processing of sensation is performed. Several components of the neurological evaluation, such as reflexes, posture, gait, and functional tasks, do not fit into distinct categories, but rather integrate sensory and motor function.

History

The history is significant in determining motor and sensory function because minor motoric and sensory dysfunction may be discernible only in the history. Motoric deficits often will present as the patient's inability to complete desired tasks. It is important to determine which tasks the patient is having difficulty performing and how he or she is attempting to perform the task. An inability to perform gross motor tasks, such as rising from a chair, suggests proximal weakness, whereas difficulty with fine motor tasks, such as preparing food, suggests a distal motor weakness (Greenberg, Aminoff, & Simon, 1993).

The patient's description of why he or she is unable to complete the tasks is also important. Symptoms such as paralysis, tremor, heaviness, clumsiness, or coordination deficits often direct the clinician toward specific neural structures.

Weakness, although suggestive of motoric dysfunction, needs to be clarified, as the patient may use this term to describe a lack of drive or enthusiasm to perform their work (Greenberg et al, 1993). Other symptoms, such as falls, also suggest that the clinician evaluate the motor domain.

Sensory deficits often present as paresthesias (abnormal spontaneous sensations), dysesthesias (unpleasant sensations elicited by normally pleasant stimuli), or numbness (an often vague sensory complaint) (Greenberg et al., 1993; Adams & Victor, 1993). Some gradual sensory losses may not be apparent to the patient, but will be expressed by clumsiness, for example, sensory ataxia, or frequent injuries, for example, from falling or burning the fingertips. When discussing sensory changes with the patient the clinician must avoid leading questions.

Formal Observation

All neurological evaluations rely strongly on the examiner's observational skills. To quantify these observations, the appearance of the patient is described. During this time it is best to observe the patient with minimal clothing. From a motoric perspective the clinician looks at posture from an anterior, posterior, and lateral view, paying particular attention to the position of the head, shoulders, pelvis, spinal curvatures, and left-right symmetry and postural deformities. From a lateral view an ideal line of gravity should pass through the ear lobe, tip of the shoulder, greater trochanter of the femur, just posterior to the patella and just anterior to the lateral malleolus (Daniels & Worthingham, 1986). A small cervical and lumbar lordosis and thoracic kyphosis should be evident (Kendall, McCreary, & Provance, 1993). In addition, the breadth and erectness of stance, neuromuscular tremors, tics, muscle fasciculations, changes in muscle bulk (atrophy or hypertrophy of the whole body, isolated limb, or isolated muscle), and any surgical scars are noted.

The status of the skin is also observed as seen in its color, shininess, and hair distribution, especially in the distal extremities. Although the skin is not part of the assessment of the sensory-motor system, the skin often reflects vascular problems which accompany sensory-motor deficits. Skin markings, such as café au lait spots, may be pathognomonic of neurological disease (DeMyer, 1994). With a sensory loss there may be scarring from the neglect of injuries that have not been perceived by the patient.

Throughout the evaluation the quality and quantity of the patient's movements are observed. Alterations in muscle tone can be seen at rest or can become activated during voluntary movement. This abnormal positioning may be continuous, such as in the neck position of spasmodic torticollis, or transient, such as when a stroke victim attempts to move an affected limb. Both the quality and quantity of movements are observed. A paucity of movements is characteristic of some conditions, such as Parkinson's disease, whereas excessive movements, such as tics, ballism, chorea, or athetotic movements, are characteristic of other conditions, such as Huntington's chorea (Greenberg et al., 1993, Adams & Victor, 1993).

If a tremor is observed it is described in terms of timing and distribution. The tremor is referred to as a physiological tremor if it is continuous, a resting tremor if it occurs during rest, an intention tremor if it takes place during active intentional movement, and a postural tremor if it arises during sustained positioning. The distribution is described in terms of the extremities involved.

Range of Motion and Muscle Tone Testing

PAUL D. HANSEN AND LYNETTE S. CHANDLER

Range of motion (ROM) testing is not classically considered as part of the evaluation of sensory-motor systems; rather, it is often categorized as a test of the musculoskeletal system, in particular, articular and musculotendinous connective tissue extensibility. While this is an appropriate classification, active and passive joint motion also provide considerable information about sensation and movement. Range assessment is divided into two components, active (patient produced) and passive (examiner performed) movement.

To assess active range of movement (AROM), the patient is asked to move from a starting position to an end range position. During the movement the range and quality of the movement are observed along with any associated movements. Active movement is constrained by what the patient is motivated, willing, and able to do (Cyriax, 1978). Motorically impaired but motivated patients may complete the requested task by substituting movements, such as leaning the trunk forward instead of using shoulder flexion when asked to reach. With these constraints acknowledged, AROM testing provides extensive information about the neurological condition of a patient. If a patient's AROM is range or quality limited, the underlying cause must be sought in terms of a change in the nervous system, musculoskeletal system, or motivation. Hence, the assessment of a patient's AROM needs to be supplemented with passive range of motion (PROM) testing, which describes the motion that is available, and a strength test (typically a manual muscle test [MMT]), which describes the amount of strength the patient possesses.

Passive range of motion is assessed by the clinician's moving the limb through its full range of motion. Initially it is moved slowly through the full range, and then repeated at a faster speed. The resistance to movement and the resistance encountered at the end range (see Cyrix, 1978) are reported.

The quality of any muscular resistance to the passive motion can be described. Clasp knife resistance initially gives substantial resistance to passive movement, but if the examiner is persistent the resistance will give and movement will be allowed. In contrast, cogwheel rigidity gives similar initial resistance, but only a limited amount of motion is allowed after the resistance gives, until the next "cog" is encountered. "Lead pipe" rigidity allows the motion to occur slowly but only with persistent force as if bending a lead pipe. Both cogwheel and lead pipe rigidity are associated with Parkinson's and other conditions of the basal ganglia. Clasp knife resistance is associated with upper motor neuron pathologies.

When handling the patient for PROM assessment the overall sensitivity of the subject to cutaneous input, such as the sensitivity seen in thalamic pain syndromes (Adams & Victor, 1993), is often observed. In addition, the excitability of the motor system to cutaneous inputs is assessed. In hyperreflexia, for example, as seen following a spinal cord injury, simply touching a limb may trigger a flexor withdrawal response.

Passive range of motion also allows the examiner to assess the stretch reflex excitability and muscle tone of the patient. Passively moving the limb stretches the muscles surrounding the joints. Through muscle spindle afferents the motor neuron is excited by the passive muscle elongation, a response that is sensitive to the velocity of the muscle stretch. A hyperreflexive muscle will respond to this passive stretch with a contraction. Occasionally the reflexive rebound will be repetitive and is termed clonus.

While muscle tone is experimentally difficult to define, clinically the extremes of high tone (hypertonia) or low tone (hypotonia) in the extremities are easy to separate. The determination of tone is based on the resistance to PROM, extensibility (resistance to PROM with a quick stretch at the end of the elongated range) and palpation (the gentle squeezing of the muscle belly) (DeMyer 1994).

Immediately following an injury of either the nervous or musculoskeletal systems PROM is not limited. For example, if the rotator cuff muscles are torn at the shoulder or the axillary nerve damaged, active motion of the shoulder is severely limited although the PROM of the shoulder is full, and often pain free. If the condition persists the articulation may develop a connective tissue contracture from disuse (Cyriax, 1978).

SENSORY TESTING

The evaluation of sensation is the most subjective component of the clinical neurological evaluation. The failure to demonstrate a sensory loss in the presence of a known pathology and the ability to find sensory changes when no organic pathology can be demonstrated speaks clearly to the weaknesses of this step in the neurological evaluation (Adams & Victor, 1993).

The cause of the weakness of the sensory evaluation lies in the organization of the sensory system. As previously discussed, in the periphery, each receptor is sensitive to a particular class of stimuli, for example, touch. In contrast, at the cerebral cortex, more abstract information is processed. This ascension of information through the sensory hierarchy has been best described in the visual system. At the receptor level, a rod in the retina is sensitive to a single photon of light, whereas in the cerebral hemisphere there are regions that react only when presented with a complex stimulus such as the form of a face (Kandel, 1995).

Viewed in this way, the sensory system progresses from the clearly identifiable stimulus to an abstract sensation as the signal travels from the bottom to the top of the neural hierarchy. In the neurological examination, we ask our patients to minimize the interactions between the inputs as they ascend through the nervous system and to report only on the simple stimulus that is presented in the periphery. Even for the most cooperative patient this is a difficult task.

When testing sensory integrity, a sufficient stimulus must be provided to cause the receptor to discharge. Without a sufficient stimulus no sensation can occur. If the clinician provides a sufficient stimulus the patient decides if the stimulus is present or absent. In this dual choice paradigm there is a 50% probability that the correct answer will be given by chance. Given that in a normal evaluation there is 100% accuracy, the patient with neuropathology will typically score less than 100%, but possibly better than 50%. The validity of the answers from a patient who is wrong 100% of the time needs to be questioned.

Classically, the evaluation of the sensory systems is according to three sensory systems: light touch, pain and temperature, and proprioception (DeMyer, 1994). The first two are tested for each peripheral nerve and spinal nerve root (Figure 2); proprioception (kinesthesia and proprioception) is assessed in each extremity, and if a deficit is suspected, for each articulation in the suspect extremity. These stimuli test the lower aspects in the sensory hierarchy, whereas higher-order sensations, such as graphesthesia and stereognosis, assess dorsal column and cortical functions.

The differentiation between light touch, as tested with a cotton wisp, and pain, as tested with the ability to perceive a sharp pin point (in contrast to a dull pressure), is done to separate the dorsal column and the anterolateral (spinothalamic) spinal pathways respectively. Although this is the classic demarcation between these two pathways, evidence suggests that neither pathway exclusively carries a set of sensations; hence, the differentiation is relative (Nolte, 1993).

The regions within each dermatome and peripheral nerve are tested for light touch as well as pain and temperature. If a deficit is found the boundaries of the area are mapped to determine whether it matches a dermatomal or peripheral nerve distribution. Often, this map by itself will describe the area specific for a dermatome, peripheral nerve, vascular (often in a distal to proximal stocking or glove pattern), or CNS injury (typically involving all of one limb or pair of limbs).

Temperature sensibility, in addition to sharp–dull testing, is a means of assessing the anterolateral tract. For the warm and cold stimuli temperatures of 45°C and 10°C, respectively, are used.

Testing of the dorsal column pathways of the spinal cord can also be assessed using vibration (128 Hz applied at a singular distal bony location as there is significant transmission of the vibration through the soft tissues) or spatial discrimination, as is needed in the recognition of figures that are drawn on the skin. Figure recognition is performed by asking the patient to identify a large figure (at least 4 cm in size), such as a letter, which is drawn on the patient's skin. If a subject is unable to identify letters, identifying the direction of a line drawn on the skin can be substituted (Adams & Victor, 1993). Both figure recognition and vibration assess dorsal column function (if the cerebral cortex is intact), but do not localize individual spinal nerve root or peripheral nerve domains. Hence, vibration and figure drawing are used to assess the integrity of the spinal cord dorsal columns and not the integrity of the peripheral nerve.

In addition to pain and touch, proprioception should be assessed. As a general term proprioception refers to the knowledge of body in space. In the neurological examination, proprioception refers to the patient's ability to know a static body position, while kinesthesia means the ability to perceive body and limb movement (DeMyer, 1994).

Testing of proprioception is performed by having the clinician passively move one of the patient's limbs to a new position. The patient then matches the contralateral limb to this position. To assess kinesthesia, the patient moves the contralateral limb while the clinician moves the ipsilateral limb. When holding the limb the clinician needs to minimize cutaneous inputs which may clue the patient as to the direction of movement. If the patient is unable to move the contralateral limb, the clinician may test proprioception by randomly moving the limb to a flexed or extended position and then asking the patient to report the direction that the limb is pointed.

Tests of higher sensory function, such as stereognosis, require the patient to identify objects by active touch based on texture, size, and shape. The location of loss for this function is believed to be the contralateral postcentral gyrus and parietal lobe (Adams & Victor, 1993).

Strength Testing

The minimal requirements needed for a muscle to contract *in vivo* are a motoneuron, an intact axon from the motoneuron to the muscle, an intact neuromotor synapse, an intact muscle, and adequate nutrition to the nerve and muscle. To control the motoneuron a variety of descending and local (spinal) excitatory and inhibitory inputs converge on the motoneuron. Testing of the motor system examines the integrity of these systems.

Each muscle's innervation can be described in terms of the peripheral nerve that innervates the muscle and the spinal segments from which that peripheral nerve arises. Extremity muscles typically are innervated by axons from multiple spinal segments that travel to the muscle in a single peripheral nerve. Commonly tested muscles and their innervation and functions are listed in Table 3. General

TABLE 3. INNERVATION OF SELECTED MUSCLES OF UPPER LIMBS

Muscle	Main root	Peripheral nerve	Main action
Upper extremity			
Supraspinatus	C4, 5, 6	Suprascapular	Abduction of arm
Infraspinatus	C5, 6	Suprascapular	External rotation of arm at shoulder
Deltoid	C5, 6	Circumflex	Abduction of arm
Biceps	C5, 6	Musculocutaneous	Elbow flexion
Brachioradialis	C5, 6	Radial	Elbow flexion
Extensor carpi radialis longus	C6, C7	Radial	Wrist extension
Flexor carpi radialis	C6, C7	Median	Wrist flexion
Extensor carpi ulnaris	C7, 8	Radial	Wrist extension
Extensor digitorum	C7, 8	Radial	Finger extension
Triceps	C6, 7, 8	Radial	Extension of forearm
Flexor carpi ulnaris	C7, 8	Ulnar	Wrist flexion
Abductor pollicis brevis	C8, T1	Median	Abduction of thumb
Opponens pollicis	C8, T1	Median	Opposition of thumb
First dorsal interosseous	C8, T1	Ulnar	Abduction of index finger
Abductor digiti minimi	C8, T1	Ulnar	Abduction of little finger
Lower extremity			
Iliopsoas	L1, 2, 3	Femoral	Hip flexion
Quadriceps femoris	L2, 3, 4	Femoral	Knee extension
Adductors	L2, 3, 4	Obturator	Adduction of thigh
Gluteus maximus	L5, S1, 2	Inferior gluteal	Hip extension
Gluteus medius and minimus, tensor fasciae latae	L5, S1	Superior gluteal	Hip abduction
Hamstrings	L5, S1, 2	Sciatic	Knee flexion
Tibialis anterior	L4, 5	Peroneal	Dorsiflexion of ankle
Extensor digitorum longus	L5, S1	Peroneal	Dorsiflexion of toes
Extensor digitorum brevis	L5, S1, 2	Peroneal	Dorsiflexion of toes
Flexor digitorum longus	S2, 3	Tibial	Plantarflexion of toes
Peronei	L5, S1, 2	Peroneal	Eversion of foot
Tibialis posterior	L4, 5	Tibial	Inversion of foot
Gastrocnemius	S1, 2	Tibial	Plantar flexion of ankle
Soleus	S1, 2	Tibial	Plantar flexion of ankle

Note: From *Clinical Neurology* (2nd ed., pp. 153–154), by D. A. Greenberg, M. J. Aminoff, and R. P. Simon, 1993, Norwalk, CT: Appleton & Lange. Copyright 1993 by Appleton & Lange. Adapted with permission.

anatomy texts have complete listing of muscle innervation and functions (for example, Williams & Warwick, 1980).

As each muscle is innervated by specific spinal segments and peripheral nerves, spinal cord and peripheral nerve injuries have characteristic motor presentations. A complete transection between the sixth and seventh cervical spinal segments (resulting in C6 tetraplegia or quadriplegia) would cause the patient to be unable to use the flexor muscles of the fingers (flexor digitorum profundus and superficialus muscles) with retention of elbow flexor strength (brachialus and biceps muscles) (see Table 3 for spinal cord innervation levels). Although the triceps muscle would retain some innervation from the 6th cervical nerve root, this nerve root does not contribute substantially to the triceps muscle, so the triceps would not be strong enough for functional use. Note that this patient would retain sensation only from the radial border of the hand up (see Figure 2).

The isolated involvement of a peripheral nerve will usually affect all of the muscles that are supplied distal to the injury site. For example, an injury to the median nerve in the carpal tunnel at the wrist will cause a paralysis of the intrinsic muscles of the radial aspect of the hand (primarily the thumb muscles) while the long flexor muscles of the wrist and fingers would be unaffected as they receive their innervation from the median nerve prior to the carpal tunnel. Specific branching of the peripheral nerves are described in general anatomy texts (for example, Williams & Warwick, 1980). Other patterns of weakness include a paralysis of pairs of limbs due to spinal cord or brain injury, or general weakness from systemic conditions such as myasthenia gravis.

In addition to the pattern of weakness, injuries of the motoneurons or peripheral nerves result in flaccid paralysis, with atrophy, fasciculations, and hyporeflexia. Injuries to neural structures involved in the corticospinal pathway cause weakness during voluntary movement, as well as spasticity, hyperreflexia, clonus, and the Babinski sign (Kessler, 1989).

To determine which muscles are strong and which are weak the primary tool of the clinical motor exam is the manual muscle test (MMT). The MMT has evolved little since it was first described at the turn of the century (Kendall et al, 1993). There are two forms of the MMT test: gross, nonspecific testing and isolated muscle testing (Palmer & Epler, 1990). A gross evaluation of strength is used as a screening examination. If there are hints of specific weakness from the screening or from the history an isolated muscle test is performed. An example of a screening strength examination is given in Table 4. An isolated MMT attempts to separate the testing of each individual muscle. Given the differences in their origins and insertions, each muscle has its own specific testing position in an isolated MMT. (See Daniels & Worthingham, 1986; Kendall et al., 1993 for specifics for testing each muscle.)

The MMT is performed by resisting the subject's movement either with gravity or gravity and manual resistance. Some authors (Daniels & Worthingham, 1986) advocate applying this resistance to an isotonic contraction, whereas other authors (Kendall et al., 1993) suggest that the resistance be applied to an isometric contraction (a break test) at midrange of muscle length. The patient's capacity to move against this resistance is then subjectively graded by the clinician on an ordinal scale (Table 5). The amount of force that a muscle needs to produce in order to receive a normal (5/5) or other grade varies with the muscle being tested and the perceived norms for the patient's sex and patient's age (Kendall et al.,

TABLE 4. EVALUATION PROCEDURES FOR MUSCLE SCREENING TESTS

Position of patient	Muscle group tested	Instruction to patient	Therapist's action
Supine	Neck and trunk flexors	1. Hold arms straight in front of body. Raise head and shoulders off table. Hold.	None
	Hip flexors	2. Keep legs straight. Raise both legs off table simultaneously. Hold.	None
	Hip abductors	3. Abduct legs to each side. Hold.	Attempt to bring legs together.
	Hip abductors	4. Keep legs together. Hold.	Attempt to separate legs.
	Hip extensors	5. Flex hips and knees, keeping soles of feet on table. Raise hips from table.	None
	Shoulder adductors	6. Bring hands together in front of chest, elbows straight. Hold.	Attempt to separate arms into horizontal abduction.
	a. Shoulder flexors and scapular upward rotators b. Shoulder extensors and scapular downward rotators c. Shoulder horizontal abductors	7. Flex shoulder to 90 degrees, elbows straight. Hold.	a. Attempt to push arms into extension. b. Attempt to push arms into flexion. c. Attempt to push arms together into horizontal adduction.
Supine or sitting	a. Shoulder abductors b. Shoulder adductors	8. Abduct shoulder to the side to shoulder level, elbows straight. Hold.	a. Attempt to push arms down to sides into shoulder adduction. b. Attempt to push arms over head into shoulder abduction.
	a. Shoulder medial rotators b. Shoulder lateral rotators c. Elbow flexors d. Elbow extensors e. Supinators	9. Hold arms at sides, elbows bent, forearms in neutral position. Hold.	a. Attempt to push arms outward into lateral rotation. b. Attempt to push arms in toward body into medial rotation. c. Attempt to push forearms toward table into elbow extension. d. Attempt to push forearms toward shoulders into elbow flexion. e. Attempt to turn palms down into pronation.

(continued)

TABLE 4. *(Continued)*

Position of patient	Muscle group tested	Instruction to patient	Therapist's action
	f. Pronators		f. Attempt to turn palms up into supination.
	g. Wrist extensors		g. Attempt to flex the wrists.
	h. Wrist flexors		h. Attempt to push palms away from body into wrist extension.
	Finger flexors	10. Squeeze my fingers. Hold.	Place index and middle fingers in patient's hands; attempt to pull fingers out.
	Finger extensors	11. Straighten fingers. Hold.	Attempt to push fingers into flexion.
	Palmar interossei	12. Adduct fingers. Hold.	Attempt to pull fingers into abduction.
	Dorsal interossei	13. Abduct fingers. Hold.	Attempt to push fingers into adduction.
	Opponens pollicis	14. Pinch my finger. Hold.	Place index finger between patient's thumb and each finger, one at a time.
Sitting	Latissimus dorsi and triceps	15. Place hands on treatment table next to hips, elbows straight, shoulders shrugged. Depress scapula by lifting buttocks off table.	None.
	Upper trapezius and levator scapulae	16. Shrug shoulder toward ears. Hold.	Push shoulders down into depression.
	Medial rotators of the hips and everters of the feet	17. Evert feet. Hold.	Push on lateral borders of each foot, into inversion and lateral rotation.
	Lateral rotators of the hips and inverters of the feet	18. Invert feet. Hold.	Push on medial border of each foot into eversion and medial rotation.
Prone	Rhomboids, middle trapezius, and posterior deltoid	19. Bend elbows level with shoulders; pinch or adduct scapulae together, raising arms from table. Hold.	Attempt to push arms down.
	Elbow and shoulder extensors	20. Begin with arms at sides, elbows straight. Raise arms off table. Hold.	Attempt to push arms down.
	Extensors of the hip, back, neck, and shoulders	21. Begin with arms at sides. Arch back, raising head, shoul-	None

TABLE 4. *(Continued)*

Position of patient	Muscle group tested	Instruction to patient	Therapist's action
		ders, arms, and legs off table simultaneously. Hold.	
Prone or sitting	a. Hamstrings	22. Flex knees. Hold.	a. Attempt to pull knees into extension.
	b. Quadriceps		b. Attempt to push knees into further flexion.
Standing	Gastrocnemius soleus	23. Stand on one leg. Rest fingers lightly on table. Rise up to tiptoes; repeat 10 times. Repeat with other leg.	None
	Dorsiflexors	24. Walk on heels for 10 steps.	None
	Hip and knee extensors	25. Do five partial deep knee bends.	None

Note. From *Clinical Assessment in Physical Therapy* (pp. 4–5) by M. L. Palmer and M. E. Epler, 1990, Philadelphia: Lippincott. Copyright 1990 by J. B. Lippincott. Reprinted with permission.

1993). In addition to strength information, the MMT is also used to assess the structural integrity of the musculotendinous unit (Table 6).

The MMT has proven to be a reliable tool. In a summary of the MMT literature Lamb (1985) concluded that exact agreement between clinicians is low (45%–60% agreement), although there is greater than 90% agreement within plus or minus one full muscle grade.

Although the MMT is a reliable tool in the clinic for the assessment of pathology (Florence et al., 1984; Lilienfeld, Jacobs, & Willis, 1954) it is not a sensitive tool. Beasley (1961) found that the MMT was unable to differentiate knee extension forces that varied by as much as 25%. In testing whether the MMT could

TABLE 5. ORDINAL GRADING OF MUSCLE STRENGTH

Grade	Description
5 Normal	Holds against strong resistance
4 Good	Holds against moderate resistance
3 Fair	Holds against gravity
	Completes range against gravity
2 Poor	Completes movement only in a gravity eliminated (horizontal) position
1 Trace	No visible motion but either the tendon becomes visible or a contraction is palpated
0 Zero	No contraction

Note. Comments can be appended to indicate that the patient is slightly stronger (s or +) or weaker (W or -) than the primary grade. Normal charting is to report grade and scale, for example, 5/5, or 3-/5. From *Muscles, Testing and Function* (pp. 84–189), by F. P. Kendall, E. K. McCreary, and P. G. Provance, 1993, Baltimore: Williams & Wilkins. Copyright 1993 by Williams & Wilkins. Adapted with permission. Also from *Muscle Testing: Techniques of Manual Examination* (5th ed., pp. 3–4), by L. Daniels and C. Worthingham, 1986, Philadelphia: Saunders. Copyright 1986 by W. B. Saunders Company. Adapted with permission.

TABLE 6. EFFECT OF INTERACTION BETWEEN STRENGTH AND PAIN DURING AN MMT

Test result	Diagnostic implication	Clinical example
Selected movement strong and painless	Normal muscle function	
Selected movement strong and painful	Minor injury of the muscle or the tissue to which the muscle attaches	Tendonitis
Selected movement weak and painless	A disruption of contractile control, either through a rupture of the muscle, or from a nervous system injury	Complete muscle rupture or injured peripheral nerve
Selected movement weak and painful	A significant injury to the muscle or the tissues to which it attaches	Incomplete tear of a muscle or bony fracture
All movements are strong and painful	Systemic inflammation of the muscles, or a hypersensitive patient	Acute phase of rheumatoid arthritis, or malingering
All movements are strong and painless	Normal muscles	Normal muscle and peripheral nerve function (although this does not exclude higher CNS disorders

differentiate between the stronger of two extremities the MMT was in agreement with more objective measures of strength in 82% of the cases (Saraniti, Gleim, Melvin, & Nicholas, 1980).

To accurately execute the movement for an isolated MMT the patient must be able to follow a command and be able to isolate the specified movement. Because of these constraints the isolated MMT is not advocated for use in patients with CNS pathology (Daniels & Worthingham, 1986; Palmer & Epler, 1990); rather, a gross MMT should be performed. Nevertheless, in patients who are able and motivated to follow the instructions and put forward a maximal effort, the manual muscle test is an excellent tool for testing the integrity of the motor neuron pool, the motor nerve, the muscle, and the tissues to which the muscle attaches.

A variety of tests have been developed to supplement or replace the MMT, for example, isokinetic tests (Cybex, Kinetron), hand-held dynamometers, cable tensiometers, and so on. While each instrument has its advocate, none have replaced the MMT in the clinic (Mayhew & Rothstein, 1985).

With the multitude of results that can be elicited, the results of the MMT are meaningful only when the positive (weakness) findings are contrasted with the negative (normal strength) results (Cyriax, 1978).

A pattern of weakness or pain that does not match an anatomical or metabolic distribution brings to mind the possibility of a psychogenic origin. If this is suspected, the patient should be reexamined at a later date and the data compared between the two examinations. Given the multitude of MMTs and sensory tests that should be tested in each patient, it is difficult for the patient to remember what his or her responses were to each test. Hence, if the patient has a fixed pathology, the pattern of muscle weakness or pain will remain constant over several days. A patient without a pathological cause for the weakness or pain will present with a changing pattern over several days or weeks. A consistent pattern in the follow-up evaluation should prompt the examiner to look once again for a physiological basis for the patients complaints.

COORDINATION

Although the functions of the cerebellum have been reinterpreted and expanded over the last decade, cerebellar function can be described simply as coordinating movement by comparing descending motor commands with ascending proprioceptive sensory information (DeMyer, 1994). The more medial (vermal) regions of the cerebellum coordinate truncal movements, whereas the more lateral (cerebellar hemispheres) regions are concerned with volitional movement, primarily of the extremities. In addition, a distinct medial component (flocculonodular lobe) of the cerebellum interacts strongly with the vestibular apparatus and nuclei.

Testing functions that are predominantly governed by the cerebellum involves the subject's performing tasks that require coordination or the accurate use of sensation to guide movement. A coordination task for the upper extremities is diadochokinesis, the performance of rapid alternating movements. This is tested by having the patients repetitively turn their hands rapidly from palm up to palm down on the thigh. Normal movement is smooth and coordinated.

Tasks that require a tight coordination between afferent and motor outflow include sliding one's heel up the contralateral shin or pointing slowly to a specific target, such as one's nose. During these movements, intention tremors, hypometria or hypermetria (undershoot and overshoot respectively), and coordination deficits can be observed.

Cerebellar deficits are seen primarily during voluntary movement. Hence, intention tremors, truncal and extremity ataxias, and ataxic gaits can be elicited only as the patient moves about.

REFLEXES

Reflex testing is done to determine whether the sensory limb, central processing, and motor limb of the reflex arc are intact. The reflexes used to access the somatomotor nervous system can be divided into the myotatic reflexes (also called tendon tap or muscle stretch reflexes) and the cutaneous reflexes.

The myotatic reflex is elicited by quickly stretching a muscle, often done clinically with a brisk tap with the reflex hammer on the muscle's tendon of insertion. This stretch stimulates the 1a sensory fiber which arises from the muscle spindle. This information then travels up the peripheral nerve and into the spinal cord where it makes a direct (monosynaptic) connection onto the motoneuron of the homonymous muscle. This motoneuron then projects its axon out through a peripheral nerve to the muscle, where it causes a brief muscle contraction (DeMyer, 1994).

Grading of the reflex is done on a scale of 0 to 4 with 0 being absent, 1 to 3 being a normal range from hyporeflexic to slightly hyperreflexic, and 4 being hyperreflexic. The most commonly tested myotatic reflexes (and their spinal nerve roots) are the biceps (C5–6), brachioradialus (C6), biceps (C7), quadriceps (L4) and achilles tendon (S1). If one of these nerve roots or the peripheral nerve is severed the reflex will be absent.

The myotatic reflex also indirectly assesses descending motor pathways. As the overall activity of the motoneuron is dependent upon multiple inputs, changes in these inputs, for example, due to a stroke of the cerebral hemisphere, may be reflected in an altered activity level of the motoneuron. Hence, a change in the descending modulation of the motoneuron may be seen in an abnormal reduction or increased myotatic reflex. Furthermore, as the descending commands can dramatically alter the excitability of the motoneuron, myotatic reflexes may be difficult to elicit in patients with normal nervous systems.

Cutaneous reflexes are polysynaptic spinal cord reflexes that utilize cutaneous sensations to elicit motor responses. The most commonly charted cutaneous reflex is the plantar response with its Babinski sign. To elicit this reflex the examiner strokes the lateral margin of the patient's sole from the heel to the base of the toes. The positive sign (Babinki's sign) of an upgoing great toe with the fanning of the other toes, is indicative of an upper motor neuron injury. The normal response is a downgoing (flexing) great toe (DeMyer, 1994).

There are a variety of other cutaneous reflexes that can be elicited. These (and the spinal nerve roots they test) include abdominal reflexes (T8-T12), cremasteric reflex (L1-2), anal wink (S2-4), and the bulbocavernous reflex (S3-4).

Gait

The examination of gait provides perhaps the clearest picture that the clinician will see of the interaction and integration within the nervous system. In terms of the sensory-motor aspects, gait provides the opportunity for the examiner to look at the impact on a highly stereotyped behavior of the motor and sensory loss discovered in the localized testing.

The function of gait is to transport the body from one location to another. While this may seem overly simplified, an individual without the motivation to move may not walk. If an impairment is present in the locomotor apparatus, a motivated patient will find a solution so that the overall function is preserved. These solutions, which often involve abnormal movements, provide a short-term fix that allows the patient to locomote; over time, however, these abnormal movements may lead to secondary problems and pain.

Gait is divided into the three distinct phases of stance, swing, and double support (see Figure 3). Each of these phases is further subdivided. Each phase and subphase has distinct functions that require specific muscles and joint movements (Pathokinesiology Service, 1993). In normal gait the lower limbs, roughly considered as the pelvis and lower extremities, provide the transport mechanism and the head, arms, and trunk are passengers.

While a thorough discussion of the analysis of gait is beyond the scope of this chapter, there are a few areas that all clinicians should be aware of. These components are best learned by observing normal individuals. Normal gait is fluid, casual, performed without vigilance, exhibits left to right symmetry, is variable in speed, and stereotyped in pattern from step to step with few extraneous movements. Variance in any of these domains suggests a potential dysfunction. The nonfluid gait, for example, is often characteristic of one of the many movement

GAIT ANALYSIS: FULL BODY

RANCHO LOS AMIGOS MEDICAL CENTER
PHYSICAL THERAPY DEPARTMENT

SENSORY AND
MOTOR
FUNCTION

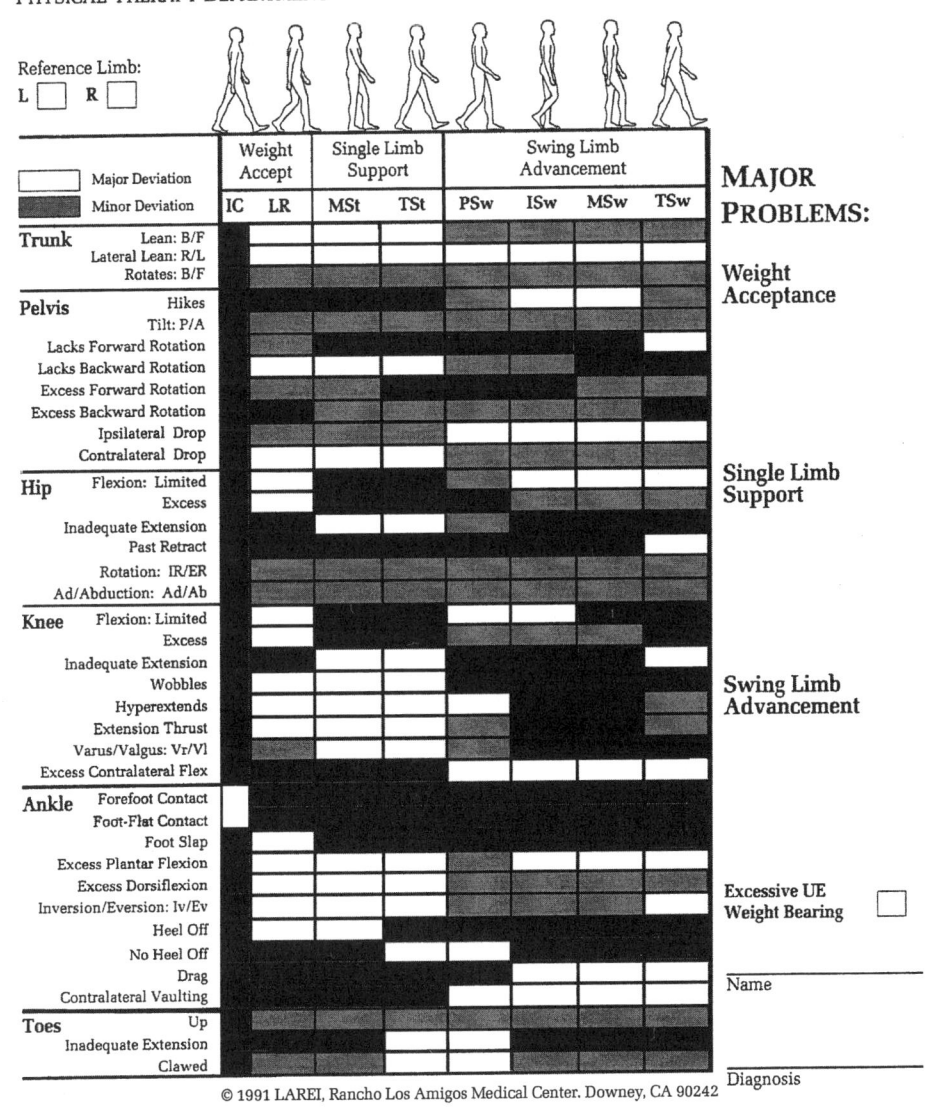

Figure 3. This gait analysis form breaks the assessment of gait into discrete phases (see top lines of chart). Each phase has specific impairments (left-hand column) which can alter function during that phase. These impairments can be critical (open box), potentially important (gray box), not important for function during that phase of gait. Abbreviations: IC = Initial Contact of the foot on the ground; LR = Loading Response as weight is transferred onto the limbs; MS = Mid Stance as the body progresses over the limb; TS = Terminal Stance as the body moves forward of the limb onto the forefoot; PSw = Pre Swing just before the limb leaves the ground; ISw = Initial Swing as the thigh advances and the foot leaves the ground; MSw = Mid Swing as the thigh continues to advance and the knee begins to extend; and TSw = Terminal Swing as the knee extends and the foot prepares for initial contact. (From *Observational Gait Analysis*, by Pathokinesiology Service, 1993, Downey, CA: Los Amigos Research and Education Institute: Author. Copyright 1991 by LAREI. Reprinted with permission.)

disorders, such as quadriplegia or stroke. A patient who uses excessive trunk motions may be using these movements to compensate for weakened lower extremity muscles that could result from a spinal cord injury or myopathy. A patient who must maintain vigilance during walking may have a sensory deficit that is being overcome through the concentration on visual inputs. The patient with an asymmetrical gait is exhibiting signs of a unilateral injury, which may range in severity from a painful bunion to a stroke.

The patient who is unable to vary speed may be exhibiting signs of Parkinson's disease. The person whose gait is nonstereotyped from step to step, or exhibits extraneous movements, may have a cerebellar injury or choreatic movement disorders such as the Huntington's disease. Thus, impairment of specific aspects of the locomotor apparatus may cause a disability of gait.

Summary

In this chapter we have provided an overview of the clinical evaluation of sensory-motor impairment, with a focus on those impairments that lead to movement disabilities. The recommended content of the exam includes history and observation, assessment of range of motion/muscle tone, sensation, strength, coordination, reflex responses, and gait analysis. Normal gait is a functional activity that requires the integration of the sensory-motor system and thus is an essential component of the suggested sensory-motor evaluation.

Patients come to a clinic because of a loss of function as reflected in pain or disability. It is the clinician's task to discover the cause of this functional loss, usually by investigating the possible presence of pathophysiology and impairment. If the functional loss is in the mobility domain, an area often treated by physical therapists, the cause of the loss can often be found in impairments of the sensory-motor system.

Assessment of a neurologically impaired patient requires a thorough neurological evaluation, of which the sensory-motor components are isolated segments. Assessment of sensory-motor impairments in isolation is not meaningful to the patient unless those impairments impact his or her disability or pain. It is important to integrate findings from the assessment at the pathophysiology and impairment levels with the assessment of function in order to address the patient's goals, usually for increased function.

Sensory-motor evaluation must be carried out in the context of a comprehensive model of assessment as suggested by the World Health Organization (1980) or the National Institutes of Health (1993). The nervous system is of such complexity that the patient with neurologic pathology should become the focus of attention of many disciplines and receive a battery of overlapping evaluations designed to pinpoint the pathophysiology, identify the impairment and strengths, assess the disabilities and abilities and environmental constraints. Treatment should then be instituted, with the patient's goals in mind, to correct or prevent progression of the pathophysiology, to correct impairments or enhance strengths, to minimize disabilities and maximize abilities, in an environment with as few limitations as possible.

References

Adams, R. D., & Victor, M. (1993). *Principles of Neurology* (5th ed.). New York: McGraw-Hill.

Beasley, W. C. (1961). Quantitative muscle testing: Principles and applications to research and clinical services. *Archives of Physical Medicine and Rehabilitation, 42,* 398-425.

Cyriax, J. (1978). *Textbook of orthopaedic medicine: Vol. 1. Diagnosis of soft tissue lesions.* London: Bailliere Tindall.

Daniels, L., & Worthingham, C. (1986). *Muscle testing: Techniques of manual examination* (5th ed.). Philadelphia: Saunders.

DeMyer, W. (1994). *Technique of the neurologic examination.* New York: McGraw-Hill, Inc.

Florence, J. M., Pandya, S., King, W. M., Robison, J. D., Signore, L. C., Wentzell, M., Province, M. A. (1984). Clinical trials in Duchenne dystrophy standardization and reliability of evaluation procedures. *Physical Therapy, 64,* 41-45.

Greenberg, D. A., Aminoff, M. J., & Simon, R. P. (1993). *Clinical Neurology* (2nd ed.) Norwalk, CT: Appleton & Lange.

Hinton, G. E., McClelland, J. L., & Rumelhart, D. E. (1986). Distributed representations. In D. E. Rumelhart & J. L. McClelland (Eds.), *Parallel distributed processing: Vol. 1. Foundations* (pp. 77-109). Cambridge, MA: MIT Press.

Kandel, E. R. (1995). Construction of the visual image. In Kandel, E. R., Schwartz, J. H., & Jessell, T. M. (Eds.), *Essentials of neural science and behavior* (pp. 387-405). Norwalk, CT: Appleton & Lange.

Kendall, F. P., McCreary, E. K., & Provance, P. G. (1993). *Muscles, testing and function.* Baltimore: Williams & Wilkins.

Kessler, E. S. (1989). The neurologic examination. In W. J. Weiner & C. G. Goetz (Eds.), *Neurology for the non-neurologist* (2nd ed., pp. 1-14). Philadelphia: Lippincott.

Lamb, R. L. (1985). Manual muscle testing. In J. M. Rothstein (Ed.), *Measurement in physical therapy* (pp. 47-56). New York: Churchill Livingstone.

Lilienfeld, A. M., Jacobs, M. & Willis, M. (1954). A study of the reproducibility of muscle testing and certain other aspects of muscle scoring. *Physical Therapy Review, 34,* 279-289.

Mayhew, T. P., & Rothstein. J. M., (1985). Measurement of muscle performance with instruments. In J. M. Rothstein (Ed.), *Measurement in physical therapy* (pp. 57-102). New York: Churchill Livingstone.

National Institutes of Health. (1993). *Research plan for the National Center of Medical Rehabilitation Research* (NIH Publication No. 93-3509). Bethesda, MD: Author.

Nolte, J. (1993). *The human brain.* St. Louis: Mosby Year Book.

Palmer, M. L. & Epler, M. E. (1990). *Clinical assessment in physical therapy.* Philadelphia: Lippincott.

Pathokmesiology Service. (1993). *Observational gait analysis.* Downy, CA: Los Amigos Research and Education Institute: Author.

Rothstein, J. M. (1985). Measurement and clinical practice: Theory and application. In J. M. Rothstein (Ed.), *Measurement in physical therapy* (pp. 1-46). New York: Churchill Livingstone.

Saraniti, A. J., Gleim, G. W., Melvin, M., & Nicholas, J. A. (1980). The relationship between subjective and objective measurements of strength. *Journal of Orthopedics and Sports Physical Therapy, 2,* 15-19.

Williams, P. L., & Warwick, R. (Eds.). (1980). *Gray's anatomy.* Philadelphia: Saunders.

World Health Organization (WHO). (1980). *International classification of impairments, disabilities, and handicaps: A manual of classification relating to the consequences of disease.* Geneva, Switzerland: Author.

9

Assessment of External Prostheses

JACQUELIN PERRY AND EDMOND AYYAPPA

Prostheses are designed to replace lost function. Historically, this term represented artificial limbs, which overcome the limitations of amputation. Today, these devices are classed as external prostheses because technical advances during the most recent decades have led to internal prostheses, which replace joints and bone segments within the limb. In this review of external prostheses, attention is focused on the management of the lower-limb amputee as the incidence is eleven times greater than the equivalent upper-extremity surgery. Following a historical review, current scientific and clinical practices are illustrated by summarizing the analysis and management of the below-knee amputee.

Progress in prosthetic development has followed advances in surgical technique, materials, engineering sophistication, and delineation of the biomechanics and physiology of gait. In ancient times amputations were urgent, life-saving measures. Surgical speed was essential as there were no anesthetics. Surviving the procedure was often doubtful because of the accompanying shock and the threats of hemorrhage and sepsis (Sanders, 1986).

HISTORICAL DEVELOPMENT

The first prostheses were simple devices designed to restore limb length. Ancient art shows that most were a wooden peg with either a hollow socket to contain the residual thigh or a trough to support the bent knee when the amputation was within the lower leg (Peltier, 1993). There is at least one report of more elaborate limbs such as an iron leg for a queen who lost a leg in battle (Vedas of India,

---

JACQUELIN PERRY   Medical Consultant, Pathokinesiology, Rancho Los Amigos Medical Center, 7601 E. Imperial Highway, Downey, California 90242.   EDMOND AYYAPPA   Veterans Affairs Medical Center, 5901 East 7th Street, Long Beach, California 90822.

Rehabilitation, edited by Goldstein and Beers. Plenum Press, New York, 1998.

3500–1800 B.C.) (Vitali, Robinson, Andrews, & Harris, 1978). The earliest museum artifact (c. 300 B.C.) was a prosthesis constructed of bronze and iron around a wooden core (Gillis, 1954). These more elaborate devices point to the added concern for reproducing anatomical form.

Whether wood or metal, prostheses were rigid supports until 1529, when Ambrose Pare applied the skill of the armor makers to design the first articulated limb, with knee, ankle, and midfoot mobility (Gillis, 1954; Sanders, 1986). These skilled artisans closely reproduced the anatomical form of the limb. Wood, leather, and metal were variably used. Further advances over the next three centuries included artificial tendons to link knee and ankle motion, waist belts and shoulder suspenders for better fixation, cushioning with rubber ankle inserts and soles, and rawhide covering of the wooden leg (A. A. Marks, 1931; G. E. Marks, 1888). This series of advances represents the culmination of clinical analyses to reproduce function and comfort. Surgical advances included hip disarticulation (1803) and Syme's amputation (1842). Following the introduction of ether anesthesia (1846) and antisepsis (1860) procedures could be more selective. During the nineteenth century there were several seemingly modern developments. Parmelee patented a suction socket for suspension of a prosthesis with a polycentric knee and an articulated foot (1863) (Sanders, 1986). Marks patented the first rubber foot (1863) (G. E. Marks, 1888). Herman substituted aluminum for steel (1868). Bigg's book on "amputations and artificial limbs" (1885) included anatomical studies on alignment (Sanders, 1986). Martin (1918) used "scientific principles of anatomy and physiology" to develop a system for fabricating artificial limbs from measurements and casts of the sound and residual limbs (Sanders, 1986, p. 24).

Scientific Assessment

All of the developments mentioned represent independent thoughtful analysis of human form and function by critical observers (Shurr & Cook, 1990). During the nineteenth century the Weber brothers (Weber, 1851) and, later, Braune and Fischer (1985) conducted experimental studies on the mechanics of gait (Steindler, 1953). Murray (1967) and Muybridge (1979) photographically documented the motions of walking. Improved instrumentation allowed other investigators to detail individual techniques of gait analysis (Elftman, 1934, 1938; Hubbard & Stetson, 1938; Patek, 1926; Schwartz, 1934; Schwartz, Trautmann, & Heath, 1936). It became evident during World War II, however, that the available artificial limbs were not meeting the needs of the amputee. In response, the National Research Council initiated a major prosthetic research program (Strong, 1970). Subsequently, the first fundamental studies of the function and fit of prostheses and their relationship to normal gait were begun in 1945 at the University of California, Berkeley (UCB) (Wilson, 1970).

Berkeley Prosthetic Project

Eberhart and Inman with the support of the National Research Council and the Veterans Administration established a team of professionals to upgrade the quality of prostheses provided to World War II veterans (Strong, 1970; Wilson,

1992). Both investigators were keenly aware of the limitations of the existing prostheses and were especially suited to the challenge of improving the situation. Eberhart, an associate professor of civil engineering, had personally experienced the limitations of existing prostheses through his relatively recent traumatic transtibial amputation. Inman, an assistant clinical professor of orthopaedic surgery, had a background of kinesiological research. They postulated that the inefficiencies of a prosthesis were reflected at the interface between the socket and the residual limb. The amputee's problems were discomfort, even frank pain, heaviness, skin irritation, and visible gait abnormalities.

Utilizing the earlier studies of human walking (Steindler, 1953; Murray, 1967) as a model, Eberhart and Inman established the University of California, Berkeley, Prosthetic Research Project. Recognizing the background that prior designers had lacked, their first objective was to develop scientific criteria for prosthetic design through a series of fundamental studies on the biomechanics of human gait (H. D. Eberhart, 1947). Their second objective was to apply these basic studies to the gait of the amputee and to prosthetic design. Comprehensive gait analysis was the result. With the aid of reflective markers (even bone pins on selected occasions), a glass walkway, multicamera photography (still, cine, and interrupted light), and imbedded forceplates, detailed three-dimensional data were obtained on the individual joint motions and weight-bearing forces occurring in gait (H. D. Eberhart, Inman, & Bressler, 1968). Force studies in the amputee were enhanced by the use of a force pylon serving as a segment of the leg. Dynamic electromyography with surface electrodes defined muscle action.

The Berkeley project specifically studied transverse rotation as this was considered to be a significant source of residual limb skin irritation (H. Eberhart, Elftman, & Inman, 1968). For more precise tracking of the joints, Steinman pins, representing the centers of rotation, were inserted into selected bony sites on the pelvis, femur, and tibia. Pin vibration and instability posed major technical challenges, limiting useful data to 12 of the 27 subjects studied (Eberhart, 1947). The researchers found two periods of synchronous rotation of the pelvis, femur, and tibia. Internal rotation occurred during the initial stance interval between heel strike (no load) and foot flat (peak load). This was followed by external rotation through the rest of stance. Motion between the femur and pelvis averaged 9 degrees. Between the tibia and femur the average arc was 8 degrees. Viewing of the limbs through a glass plate walkway allowed measurement of tibial rotation on the foot, internally during initial stance and then externally after mid stance for an average arc of 15 degrees. The ankle pin (malleolar yoke) was more difficult to maintain, limiting the data to 3 subjects. It was concluded, however, that transverse rotation was a significant function and should be provided by a torque absorber unit at the ankle.

The initial biomechanical gait studies of normal ankle function by the UCB project identified significant plantar flexion, dorsiflexion, and eversion and inversion motions and moments as the limb rotated forward over the supporting foot (Bresler & Berry, 1950). A high-impact force with floor contact also was defined. At that time the prosthetic equivalent was an articulated ankle with a wooden foot. Some operated as a simple mechanical hinge, while others included motion in all three planes using various rubber bumper designs to restrain motion and lessen impact (E. M. Wagner & Catranis, 1968). The attention of the Berkeley effort focused on the analysis of the different types of mechanical ankle and some new designs were contributed. All styles, however, were heavy and the rubber bumper systems remained functionally inadequate.

From the project's many studies the kinematics and kinetics of normal locomotion were defined. These included the paths of motion at each joint, segment and center of gravity displacement patterns, timing and relative intensity of the major muscle groups, and the ground reaction forces (H. D. Eberhart, Inman, & Bressler, 1968). The Berkeley prosthetic team expanded knowledge to include the shear forces that occur as the body rolls forward over the supporting limb.

ABOVE-KNEE PROSTHETIC ASSESSMENT

Study of the above-knee amputee by the Berkeley group initially emphasized the function of the unilateral amputee as the problems of this group appeared more critical at that time (H. Eberhart, Elftman, & Inman, 1968). Motion analysis showed a fully extended knee starting in terminal swing and continuing through stance. The inadequate ankle plantar flexion that followed heel strike and threatened knee stability was attributed to dependence on an ankle bumper in place of the lost pretibial muscle control. Active ipsilateral thigh control and postural adaptations by the sound limb and trunk were identified as the variable mechanisms used by individual amputee to assure knee extension stability. Rotation of the fully extended limb across the ankle prior to heel rise caused a maximum rise of the hip (and center of gravity) which was interpreted as vaulting. Compensatory actions by the sound limb were identified. In swing, the inability of the prosthetic foot to generate a propelling force to initiate limb advancement was interpreted as a need to restrict prosthesis weight so that the work of the hip flexors would not be excessive. The amputee also demonstrated rapid hip extension in terminal swing to use tibial inertia as a means of completing knee extension in preparation for stance (Radcliffe, 1970c). These findings of excessive knee extension in stance and excessive hip action in swing formed the basis for others to design more sophisticated knee joints to replace the then-dominant single-axis constant friction joint.

The biomechanical response to the problem of residual limb discomfort was twofold. Torque absorbers were designed, but the final solution was the combination of improved socket design, simplified limb hygiene (Levy, 1970), and surgical preservation of muscle function by myodesis when possible, in addition to more normal joint mechanics. Anatomical contouring and total contact were introduced. For the above-knee amputee, the UCB project developed the quadrilateral socket, which emphasized total contact, reduced anterior-posterior dimension, and theorized skeletal weight-bearing transfer through the ischial tuberosity (Lehneis, 1984, 1985; Radcliffe, 1970c). Subsequent developers have contributed many design innovations based on amputee feedback as well as prosthetist, engineer, and physician input. Most notable are the narrow M-L designs of CAT–CAM and NSNA (Lehneis, 1985; Lehneis, Chu, & Adelglass, 1984; Long, 1975, 1985; Sabolich, 1985).

BELOW-KNEE PROSTHETIC ASSESSMENT

In 1957 the Berkeley project was specifically commissioned to reconsider the below-knee (BK) prosthesis and an advisory conference was held (Wilson, 1969). The basic BK prosthesis was attached to the limb with a thigh lacer that included steel knee joints and a foot using some type of articulated ankle. Detailed review of the normal and BK amputee gait data resulted in a totally new approach that led to two developments: the PTB (patellar-tendon bearing) prosthesis and the SACH foot (solid ankle, cushion heel) (Ayyappa, 1994; Radcliffe & Foort, 1961; Wilson, 1970). The improved BK socket was a PTB design which closely followed the contours of the proximal tibia (Foort, 1970). The PTB prosthesis replaced the thigh lacer with supracondylar fixation, again using the advantages of anatomical contour and total contact. The SACH foot has a solid wood (or plastic) keel imbedded in the proximal portion of a stiff foam foot that allowed a slight degree of forefoot flexibility while the cushion heel lessens the impact of initial floor contact and through vertical compression was thought to simulate articulated ankle planter flexion. Material compressibility replaced the mechanical joint of the single-axis foot. Foot contour, alignment, and elasticity facilitated progression onto the forefoot. This design followed careful analysis of the essential degrees of mobility. The SACH foot became the standard terminal device for most amputees, though feet with multidirectional articulated ankles continue to be preferred for a few situations, primarily those involving uneven terrain.

The era of the University of California, Berkeley, prosthetic project represented the most concentrated period of prosthetic advancement. The comprehensiveness of the basic gait studies and their application to the biomechanics of the different prosthetic needs for the above-knee (Radcliffe, 1970c), below-knee (Radcliffe, 1970a), and Syme (Radcliffe, 1970b) amputations established design criteria that still are appropriate today. Contributions of subsequent investigators continue to be largely based on the Berkeley criteria.

Extensive calculations and interpretations were required to relate normal and amputee gait data to the problems of prosthetic design. Data reduction was a slow process, however, as all motion measurements had to be performed by hand. There were no automated film analyzers to identify the motion patterns, nor computers for rapid data processing. Hence the number of subjects studied was limited. The project also had the additional depth of considering all three planes of motion, in contrast to prior studies which analyzed only the sagittal plane of gait progression (Inman, Ralston, & Todd, 1981). Subsequent investigators with the aid of more advanced instrumentation have replicated and expanded the Eberhart–Inman work but have not found it in error.

MODERNIZATION OF GAIT ANALYSIS

Introduction of the Berkeley gait laboratory capability into other academic and clinical centers was slow as the techniques available at that time were very labor intensive and consequently costly. Murray adopted the Berkeley technique of interrupted light photography and manual calculation of joint motion to develop a much larger data base of normal gait which became the basic reference data (Murray, Drought, & Kory, 1964). One of her clinical studies was the gait of above-knee amputees using a constant friction knee joint, which demonstrated the absence of knee flexion in stance and a walking speed 66% of normal (Murray, Sepic, Gardner, & Mollinger, 1980).

Efforts to automate motion analysis began with Karpovich's introduction of a single axis goniometer for the ankle and the knee (Karpovich & Karpovich, 1959;

Tipton & Karpovich, 1965). This concept was advanced to a three-dimensional unit by Chou (Kettelkamp, Johnson, Smidt, Chao, & Walker, 1970) and a total limb system by Lamoreux (1981). (Zuniga, Leavitt, Calvert, Canzonari, and Peterson (1972) combined the use of a knee electrogoniometer with footswitches to document the foot-floor contact patterns and walking speeds as well as knee motion. His study of above-knee amputee gait confirmed the lack of knee flexion in stance and displayed a foot support pattern characterized by a marked reduction in footflat time leading to a 76% increase in forefoot support. Sutherland and Hagey (1972) introduced the first step of automated photographic motion analysis with a technique that simplified the measurement of joint angles. With the use of motion picture film a series of limb postures occurring during the gait cycle were stored on individual frames. The x, y coordinates of each joint marker were defined with a Vangard motion analyzer rather than by manual measurement. Further automation with an ultrasound system provided instantaneous marker location and transmission of coordinates to a computer, which created motion graphs. The remaining disadvantage of having to manually designate the markers on each frame was overcome by the introduction of remote optical sensing using cameras which record the marker locations digitally. Two types of marker systems have been developed. The optoelectrical technique (Selspot) uses electrically activated light-emitting diodes which flash in a predetermined sequence by multiplex control (Woltring & Marsolais, 1980). The other approach (Vicon) uses multiple cameras and passive infrared retroreflecting markers (Andrews & Jarrett, 1976). Both techniques have been modified and embellished by subsequent developers. Today there are several commercially available systems with computer programs that define motion occurring in the three basic planes as well as intermediary views.

The designation of muscle function by dynamic electromyography has been refined by advances in electrode design and signal recording. Surface electrodes now contain common-mode-rejection electronics to provide cleaner signals and computer processing permits more discrete signals (Basmajian & Deluca, 1985). Intramuscular wire electrodes now are more flexible, following the introduction of 50-micron stainless steel alloy wire (Basmajian & Stecko, 1962). The advantage of the wire electrodes is the ability to accurately differentiate the action of adjacent muscles by exclusion of crosstalk (Perry, 1992). Surface recording, however, is more convenient. An EMG analyzer is available which can define the phasing and relative intensity of the muscle action during gait (Perry, Bontrager, Bogey, Gronley, & Barnes, 1993).

Ground reaction forces continue to be used to identify the impact of limb loading. In addition, moment calculations, initially interpreted as demands on the limb during walking (Bresler & Berry, 1953), have been further refined as quantified indicators of muscle control. For this purpose instantaneous vectors with corrections for inertia and gravity are related to the corresponding kinetics. By including the rates of moment generation "power" also is determined (Winter, 1983). These techniques have been presented as a means of identifying muscle function without using dynamic electromyography (Czerniecki, Gitter, & Munro, 1987). When control of all limb joints is in normal dynamic balance, the power calculations are compatible with the muscle action implied by EMG (Winter, 1983). Such is not the case with below-knee amputee gait, however, as reduced knee motion and secondarily decreased power calculations found during the mechanics of weight acceptance actually are accompanied by increased muscle action (Torburn, Perry, Ayyappa, & Shanfield, 1990). It appears that the prosthetic foot, as a compromise between material optimization and function, introduces some artifacts.

CURRENT SCIENTIFIC ASSESSMENT OF BELOW-KNEE PROSTHETICS

Today the major clinical concern is the below-knee (trans-tibial) amputee. Current gait analysis studies primarily focus on comparing the effectiveness of different types of prosthetic feet. Prosthetic foot designers attempt to passively reproduce, by material quality, the normal dynamic functional balance between mobility and stability. The greatest difference is between the relatively rigid SACH foot and the mobile hinge of a single-axis foot (Goh, Solomonidis, Spence, & Paul, 1984). Motion analysis showed the articulated ankle improved weight acceptance stability by providing significant plantar flexion, allowing an earlier foot flat posture (Doane & Holt, 1983). However, the arc of knee flexion and the timing and intensity of the quadriceps did not differ (Culham, Peat, & Newell, 1986). Stance limb progression was aided by the mobile ankle in two ways. Footswitch analysis showed the mobile ankle had a longer period of single limb stance while total stance was shorter. In addition, there was more prolonged hamstring action with the SACH foot; this implies that a forward lean was used to improve progression over the less yielding foot. These functional advantages of a mobile prosthetic foot would be significant in a marginal walker, but the single-axis design is heavier and less durable.

Recently the use of new energy-storing materials has been applied to prosthetic foot design. The initial objective was to facilitate running as the presence of normal knee control gives the BK amputees considerable functional potential (Burgess, Hittenberger, Forsgren, & Lindh, 1983). All of the foot designs emphasize controlled dorsiflexion mobility for greater "push-off." It is assumed that these new dynamic, elastic-response prosthetic feet, as they are called, would be advantageous for the average walker, in particular, by reducing the greater than normal energy cost currently experienced by amputees (Waters, Perry, Antonelli, & Hislop, 1976). The most mobile design is the long-shafted Flex foot, yet it has not lessened the muscular demands of weight acceptance. The simulated "ankle plantar flexion" during limb loading by the FLEX foot, (4°) is no better than the SACH (6°) foot. Both are markedly less than the normal controlled yet rapid ankle plantar flexion of 12 degrees used to reduce the propulsive effect of the heel rocker by allowing early forefoot contact. Both prosthetic feet cause a significant delay in attaining the stability of foot flat (J. Wagner, Sienko, Supan, & Barth, 1987).

The intact limb uses a modest arc (15°–20°) of knee flexion to absorb the shock of floor contact (Perry, 1992). Prosthetic feet rely on a cushion heel. The time needed for adequate cushion compression delays the drop of the forefoot to the floor. This perpetuates an unsteady, heel-only source of support that requires increased active muscular control of the knee and hip to assure weight-bearing stability. This subtle source of instability is obscured by two findings. Displacement of the customary heel marker gives a false reading of plantar flexion (Torburn et al., 1990). Heel cushion compression delays the rate of initial tibial advancement, resulting in reduced weight acceptance knee flexion for the BK amputee compared to normal. This difference is reflected in calculations of subnormal moments and powers (Barth, Schumacher, & Sienko Thomas, 1992; Gitter, Czerniecki, & DeGroot, 1991). These findings have been interpreted as a sign of reduced muscle demand and a corresponding conservation of energy has been attributed to the dynamic elastic-response feet. Direct EMG recordings, however, show significantly increased action by both the quadriceps and hip extensors (Torburn et al., 1990). In addition, energy cost measurements are significantly higher than normal for all of the feet tested (Powers, Torburn, Perry, & Ayyappa, 1994). The source of this discrepancy appears to be the mechanics of the prosthetic hindfoot, described earlier. Further investigation is needed.

Progression over the supporting foot, however, is greatly facilitated by the elastic dorsiflexion of the Flex foot (20°) compared to the SACH foot (11°) (Torburn et al., 1990; J. Wagner et al., 1987). Unloading the foot in the subsequent double support period provides a rapid elastic recoil to prepare the limb for swing. The normal foot has a 30° recoil arc (from 10° dorsiflexion to 20° plantar flexion). The combination of the terminal stance dorsiflexion torque and the arcs of rapid motion is interpreted by power calculations as the foot's "push-off" capability (Barth et al., 1992; Gitter et al., 1991). Weight transfer to the other limb also reflects prosthetic foot mobility. The vertical ground reaction force on the sound limb was found to be greater than normal with the SACH and all the prosthetic foot designs except the Flex foot (Torburn et al., 1990; J. Wagner et al., 1987).

The inconsistencies in the interpretations of quantified gait data reflect the complexities of both normal and prosthetic gait. These will be resolved as new analytic techniques are developed that specifically address the areas of conflict.

Instrumented gait analysis provides the basic principles for prosthetic design and patient management. It also allows objective measurement of the results. Between these two extremes are the events of daily care. Effectiveness in this area relies on the clinical techniques of assessment.

Clinical Assessment

Acceptance of the prosthesis is the ultimate requirement of a successful rehabilitative outcome for the amputee. The determinant may represent a critical balance between cosmetics and function even though these two characteristics occasionally compete. A prime example of conflict is the use of a locked knee on an above-knee amputee's prosthesis. The cosmetic detraction of a stiff-knee gait has been traded for weight-bearing instability.

Although function generally precludes cosmetics, there are some exceptions. The perceptions and expectations of patients are enormously variable. Cosmetics, unlike function, is primarily an art, the standards of which, lie in the eye of the beholder, the patient.

Function can be more objectively assessed. The outcome is dependent upon how well the qualities of the prosthesis address the specific physiological condition of the patient. There are three primary factors:

1. Socket interface shape: contour, reliefs, materials
2. Componentry: design and selection
3. Alignment: relationships of components and socket

The details of these criteria differ with the level of the amputation. Since below-knee amputations represent the majority today (Bresler & Berry, 1953), this chapter focuses on that group.

Past custom designated the terminal segment of an amputated limb as a "stump." Recent amputees who are adjusting to a new body image, however, often are extremely offended by the use of the word "stump." Hence the phrase "residual limb" has been adopted. It is one step in assisting the amputee to establish an improved self-image.

Socket Interface Shape

The socket is the junction between the residual limb and the prosthesis. At this interface the forces of walking are exchanged. Between the skeletal structure that bears body weight and the prosthesis that reacts with the ground are the soft tissues and skin of the limb. The contour and material qualities of the socket must intimately match the qualities of the limb to assure a noninjurious transmission of the ever-changing forces of gait. Consequently the basic determinants of the socket interface shape are the characteristics of the amputated limb. A transparent test socket has become standard practice as a fitting evaluation tool for amputees at all levels (Ayyappa, 1994; Hamontree & Snelson, 1973; Quigley, 1985).

Measurement of the residual limb is the first phase of assessment. To determine residual limb length of a below-knee amputee the natural tendency is to make the measurement from the medial tibial plateau to the distal end, but this value does not relate to the prosthesis. By contrast, determining the length from mid-patella-tendon (half of the distance from the inferior border of the patella to the tibial tubercle) has prosthetic significance as the mid-tendon landmark is clearly defined on the socket wall. Thus, this latter measurement has relevance even when the patient is not present. The difference in length from the mid-tendon and medial tibial plateau landmarks is less than a quarter of an inch, but it can be a factor in fitting accuracy. The optimum residual limb length generally is considered to be 8 inches or 20 cm. For a child or young adult the landmark is the distal border of the gastrocnemius.

The quality of the soft tissues is judged by limb shape (conical, cylindrical, or bulbous), thickness of the distal padding, location of the suture line, general subcutaneous tissue quality (light or heavy), and atrophy of the residual limb or thigh (a common finding with thigh lacers). There is close inspection of the skin for discoloration, abrasions, boils, edema, pressure points, and healed and unhealed scars. Evidence of bursae, spurs, redundant tissue, and muscle bunching also is sought. In assessing pain it is particularly important to differentiate phantom pain from symptoms induced by the prosthesis.

Any degree of knee contracture is carefully measured. It is a general rule that a below-knee amputee with a knee flexion contracture greater than 25 degrees cannot ambulate on a conventional patella-tendon–bearing prosthesis. Design variants such as a bent-knee prosthesis might be appropriate but carry significant cosmetic and functional disadvantages.

Socket contact and contour adequacy is judged from the health of the skin and underlying soft tissues and absence of prosthetically induced pain. Following the introduction of the patella-tendon–bearing prosthesis and its popliteal counter pressure for stability (Foort, 1970; Radcliffe & Foort, 1961; Wilson, 1969), there has been a progressive drive toward total contact (Knapp & Cummings, 1992; Staros & Goralnik, 1981). Today the term is "total surface bearing" (TSB). Moderated contouring is still used about the patellar tendon, pretibial muscle apex, medial condylar flare, and popliteal area to enhance weight-bearing stability (Breakey, 1973; Quigley & Wilson, 1975).

Socket interface material assessment is based on the amount of cushioning the residual limb tissues require for comfort and skin health. An unlined, "hard socket" is appropriate for the posttrauma amputee with good tissue integrity as it provides superior hygiene, excellent proprioception, and a streamlined, bulkless silhouette. The hard socket, however, is unforgiving of any residual limb problems and therefore is limited to posttrauma amputees with optimal tissue coverage. The "soft" sockets are generally the interface of choice, especially for dysvascular patients. These choices include liners of Kemblo, Spenco, silicone gel, air cushion, or a flexible socket design. A recent development of significance is the TEC liner, which provides 3/8" of gelatinous material against the skin. Criteria for selection are shear or pressure sensitivity, ulceration history, excessive perspiration, pain, durability, and fabrication ease.

Fabrication of the socket has always been a subtle art with effectiveness varying among prosthetists. Recently computerized socket designs using CAD–CAM techniques have been introduced. Assessment of this technique's effectiveness is based on accuracy of fit, turn-around time, availability to the practicing prosthetists, and cost. Currently, there is greater use of the Automated Fabrication of Mobility Aids (AFMA) system in the developing countries, where service is scant, than in the United States. Superiority over the hand-casted and modified socket of an experienced prosthetist has not been confirmed.

Component Design and Selection

Assessment of componentry relates primarily to the choice of shaft pylon, suspension system, and prosthetic foot. To date these choices are primarily subjective.

Historically, the limb shaft was a willow wood shell covered with highly varnished parchment leather (A. A. Marks, 1931; G. E. Marks, 1888). Today there are two approaches. Most common is the continuation of the exoskeletal design with the wood largely replaced by rigid polyurethane foam. Cosmetics are preserved by shaping the leg to match the sound limb. The exoskeletal limb is considered to be more durable, less vulnerable to water damage, and to cost less initially. An alternate approach is an endoskeletal design. There is a central supporting structure with a cosmetic foam cover. The advantages are adjustability throughout the process, component interchangeability, lightness, and improved cosmetics, but there also is a limit on subject weight. For special situations there is the choice of ultralight, heavy-duty, or unique sport or occupational designs.

Suspension systems for the below-knee prosthesis contribute to knee control as well as prosthetic fixation as they all extend at least to the supracondylar area. The selection criteria relate largely to the limb's anatomy such as bony prominences, obesity, ligamentous laxity, and skin quality. The choices are a supracondylar cuff, supracondylar wedges, an elasticized cloth suspension sleeve, and a thigh-lacer with knee joints. The thigh-lacer provides maximum medial-lateral stability and transfers above the knee in cases of chronic weight-bearing ulcerations.

Prosthetic foot selection has been complicated by the introduction of the dynamic, elastic-response feet (DER). Rather than having a choice only between the stable SACH foot and one with an articulated ankle, there are now a great many designs. As discussed in the gait analysis section all of the new designs seek to improve forward roll over the supporting foot (i.e., "push-off"). DER foot designs vary as to the site of flexibility. These include forefoot, midfoot, ankle, and tibial shaft. Except for the dominant flexibility of the Flex foot, no functional differences have been documented. There also is the multiaxial ankle joint, which adds a degree of eversion–inversion and sometimes rotation. As a result, the considerations in selecting a prosthetic foot are cost, weight, durability, activity level of the amputee, and experience of the prosthetist (G. E. Marks, 1888; Michael, 1987).

Alignment

The positional relationship between the socket and prosthetic foot is critical, yet also highly subjective. The goal is to place the knee in a stable (minimally flexed) weight-bearing posture without causing hyperextension in late stance and to have the lower limb follow a normal path of motion in swing. Static or "bench alignment" uses the subcutaneous crest of the tibia (tibial blade) to indicate vertical alignment of the residual limb. In the sagittal plane this landmark, at its origin at the tibial tubercle, is angled about 5 degrees forward of the perpendicular to the tibial plateau that serves as the supporting surface for the knee joint. Hence, if one aligns the socket by the tibial blade, the limb is set with the tibia tilted slightly forward to avoid a backward thrust as the amputee walks. The exact angle still is being debated. Final positioning of the limb is determined by observational analysis of the subject's gait and feedback from the amputee. Smoothness of the gait, medial-lateral verticality of the shank, the absence of abnormal motions in swing such as a whip, and an erect trunk posture are the observational criteria used for assessment. Comfort and ease of walking are the amputee's criteria. A 1-subject study of repeated realignment by an experienced prosthetist over a 2-year period documented inconsistencies in the alignment accepted as being good (Hannah & Morrison, 1984; Zahedi, Spence, Solomonidis, & Paul, 1986), a decision confirmed each time by three other prosthetists. It was concluded that there was a range of alignment variations that were satisfactory to both the prosthetist and amputee.

In alignment of the coronal plane, there is an attempt to mimic the slight varus moment seen at the knee during midstance of normal human locomotion. This moment, usually, is achieved by a slight medial inset of the prosthetic foot relative to the socket, taking advantage of the good weight-bearing areas in the proximal-medial and lateral-distal regions of the residual limb. Pressure in these areas is well tolerated and avoids the complications of excessive pressure on the cut end of the tibia (medial-distal) and the head of the fibula (lateral-proximal). Although observational analysis is appropriate for general clinical care, the alignment of amputees with complex fitting problems is best resolved through objective confirmation using forceplate and motion data.

Conclusion

Instrumented gait analysis provides the most objective assessment of prosthetic function. This information serves as guidelines for both prosthetic design and clinical application. The daily practice of amputee management, however, depends on the subjective skills of the prosthetist. Thus, it is essential for optimum advancement of prosthetic effectiveness that researchers and clinicians work closely together.

References

Andrews, B. J., & Jarrett, M. O. (1976). *On-line kinematic data acquisition.* Glasgow, Scotland: University of Strathclyde.

Ayyappa, E. (1994). *Prosthetic desk reference: Prosthetic program manual.* Dominguez Hills, CA: California State University–Dominguez Hills.

Barth, D. G., Schumacher, L., & Sienko Thomas, S. (1992). Gait analysis and energy cost of below-knee amputees wearing six different prosthetic feet. *Journal of Prosthetics and Orthotics, 4*(2), 63-75.

Basmajian, J. V., & Deluca, C. J. (1985). *Muscles alive: Their functions revealed by electromyography.* Baltimore: Williams & Wilkins.

Basmajian, J. V., & Stecko, G. A. (1962). A new bipolar indwelling electrode for electromyography. *Journal of Applied Physiology, 17,* 849.

Braune, W., & Fischer, O. (1985). *On the centre of gravity of the human body as related to the equipment of the German infantry soldier.* Berlin, Germany: Springer-Verlag.

Breakey, J. W. (1973). Criteria for use of supracondylar and supracondylar-suprapatellar suspension for below-knee prostheses. *Orthotics and Prosthetics, 27,* 14-18.

Bresler, B., & Berry, F. R. (1950). *Energy characteristics of normal and prosthetic ankle joints.* (Prosthetic Devices Research Project, National Research Council, Series 11[22]). Berkeley: University of California Institute of Engineering Research.

Bresler, B., & Berry, F. R. (1953). *Energy and power in the leg during normal level walking.* (Prosthetic Devices Research Project, National Research Council). Berkeley: University of California.

Burgess, E. M., Hittenberger, D. A., Forsgren, S. M., & Lindh, D. V. (1983). The Seattle prosthetic foot—a design for active sports: Preliminary studies. *Orthotics and Prosthetics, 37*(1), 25-31.

Culham, E. G., Peat, M., & Newell, E. (1986). Below-knee amputation: A comparison of the effect of the SACH foot and single axis foot on electromyographic patterns during locomotion. *Prosthetics Orthotics International, 10,* 15-22.

Czerniecki, J. M., Gitter, A., & Munro, C. F. (1987). Muscular power output characteristics of amputee running gait. *Archives of Physical Medicine and Rehabilitation, 68,* 637.

Doane, N. E., & Holt, L. E. (1983). A comparison of the SACH and single axis foot in the gait of unilateral below-knee amputees. *Prosthetics and Orthotics International, 7,* 33-36.

Eberhart, H., Elftman, H., & Inman, V. (1968). The locomotor mechanism of the amputee. In P. Klopsteg & P. Wilson (Eds.), *Human limbs and their substitutes* (pp. 472-480). New York: Hafner.

Eberhart, H. D. (1947). Prosthetic device research project. *Subcontractor's report on fundamental studies on human locomotion and other information related to the design of artificial limbs* (Contract No. VAm 21223). Washington, DC: National Research Council, Committee on Artificial Limbs.

Eberhart, H. D., Inman, V. T., & Bressler, B. (1968). The principal elements in human locomotion. In P. E. Klopsteg & P. D. Wilson (Eds.), *Human limbs and their substitutes* (pp. 437-471). New York: Hafner.

Elftman, H. (1934). A cinematic study of the distribution of pressure in the human foot. *The Anatomical Record, 59*(4), 481-491.

Elftman, H. (1938). The measurement of the external force in walking. *Science, 88,* 152-153.

Foort, J. (1970). The patellar-tendon-bearing prosthesis for below-knee amputees: A critical review of technique and criteria. In Committee on Prosthetic Research and Development, *Selected articles from Artificial Limbs (1/54-2/66)* (pp. 353-362). Huntington, NY: Krieger.

Gillis, L. (1954). *Amputations.* New York: Grune & Stratton.

Gitter, A., Czerniecki, J. M., & DeGroot, D. M. (1991). Biomechanical analysis of the influence of prosthetic feet on below-knee amputee walking. *American Journal of Physical Medicine and Rehabilitation, 70,* 142-148.

Goh, J. C. H., Solomonidis, S. E., Spence, W. D., & Paul, J. P. (1984). Biomechanical evaluation of SACH and uniaxial feet. *Prosthetics and Orthotics International, 8,* 147-154.

Hamontree, S., & Snelson, R. (1973). The use of check sockets in lower limb prosthetics. *Orthotics and Prosthetics, 27,* 30-33.

Hannah, R. E., & Morrison, J. B. (1984). Prostheses alignment: Effect on gait of persons with below-knee amputations. *Archives of Physical Medicine and Rehabilitation, 65,* 159-162.

Hubbard, A. W., & Stetson, R. H. (1938). An experimental analysis of human locomotion. *American Journal of Physiology, 124,* 300-314.

Inman, V. T., Ralston, H. J., & Todd, F. (1981). *Human walking.* Baltimore: Williams & Wilkins.

Karpovich, P. V., & Karpovich, G. P. (1959). Electrogoniometer: A new device for study of joints in action. *Federation Proceedings, 18,* 79.

Kettelkamp, D. B., Johnson, R. J., Smidt, G. L., Chao, E. Y., & Walker, M. (1970). An electrogoniometric study of knee motion in normal gait. *Journal of Bone and Joint Surgery, 52A,* 775-790.

Knapp, S., & Cummings, D. (1992). Prosthetic management. In J. H. Bowker & J. W. Michael (Eds.), *Atlas of limb prosthetics* (pp. 453-478). St. Louis, MO: Mosby Year Book.

Lamoreux, L. (1981). Exoskeleton goniometry. *Bulletin of Prosthetics Research 18*(1), 288-290.

Lehneis, H. R. (1984). Evolution of the AK socket. *Clinical Prosthetics and Orthotics, 8,* 11.

Lehneis, H. R. (1985). Beyond the quadrilateral. *Clinical Prosthetics and Orthotics, 10,* 6-8.

Lehneis, H. R., Chu, D. S., & Adelglass, H. (1984). Flexible prosthetic socket techniques. *Clinical Prosthetics and Orthotics, 8,* 6-8.

Levy, S. W. (1970). The skin problems of the lower extremity amputee. In Committee on Prosthetic Research and Development, *Selected articles from* Artificial Limbs *(1/54-2/66).* Huntington, NY: Krieger.

Long, I. A. (1975). Allowing normal adduction of the femur in above knee amputations. *Orthotics and Prosthetics, 29,* 53-54.

Long, I. A. (1985). Normal shape-normal alignment (NSNA) above-knee prosthesis. *Clinical Prosthetics and Orthotics, 10,* 9-14.

Marks, A. A. (1931). *Manual of artificial limbs.* New York: Author.

Marks, G. E. (1888). *A treatise on artificial limbs with rubber hands and feet.* New York: A. A. Marks.

Michael, J. (1987). Energy storing feet: A clinical comparison. *Clinical Prosthetics and Orthotics, 11,* 154-168.

Murray, M. P. (1967). Gait as a total pattern of movement. *American Journal of Physical Medicine, 46,* 290-332.

Murray, M. P., Drought, A. B., & Kory, R. C. (1964). Walking patterns of normal men. *Journal of Bone and Joint Surgery, 46A,* 335-360.

Murray, M. P., Sepic, S. B., Gardner, G. M., & Mollinger, L. A. (1980). Gait patterns of above-knee amputees using constant-friction knee components. *Bulletin of Prosthetics Research, 17,* 35-46.

Muybridge, E. (1979). *Muybridge's complete human and animal locomotion.* New York: Dover.

Patek, S. D. (1926). The angle of gait in women. *American Journal of Anthropology, 9,* 273-291.

Peltier, L. F. (1993). *Orthopedics: A history and iconography.* San Francisco: Norman.

Perry, J. (1992). *Gait analysis, normal and pathological function.* Thorofare, NJ: Slack.

Perry, J., Bontrager, E. L., Bogey, R. A., Gronley, J. K., & Barnes, L. A. (1993). The Rancho EMG Analyzer: A computerized system for gait analysis. *Journal of Biomedical Engineering, 15,* 487-496.

Powers, C. M., Torburn, L., Perry, J., & Ayyappa, E. (1994). Influence of prosthetic foot design on sound limb loading in adults with unilateral below-knee amputations. *Archives of Physical Medicine and Rehabilitation, 75,* 825-829.

Quigley, M. J. (1985). The role of test socket procedures in today's prosthetic practice. *Clinical Prosthetics and Orthotics, 10,* 11-12.

Quigley, M. J., & Wilson, A. B. (1975). An evaluation of three casting techniques for patellar tendon bearing prostheses. *Orthotics and Prosthetics, 29,* 21-32.

Radcliffe, C. W. (1970a). The biomechanics of below-knee prostheses in normal bipedal walking. In Committee on Prosthetic Research and Development, *Selected articles from* Artificial Limbs *(1/54-2/66)* (pp. 295-303). Huntington, NY: Krieger.

Radcliffe, C. W. (1970b). The biomechanics of the Syme prosthesis. In Committee on Prosthetic Research and Development. *Selected articles from* Artificial Limbs *(1/54-2/66)* (pp. 273-282). Huntington, NY: Krieger.

Radcliffe, C. W. (1970c). Functional considerations in the fitting of above knee prostheses. In Committee on Prosthetic Research and Development, *Selected articles from* Artificial Limbs *(1/54-2/66)* (pp. 5-30). Huntington, NY: Krieger.

Radcliffe, C. W., & Foort, J. (1961). *The patellar tendon bearing below knee prosthesis.* Berkeley: University of California Biomechanics Laboratory.

Sabolich, J. (1985). Contoured adducted trochanteric-controlled alignment method (CAT-CAM): Introduction and basic principles. *Clinical Prosthetics and Orthotics, 9*(4), 15-26.

Sanders, G. T. (1986). *Lower limb amputations: A guide to rehabilitation.* Philadelphia: Davis.

Schwartz, R. P. (1934). Kinetics of human gait. *Journal of Bone and Joint Surgery, 16,* 334-350.

Schwartz, R. P., Trautmann, O., & Heath, A. L. (1936). Gait and muscle function recorded by the electrobasograph. *Journal of Bone and Joint Surgery, 18*(2), 445-454.

Shurr, D. G. & Cook, T. M. (1990). *Prosthetics and orthotics.* Norwalk, CT: Appleton Lange.

Staros, A., & Goralnik, B. (1981). Lower limb prosthetic systems. In American Academy of Orthopaedic Surgeons (Ed.), *Atlas of limb prosthetics* (pp. 279-280). St. Louis, MO: Mosby.

Steindler, A. (1953). Historic review of the studies and investigations made in relation to human gait. *Journal of Bone and Joint Surgery, 35A,* 540-542.

Strong, F. S. (1970). Artificial limbs—today and tomorrow. In Committee on Prosthetic Research and Development, *Selected Articles from* Artificial Limbs *(1/54-2/66)* (pp. 1-2). Huntington, NY: Krieger.

Sutherland, D. H., & Hagy, J. L. (1972). Measurement of gait movements from motion picture film. *Journal of Bone and Joint Surgery, 54A,* 787-797.

Tipton, C. M., & Karpovich, P. V. (1965). Electrogoniometric records of knee and ankle movements in pathologic gaits. *Archives of Physical Medicine and Rehabilitation, 46,* 267-272.

Torburn, L., Perry, J., Ayyappa, E., & Shanfield, S. L. (1990). Below knee amputee gait with dynamic elastic response prosthetic feet: A pilot study. *Journal of Rehabilitation Research and Development, 27*(4), 369-384.

Vitali, M., Robinson, K. P., Andrews, B. G., & Harris, E. E. (1978). *Amputations and prostheses.* London: Bailliere Tindall.

Wagner, E. M., & Catranis, J. G. (1968). New developments in lower-extremity prostheses. In P. E. Klopsteg & P. D. Wilson (Eds.), *Human limbs and their substitutes* (pp. 481-616). New York: Hafner.

Wagner, J., Sienko, S., Supan, T., & Barth, D. (1987). Motion analysis of SACH vs. Flex-Foot in moderately active below-knee amputees. *Clinical Prosthetics and Orthotics, 11,* 55-62.

Waters, R. L., Perry, J., Antonelli, D., & Hislop, H. (1976). Energy cost of walking of amputees: The influence of level of amputation. *Journal of Bone and Joint Surgery, 58A,* 42-46.

Weber, E. F. (1851). *Ueber die Langenverhaltnisse der Fleischfasern der Muskeln im Allgemeinen.* Berichte uber die Verhandlung der Koniglich Sachsischen Gesellschact der Wissenschaften zu Leipzig, mathematisch-physische classe. Jahrgang 1849. Leipzig, Weidmansche Buchhandling, Sitzung Am 16, August 1851.

Wilson, A. B. (1969). Recent advances in below-knee prosthetics. *Artificial Limbs, 13,* 1-12.

Wilson, A. B. (1970). The prosthetic and orthotic programs. *Artificial Limbs, 14,* 1-18.

Wilson, A. B. (1992). History of amputation surgery and prosthetics. In J. H. Bowker & J. W. Michael (Eds.), *Atlas of limb prosthetics: Surgical, prosthetic and rehabilitation principles* (pp. 3-15). St. Louis, MO: Mosby Year Book.

Winter, D. A. (1983). Energy generation and absorption at the ankle and knee during fast, natural, and slow cadences. *Clinical Orthopaedics and Related Research, 175,* 147-154.

Woltring, H. J., & Marsolais, E. B. (1980). Optoelectric (Selspot) gait measurement in two- and three-dimensional space: A preliminary report. *Bulletin of Prosthetics Research, 17,* 46-52.

Zahedi, M. S., Spence, W. D., Solomonidis, S. E., & Paul, J. P. (1986). Alignment of lower-limb prostheses. *Journal of Rehabilitation Research Development, 23,* 2-19.

Zuniga, E. N., Leavitt, L. A., Calvert, J. C., Canzonari, J., & Peterson, C. R. (1972). Gait patterns in above-knee amputees. *Archives of Physical Medicine and Rehabilitation, 53,* 373-381.

IV

Clinical Considerations

10

Rehabilitation Assessment and Planning for Head Trauma

ANNE-LISE CHRISTENSEN AND THOMAS W. TEASDALE

Introduction

As documented elsewhere in this book, the physical, psychological, and social consequences of traumatic brain injury have long been the subject of close investigation. Pioneering work with head-injured war victims was done by Goldstein in Germany and Luria in the then Soviet Union. However, until recently the view has tended to predominate, particularly in the medical community, that once an injury has occurred, little can be done. Ramon Y Cajal's celebrated doctrine, formulated in the early decades of the century, that nerve cells, unlike those of many other of the body's organs, cannot regenerate probably contributed much to the pessimism. An injured brain cannot repair itself in the way that an injured muscle may be able to do. Thus, it is believed that the functions subserved by any particular set of nerve cells are lost with the death of those nerve cells. Although it has been recognized that in fact a degree of recovery of physical and psychological function does occur, nonetheless this is expected typically to be only partial, reaching a plateau within the first year or so after injury.

Assistance in recovery has been confined largely to physical rehabilitation, such as the retraining of the use of limbs, which are themselves often directly damaged in whatever event led to the head injury. Psychological and social rehabilitation were not thought of as attainable objectives. Thus, rehabilitation was primarily

---

ANNE-LISE CHRISTENSEN Center for Rehabilitation of Brain Injury, University of Copenhagen, Amager 88 Njalsgade, DK-2300, Copenhagen S, Denmark.   THOMAS W. TEASDALE Psychological Laboratory, University of Copenhagen, Amager 88 Njalsgade, DK-2300, Copenhagen S, Denmark.

Rehabilitation, edited by Goldstein and Beers. Plenum Press, New York, 1998.

viewed as the province of physiotherapists and perhaps occupational therapists. Psychologists were employed only for assessment and diagnostic purposes.

This gloomy view has been especially regrettable for several reasons. First, improvements in medical care in the acute phase is producing a steady increase in the rate of survival from head injuries which earlier would have been fatal. Second, there is little evidence to suggest that survivors of head injuries have thereafter any elevated mortality rate; thus they may live many years after the injury. Third, they are predominately young, often children or teenagers.

Recent decades have witnessed a growing optimism with regard to the prospects for psychosocial rehabilitation following brain injury. There is emerging evidence that there can be a degree of restoration of neural pathways (LeVere, 1988), and that regeneration can take place in some circumstances. Furthermore, there has been some beginning success in the repair of injured brains using fetal brain tissue (Stein, Glasier, & Hoffman, 1994). More important, however, is the pragmatic demonstration that rehabilitation programs, particularly those based on neuropsychological principles and addressing the totality of the individual patient's condition and needs, can be effective.

In Israel, rehabilitation work has begun by, among others, Ben-Yishay (Ben-Yishay & Gold, 1990). working, as did Goldstein and Luria, with war victims, in this case soldiers injured during the 1967 six-day war. Ben-Yishay subsequently successfully transferred his program to New York University under the sponsorship of Leonard Diller. The work of Prigatano (1986) added the important dimension of psychotherapy to the rehabilitation process. At the same time, the growing concerns of the head-injured, their relatives, and the relevant professionals led to the establishment of the National Head Injury Foundation in the United States in 1978. Outcome studies of rehabilitation programs in the United States have recently been reviewed by Cope (1994).

Although rehabilitation in the early phases after injury has been developing particularly in Germany, it is generally true that postacute rehabilitation in Europe has lagged behind the United States. The gap is now, however, closing rapidly. In what follows, we describe the rehabilitation program implemented at our own center, which was one of the first to be established in Europe.

In addition to the influences previously mentioned, our program has been particularly inspired by the work of Luria (1980). Luria was influenced by psychoanalysis, and this could partially account for his humanistic approach, in that he preferred to view cognition in the context of individuality rather than viewing cognitive functions in isolation, as Western neuropsychology has tended to do. Thus, Luria considered the brain as a system of greater complexity than is susceptible to positivist neuroscientific analysis. The scientific approach can, however, in Luria's view, provide a framework which a more humanistic psychology, concerned with the individual in context, can then build upon. Luria's approach seems to encompass a resolution between the biomedical/technological view of the brain and thus of head injury, and the social/cultural view.

This approach brings into relief the particular contributions of the psychologist to the rehabilitation process. It serves to emphasize that rehabilitation must strive to restore to patients not only everyday skills and abilities, however important these ADL functions are. It must also strive to restore the patient's sense of identity, satisfaction, and meaning of life.

Assessment

Within our program, the initial assessment of the brain-injured patient is undertaken following a preliminary screening of the referral documents, especially including medical records. Our center accepts adults who have sustained an acute brain injury. Somewhat more than half of our patients have suffered head trauma, most often because of traffic accidents (the largest remaining group being stroke patients).

The assessment proper is preceded by an interview that undertakes a global evaluation of what Stein has called the "context" of the injury (Stein et al., 1994), namely, the premorbid history and personality, the patient's social environment and resources, and, importantly, his or her present emotional state and degree of insight into the condition.

The main assessment is the Luria Neuropsychological Investigation (Christensen, 1975). This procedure comprises a systematically arranged series of tests, mainly devised by Luria himself, and organized in keeping with his neuropsychological theory.

Many psychometric test procedures are rigid in their administration, quantitative in their scoring, and norm-based in their interpretation. By contrast, the LNI is flexible and adaptive, largely qualitative in its scoring, and its interpretation is based on hypothesis formation. In this process, the picture of symptoms is considered in its totality and conceptualized in terms of a syndrome in which the body of evidence is consistent.

The adaptiveness of the LNI is particularly effective in working with brain-injured patients. It allows the tester to avoid relentless testing of functions which are in fact well preserved and to concentrate instead on exploring areas of reduced functioning before the patient becomes fatigued, which would affect performance on virtually any test.

Our assessment procedure also includes some standardized psychological tests (including the Rorschach and elements from the Wechsler Memory Scale), however, for purposes of comparison with other centers and for research. It also includes the gathering of questionnaire data concerning social and employment circumstances, emotion, mood, and personality.

Although we have hitherto employed widely used questionnaires such as the SCL-90-R (Derogatis, 1983) and Katz Social Adjustment Scale (Hogarty and Katz, 1971), it has been our experience, and the experience of some others (Lezak, 1995) that these may not be ideal for brain trauma patients. In particular the SCL-90, with scales for conventional psychopathology, can yield misleading profiles for such patients. We are therefore currently involved in a European collaborative effort to develop a questionnaire tailored to brain-injured patients and close relatives, that is suitable for assessing both postinjury status and changes subsequent to rehabilitation (Teasdale et al., 1997).

There are also language therapist and physiotherapist assessments. The former is concerned with aphasia and other communication difficulties. The latter is particularly detailed with respect to motor deficits and activities of daily living. All of these sources of information are combined with earlier neurological and other medical records to form an overall picture of the patient's strengths and weaknesses.

The LNI plays the pivotal role in the planning of the subsequent rehabilitation training, since it reveals the understanding and insight that the patient can gain. The interpretation of the LNI is performed in accordance with Luria's hierarchical principles of fixed systems, whereby the loss of elements results in the necessity to create new systems.

The Program

Our program admits postacute adult patients, preferably as early as possible after injury, in groups of between 12 and 15. They attend our Center for 4 days per week, about 6 hours per day, for 4 to 5 months.

The Center has closely collaborating multidisciplinary staff, comprising clinical neuropsychologists, language therapists, a special education teacher, and physiotherapists. As a recent addition, we also have a consultant psychiatrist. Our location at a university is in some ways particularly fitting since the Center, too, is a learning environment where the progress of participants is very much dependent on their own efforts. Indeed, in recognition of the educational setting, we refer to the people entering our program as "students" rather than "patients" although we use the latter term in what follows.

After the start of the program, each patient is allocated a psychologist who bears overall responsibility for the progress of the patient through the program and continuing thereafter and who is therefore termed the primary therapist. In a sense, the psychologist thus fulfills much of the role described by McMillan and colleagues (1988) as the case manager but with the important qualification that this role is performed in conjunction with the building up of a therapeutic alliance between the patient and the psychologists concerned. This alliance process may begin with the Luria Neuropsychological Investigation during the initial assessment. The psychologist will thus have several individual contact hours with the patient through the week. These hours can include both cognitive training, of a form geared to the nature of the specific dysfunctions, and psychotherapeutic sessions. It is worth noting that the allocation of patients is made on the basis of the psychologist's experience and qualification and, in some cases, degree of prior encounter with the patient through the referral and initial evaluation process. The allocation is made following discussions in which all therapists take part. Changes of patient–psychologist pairings during the program are not encouraged and occur only rarely.

The content of the program has an overall structure influenced by the work of Ben-Yishay and Prigatano mentioned previously, particularly their holistic approach. Uniquely for our program, however, each day begins with a morning meeting that all of the patients attend, together with at least two staff members, and that has a very specific agenda. Following Danish educational tradition, this commences with singing a traditional song. The meeting also includes a cognitively demanding sequence of light gymnastics and the selection and presentation of current news items from television or newspapers. Responsibility for chairing the meeting, leading the gymnastics, presenting the news items, researching and presenting a lexicon time, and taking minutes for the meeting is assigned to five of the patients at a time on a rotating weekly schedule. The meetings encourage an energetic motivation for the day's subsequent activities and they can be seen in Luria's terms as activating cortical tonus to attention and arousal. Orientation toward the immediate and the outside world are involved, but the meetings also cover dimensions of interaction, mutual stimulation, and emotional support.

Another important element of our program is group psychotherapeutic sessions in which the patients attain insight, under the guidance of psychologists, to help and support each other in articulating and working through the emotional and social difficulties their injury has created for them and with which they have usually not yet dealt. We need to promote not the mechanisms that in neurotic patients are termed defenses but, rather, to use Goldstein's terminology, "protective mechanisms." The purpose of participating in the group psychotherapeutic sessions is, however, not only to articulate the patients' individual feelings and experiences, but to bring them out of the emotional isolation and introspection into which they so often have fallen and to give them an awareness of and insight into the problems and difficulties of those around them. Group psychotherapy has through time been conducted according to different models. In the early days the center had a list of prearranged topics (e.g., own or family reactions to the injury were systematically worked through; Prigatano, 1986). In recent years, the sessions have tended to be more free-flowing and more responsive to the immediate concerns of the patients themselves.

Cognitive training sessions can be one-to-one cooperations between the patient and his or her primary therapist or they can include several patients who are required to work together on the solution of a problem. They frequently focus upon whatever highly practiced automatic skills can be reactivated in the patients. Some use is made of computerized training programs that can confront the patients with cognitively challenging tasks, for example, strategy or abstract reasoning, and can automate the recording of their responses. It is important to emphasize, however, that use of computers in this way is individually coordinated with the other elements of the rehabilitation program; it is also done in collaboration with the therapist who is thereby able to observe the strategies adapted by the patient and to give him or her all-important feedback on these strategies.

Most of our groups have a few patients who are aphasic to some degree (although aphasia is more common among the stroke patients than among head trauma patients) and these receive special training from our language therapist. For those with dysarthrias and respiratory problems, we have a consultant voice therapist.

Some of our teenage brain-injured patients had not completed their formal education at the time of their injury and may require help from our special education teacher in the subjects of basic mathematics and Danish language.

The program also contains elements of physical training. This should not be confused with physiotherapy in the traditional sense. In part, we are concerned for improving the physical condition of our patients which, following perhaps months of very little exertion, is often lamentable. But the sessions also have strong cognitive elements as the patients monitor their own progress through the programmed series of exercises. Group interaction is also fostered during these sessions. Since the university lacks facilities for physical training this takes place at a fitness club at a nearby hotel.

In addition, our physiotherapists also work in areas that elsewhere might be the province of occupational therapists. Thus, for instance, a cooking class is arranged for patients who need to learn to prepare their own meals. The use of public transportation is also trained when appropriate. Many patients have become averse to using public transportation, perhaps owing to difficulties created by hemipareses, spatial disorientation, or memory problems. Patients may also be retrained to cycle, and indeed those with severe hemipareses may be introduced to the possibility of using a tricycle. We also collaborate with a governmental unit that retrains driving skills for handicapped people.

The fact that our patients enter the program in groups and are thereafter in daily contact with each other, as well as with the Center's staff, for 4 to 5 months promotes a kindred spirit among them, which can be very supportive socially, and provides a generally warm and hospitable atmosphere at the Center.

A very important element in our program is our contact with the close relatives of the patients. From the very beginning of the program we encourage these relatives, often the parents of our younger patients or the spouses of the older ones, to join fortnightly meetings held in the late afternoon after the patients themselves have left for the day. The meetings allow the relatives to discuss and share the concerns that they themselves have experienced, such as the alarm and the grief at the time of the initial injury, perhaps the relief and exultation when the patient emerged from a coma, and, all too often, their disappointment and frustration as the early recovery halted too far short of a complete return to the condition prior to injury. We frequently discover that this disappointment and frustration has been engendered by inadequate and misleading information given to the relatives at the hospital. The meetings also afford us an invaluable opportunity to obtain an insight on the patient in his or her home environment and to gain firsthand information about whatever problems there might be, practical, interpersonal, emotional, etc. In addition, the meetings foster collaboration between the relatives and the Center's staff.

During the 4- to 5-month period, as the patient becomes ready, concern is gradually switched from a specific training of cognitive and social skills and resolving emotional and motivational problems toward a concern for the reentry of the patient into the community. A first consideration is naturally education and employment. Where appropriate, contact is established with the patient's former workplace or with other potential employers, although often the patients return to a less demanding form of employment than they had had preinjury. Second, because many of our younger patients may still be living with their parents as a consequence of their greater dependence following injury, we attempt to help them establish themselves in independent homes. Patients who have needed home assistance are trained to reduce this dependency as much as possible.

Another very important area of concern is promoting leisure-time activities and hobbies, in particular those that encourage social activities, in order to counter isolation and withdrawal. Examples could include amateur dramatics and choir singing, ornithological society membership and even political activity. Some activities, including a sailing project and a chess club, have been sponsored by the Center itself, and students are welcome to continue physical training at the fitness club for one afternoon a week if they so wish.

A vital phase of the Center's program is our sustained contact with the patients after the completion of the 4 to 5 months of daily attendance at the Center. Regular meetings are arranged with the group and hours with the primary psychologists take place according to individual needs for some patients continuing training at the Center; for instance, an hour per week may be arranged. This tends to be particularly true for the aphasics. On occasion, fortnightly meetings are held for a small group of aphasics who with the support of a psychologist discuss topics of common interest such as current affairs.

Continued contact with the patients while they are returning to work or to an educational program is often of crucial importance. We have found it advantageous to talk to employers in order to give them realistic expectations about what the patient will be able to accomplish and how he or she can be helped in the working environment. Thereafter the situation needs to be monitored so that emerging problems can be dealt with before they develop to the level at which the patient abandons the job.

The Center is at present collaborating with the Ministry for Social Affairs on a project with the aim of establishing the optimal conditions for work reentry for brain-injured people. In this project a brain-injured person returning to a work environment will be supported at that work by a "coach." This coach will be a person familiar with the job in question and he or she will be supervised by a professional from the Copenhagen Center who is familiar with the brain-injured person. Both coaching and supervision will continue for an as yet indefinite period during which time the employer will be reimbursed for wages paid to the coach and to the patient.

In general, while not wishing to encourage overdependence, our Center has an open-door policy toward former patients whereby all are welcome to contact us when they feel they have need for us. Although the Center has now accumulated more than 200 former patients the load from this contact has not yet become overwhelming.

We have published a number of reports concerning the outcome following rehabilitation for our patients. In general, we have found encouraging rates of returning to a working or educational environment, greater degrees of independent living, and an enriched use of leisure time with more involvement in pastimes, sports, and hobbies (Christensen et. al., 1992; Teasdale and Christensen, 1994). For most of our analyses we have grouped all patients together, but for one report we have focused on a comparison of head-injury patients ($n = 22$) with stroke patients ($n = 44$) (Teasdale, Christensen, and Pinner, 1993). The patients in this study were among the first to complete the rehabilitation program in the years 1985 through 1987.

Compared to the stroke patients, our-head injury patients were somewhat younger (mean = 27 years. s.a. = 9 years) and were more preponderantly male (73%). Fewer than 5% of them had not been in coma and most had been in coma for more than one week. The mean duration of hospitalization was 138 days (s.a. = 96 days), and on average they were a little over 3 years postinjury at the time they entered our program.

It is worth mentioning that the time between injury and entry into the program is now much shorter than for these patients just mentioned. In part, this is a reflection of the increasing acceptance among the medical community in Denmark of the value of postacute neuropsychological rehabilitation. In the early years of the Center's existence, the view predominated that it was better to postpone any rehabilitation until all "spontaneous" recovery had ceased. It is now a widespread demand that rehabilitation be a natural continuation as soon as possible after patients have become medically stable.

Among our head-injury patients, about one third had been living in marital or other partner relationships at the time of the injury. About half of these relationships had ended by the time they entered the program, in most cases due directly or indirectly to the injury itself and the strain that this had placed on both the patient and his or her partner. By the time of a follow-up 1 year after completing the program, the proportion in partner relationship had again risen to about one third. Although this has not been a controlled study, there is good clinical evidence to suggest the restored proportion is in part due to the regaining of social skills by the patients in the course of the rehabilitation program.

Whereas over one third of the head-injury patients had required home services (e.g., home help, district nurse visits), this proportion was reduced to under 15% by the 1-year follow-up. This seems likely to have been due to the program's promotion of independence and self-sufficiency.

Almost none of the patients were employed or were pursuing an education at the time they entered the program. Indeed this would have been an exclusion criterion, since the program itself requires a full-time commitment. A few of the patients could return to their original work or education on completion of the rehabilitation program, but only if the rehabilitation itself was successful. About one third of the others had attempted to return to their work or education prior to entering the program and had been unsuccessful. More than 30% had returned to competitive employment or a full-time education by the 1-year follow-up, but a further 30% were engaged in sheltered or otherwise state-supported work and training schemes. These percentages stand in sharp contrast to the much lower figures reported in Denmark by Thomsen (1992) for a group of head-injured patients who did not receive postacute psychosocial rehabilitation. As stated previously, the program increasingly focuses on the issue of occupation and education toward the end of the 4- to 5-month period, and the Center's staff, in particular the primary therapists, are typically actively involved in this aspect of the patient's life throughout the following months.

The proportion of head-injury patients who regularly engaged in leisure activities that took them out of their own home and into the company of others fell from about two-thirds prior to injury to about one-quarter at the time of entering the program. By the time of the 1-year follow-up, the figure had risen to 55%, and the number of actual hours per week involved has risen again to its preinjury level (about 10 hours).

In summary, one may conclude that according to a set of comparatively objective and certainly highly valid outcome measures, patients have at the time of 1-year follow-up a more fulfilled and active life than prior to entering the program.

A program as extensive as ours cannot be provided cheaply; it costs the equivalent of about $34,000 per patient. Evidence suggests, however, that even when evaluated in purely financial terms, it is worthwhile. An independently conducted follow-up study of the Center's patients concluded that the public sector financial savings from reduced requirements for health and social service provisions among patients completing our program lead to a recouping of the cost of the program within 3 years (Mehlbye & Larsen, 1994).

The Future

Rehabilitation experienced an explosive growth particularly in the United States during the 1970s and 1980s. The last few years have, however, seen signs that this growth may reverse itself. If this is to be prevented, and the early progress is to continue, then rehabilitation professionals must rethink their goals and the objectives by which those goals are to be achieved.

Evaluation of the effectiveness of rehabilitation will continue to be a priority. It may no longer be feasible to conduct controlled studies in which potential patients are assigned to rehabilitation or no rehabilitation at random. The total withholding of rehabilitation solely for the purpose of research is generally considered unethical. A more realistic and a more relevant goal is the comparison of rehabilitation programs. The holistic program we have described and many others like it are expensive and time-consuming. One might reasonably ask whether the programs' (generally encouraging) results could be achieved with programs involving less time and effort. Even though our own clinical experience suggests that this is not the case, we are currently involved in proposals for a multicenter collaborative study that will compare outcomes of holistic programs with less elaborated ones.

A second consideration for the future must be promoting a greater awareness of, and provision for, children who sustain brain injury, young brain trauma patients. Despite the increasing acceptance of the need for postacute rehabilitation of adult patients, it regrettably remains true that younger patients are expected to be able to return to the normal school environment once they have become medically stable. This expectation is perhaps fostered by the view that the young brain is more resilient to the effects of injury and more readily recovers from it. There is cumulating evidence that this expectation is erroneous, but children obviously cannot be introduced into rehabilitation programs designed for adults. It is a positive development that programs for children are beginning to appear (Prigatano, 1993).

A third trend for the future would be the extension of rehabilitation to the very severely injured. Most postacute programs, including our own, cater to moderate to severe levels of injury, in which patients must be capable of independent mobility and have at least basic communication skills. Until recently, too little has been done for the very severely injured, including those in a persistent vegetative state, beyond attending to basic physical needs. Evidence is now emerging to suggest that with appropriate and intensive interventions, the level of functioning of such patients can in some cases be considerably improved (Schonle, 1996).

References

Ben-Yishay, Y. & Gold, J. (1990). Therapeutic milieu approach to neuropsychological rehabilitation. In Rodger Llewellyn Wood (Ed.), *Neurobehavioural sequelae of traumatic brain injury* (pp. 194-215). London: Taylor & Francis.

Christensen, A-L. (1975). *Luria's neuropsychological investigation*. Manual test materials. New York: Spectrum.

Christensen, A. L., Pinner, E. M., Moller Pedersen, P., Teasdale, T. W., & Trexler, L. E. (1992). Psychosocial outcome following individualized neuropsychological rehabilitation of brain damage. *Acta Neurologica Scandinavica, 85,* 32-38.

Cope, N. D. (1994). Traumatic brain-injury rehabilitation outcomes in the United States. In A-L. Christensen & B. P. Uzzell (Eds.)., *Brain injury and neuropsychological rehabilitation: International perspectives* (pp. 201-220). Hillsdale, NJ: Erlbaum.

Derogatis, L. R. (1983). *SCL-90-R. Administration, scoring & procedures manual-II.* (2nd ed.). Towson, MD: Clinical Psychometric Research.

Hogarty, G. E., Skatz, M. M. (1971). Norms of adjustment and social behavior. *Archives of General Psychiatry, 25,* 470-480.

LeVere, T. E. (1988). Neural system imbalances and the consequence of large brain injuries. In S. Finger, T. E. LeVere, C. R. Almli, & D. G. Stein (Eds.), *Brain injury and recovery: Theoretical and controversial issues* (pp. 15-26). New York: Plenum.

Lezak, M. D. (1995). *Neuropsychological assessment* (3rd ed.). New York: Oxford University Press.

Luria, A. R. (1980). *Higher cortical functions in man.* New York: Basic Books.

McMillan, T. M., Greenwood, R. J., Morris, J. R., & Brooks, D. N. (1988). An introduction to the concept of head injury case management with respect to the need for service provision. *Clinical Rehabilitation, 2,* 319-322.

Mehlbye, J., & Larsen, A. (1994). Psychosocial outcome in Denmark. In A-L. Christensen & B. P. Uzzell (Eds.). *Brain injury and neuropsychological rehabilitation: International perspectives* (pp. 235-244). Hillsdale, NJ: Erlbaum.

Prigatano, G. P. (1986). *Neuropsychological rehabilitation after brain injury.* Baltimore: Johns Hopkins University Press.

Prigatano, G. P. (1993). Issues in the neuropsychological rehabilitation of children with brain dysfunction. Editorial: Toward developing neuropsychologically oriented rehabilitation programs for children. *Neuropsychological Rehabilitation, 3,* 297-299.

Schonle, P. W. (1996). Rehabilitation in patients with traumatic brain injury. *Nervenheilkunde, 15,* 220-225.

Stein, D. G., Glasier, M. M., & Hoffman, S. W. (1994). Pharmacological treatments for brain-injury repair: Progress and prognosis. In A-L. Christensen & B. P. Uzzell (Eds.), *Brain injury and neuropsychological rehabilitation: International perspectives* (pp. 17-40). Hillsdale, NJ: Erlbaum.

Teasdale, T. W. (1994). Psychosocial outcome in Denmark. In A-L. Christensen & B. P. Uzzell (Eds.). *Brain injury and neuropsychological rehabilitation: International perspectives* (pp. 235-244). Hillsdale, NJ: Erlbaum.

Teasdale, T. W., & Christensen, A. (1994). Psychosocial outcome in Denmark. In A. Christensen & B. P. Uzzell (Eds.), *Brain injury and neuropsychological rehabilitation: International perspectives* (pp. 235-244). Hillsdale, NJ: Erlbaum.

Teasdale, T. W., Christensen, A-L., & Pinner, E. M. (1993). Psychosocial rehabilitation of cranial trauma and stroke patients. *Brain Injury, 6,* 21-28.

Teasdale, T. W., Chrsitensen, A., Willmes, K., Deloche, G., Braga, L., Stachowiak, F. J., Vendrell, J. M., Castro-Caldas, A., Laaksonen, R. K., & Leclercq, M. (1997). Subjective experience in brain-injured patients and their close relatives: A European Brain Injury Questionnaire study. *Brain Injury, 11,* 543-563.

Thomsen, I. V. (1992). Late psycho-social outcome in severe traumatic brain injury: Preliminary results of a third follow-up study after 20 years. *Scandinavian Journal of Rehabilitation Medicine, 20* (Suppl.), 142-152.

11

Neuropsychological Assessment for Planning Cognitive Interventions

WILLIAM DREW GOUVIER, JUDITH R. O'JILE, AND LAURIE M. RYAN

INTRODUCTION

Neuropsychological assessment has changed in focus over the last decade from an emphasis on determining lesion localization to delineation of brain–behavior relationships and the establishment of functional strengths and weaknesses. This understanding of brain–behavior relationships and the pattern of strengths and deficits is essential to planning rehabilitation of the brain-damaged patient. In this chapter, the relationship between rehabilitation of cognitive deficits and neuropsychological assessment will be discussed.

Chelune and Moehle (1986) described this evolution as progressing through neuropsychology's initial "static phase" from 1945 through the mid-1960s. At this time, the primary focus of concern was to establish optimal diagnostic hit rates for neuropsychological tests. Attention was directed toward the establishment of decision-making rules to permit neuropsychological assessment to achieve optimal diagnostic accuracy. From the period beginning in the 1960s, neuropsychology entered its second phase, referred to by Rourke (1982) as the "cognitive phase." Rourke characterized this as "an attempt to understand the psychological tests and measures being employed in attempts to establish brain–behavior relationships" (p. 2). Neuropsychological assessment conducted from the zeitgeist of the cognitive phase focused on task analysis. In an attempt to go beyond simple diagnostic statements, great efforts were made to become descriptive in the diagnostic report. The goal was to identify the cognitive-behavioral deficits associated with a particular brain lesion and to be able to make a profile describing the patient's pattern of neuropsychological strengths and weaknesses. From the springboard of the cognitive phase of neuropsychological assessment, investigators entered directly what Chelune and Moehle (1986) referred to as the "dynamic phase." In this phase, neuropsychologists are being called on to make not just diagnostic evaluations and descriptive statements, but to answer prescriptive questions such as, Is the patient ready to work? Can the patient return to driving? Is the patient suited better for a vocational program or for a college program in his or her rehabilitation? To answer these questions, neuropsychological assessment procedures need to go beyond those used solely for purposes of lesion localization and establishing the presence or absence of brain damage. Indeed, a neuropsychological assessment designed to answer these questions needs to follow a progression of identifying deficits in ability in the criterion environment. It should encompass a broad-brush survey of neuropsychological functioning, fractionating out the nature of the deficits that are observed in this survey. That fractionated assessment should identify ways those deficits might be countered in a cognitive intervention to allow the patient to relearn, recover, or compensate for lost neuropsychological abilities. This assessment should also evaluate the impact of that intervention on the abilities themselves (Level 1) and on closely related abilities (as identified in the Level 3 fractionating assessment), and determine that the intervention has had some broad impact on neurocognitive functioning (Level 2). Finally, it should establish that the intervention has had some impact on alleviating the real-world deficits that originally brought the patient to the attention of a neuropsychological practitioner.

LEVEL 1 ASSESSMENT: THE CRITERION ENVIRONMENT OF THE REAL WORLD

Brain damage causes limitations in learning, memory, adaptive abilities, ambulation skills, and a host of cognitive abilities necessary to support the activities of daily living demanded by the real world. The reader is referred to the Chelune and Moehle (1986) chapter entitled "Neuropsychological Assessment and Everyday Functioning" and to the recent volume by Sbordone and Long (1996) entitled *The Ecological Validity of Neuropsychological Tests* for a detailed discussion of the relationship between neuropsychological assessment and the criterion environment and an elaboration on the importance of assessment of functioning in that criterion environment to establish the ultimate validity of the clinical assessment and intervention attempts. A thorough neuropsychological evaluation is the most comprehensive of all approaches to psychological testing and is careful to tap each of the three legs of the biopsychosocial triad; such assessment necessarily involves more than the administration of psychological tests. The clinical database used in a thorough neuropsychological assessment includes evaluation of functioning in the criterion environment. This should be established by identification of demographic and academic status, occupational history, patients' and relatives' reports of disability in the present environment, behavioral observations, and psychologists' analysis of observations related by patients, relatives, and others in the criterion environment. In addition to this information, further data regarding patients' level functioning can be gathered from rating forms that patients' families can complete. Such instruments include the Relatives' Assessment of Patient Functioning Inventory (Chelune, Heaton, & Lehman, 1986) and Katz Adjustment Scale: Relatives' Form (Katz & Lyerly, 1963). The Iowa Collateral Interview (Roberts & Varney, unpublished manuscript) and the Portland Adaptability Scale (Lezak, 1987) are structured interviews that can be used to supplement the rating forms. Personality and emotional functioning data are gathered on a current level and on a retrospective level to establish whether personality or emotional problems identified in the present assessment represent newly acquired deficits or long-standing premorbid ones. Finally, of course, a survey of neuropsychological functioning itself is performed.

The survey of psychosocial background is vital, because the clinical picture appears against the background of premorbid psychosocial functioning. Brain injury and recovery of function can be influenced by factors such as cerebral dominance, premorbid intellectual ability, age, social support, motivation, and premorbid mental stability (Golden, 1978; Heilman & Valenstein, 1979).

Factors that contribute to successful outcome include premorbid economic and intellectual accomplishments, right-handedness, younger age, and good premorbid emotional adjustment (Golden, 1978; Keenan & Brassell, 1974). Brain injury can also influence adjustment to the family unit (Lezak, 1978), and there is evidence to suggest that the family's ability to cope with the patient's changes may be a key factor in how emotionally disabling the patient's physical and intellectual deficits will be (McKinlay & Brooks, 1984).

The validity of Level 1 assessments of the patient's ability to function in the criterion environment may be threatened in several ways. These must be recognized and taken into account in interpreting the results of any such assessment. Some of these threats include problems with reliance on reports of the patient's functioning by family, friends, and caregivers, especially since these individuals may be adversely affected by the patient's injuries as well (Novack, Bergquist, Bennett, & Gouvier, 1991) and their perception of the patient's capacity to function may be somewhat biased. Another problem threatening the validity of criterion environment assessment is the presence of anosognosia. It is not unusual for an obtunded quadraparetic patient in the rehabilitation center to blurt out, "I could walk just fine if all these people would let go of me!" Patients are often very poor judges of their own capacity to function, and even patients who demonstrate rather good insight for some aspects of their disability may remain oblivious to other aspects (Gouvier, Cottam, Webster, Beissell, & Wofford, 1984). A final threat to the internal validity of ecological criterion environment assessment is the unstandardized and often unstructured behavioral observations by the neuropsychologists. Nevertheless, the patient's behavior in the waiting room and in interacting with his or her family or caregivers provides an important additional perspective on the patient's functional abilities in the everyday world. It should be noted that everyday functioning assessment is as close as the profession of neuropsychology can come to a survey of consumer satisfaction. It is important in our service to the public that we at least attempt to gather data that reflects the impact and significance of the efforts that we ask of the patient in pursuit of a course of neuropsychological assessment and treatment for cognitive disabilities.

LEVEL 2 ASSESSMENT: THE NEUROPSYCHOLOGICAL BATTERY

For purposes of treatment planning, the initial neuropsychological evaluations should be comprehensive, assessing an array of cognitive abilities. Significant deficits may be completely overlooked if an assessment is too narrowly focused. Similarly, inadequate assessment of abilities may result in misleading findings because the performance of some cognitive abilities is dependent on the integrity of other processes. A comprehensive approach acknowledges the extreme complexity of human cognitive functions (Cullum, Kuck, & Ruff, 1987). The basic areas of cognitive function that should generally be assessed in the brain-injured patient include arousal, attention, and concentration; sensory, perceptual, and motor functioning; visuospatial, visuoperceptual, and visuoconstructional abilities; language functions; executive functions; intellectual ability; academic achievement; learning and memory; and reasoning and problem solving. Many neuropsychological measures require a number of underlying abilities for successful completion of the task (Cullum et al., 1987). As a result it should be noted that any categorization of neuropsychological tests and functions is arbitrary. Many measures could fit several categories, and the following list should be considered from that viewpoint. The following tests have been chosen as exemplars of specific categories, but there are others not listed that would have been equally appropriate.

AROUSAL, ATTENTION, AND CONCENTRATION. "Many cognitive deficits result from reduced arousal and attentional abilities that provide the background tone upon which higher cortical functions are based. Hence, reduced ability to behave effectively may represent impairment in levels of arousal and alertness, rather than impaired higher cortical functions per se" (Gouvier, Webster, & Blanton, 1986, p. 288). For example, deficits on a measure of verbal recall may be the result of decreased attention rather than a true memory impairment. Because of the important role of these functions in other neuropsychological abilities and many daily tasks, assessment of attention/concentration abilities is crucial in the neuropsychological evaluation of the brain-injured patient (Cullum et al., 1987).

Basic attentional abilities can be assessed clinically during an interview, observation, and tasks that include digit repetition (digit span), number or letter cancellation tasks, and serial addition tasks (Cullum et al., 1987). More subtle deficits may not be apparent on highly structured neuropsychological measures but can contribute to lower frustration tolerance and associated agitation (Mapou, 1992). This can emerge under conditions of stress, change, or ambiguity, such as in a complex rehabilitation task. Examples of tests that quantify attentional problems are the Paced Auditory Serial Addition Test (PASAT; Gronwall & Wrightson, 1974) and the Symbol Digit Modalities Test (SDMT; Smith, 1973).

SENSORY, PERCEPTUAL AND MOTOR FUNCTIONING. It is essential to perform screening evaluations of sensory and perceptual functions. Sensory testing will indicate whether referral for thorough optometric or audiologic evaluations is indicated. This is important because many weaknesses that are identified as cognitive deficits might actually represent normal processing of distorted sensory information (the computer programmer's "garbage in–garbage out" analogy applies here) (Gouvier, 1987). Perceptual testing is important because information regarding preferred and nonpreferred channels of information and accurate versus inaccurate or absent information access (Gouvier, 1982) can be derived. Frequently used measures include the Reitan–Kløve Sensory-Perceptual Examination, Speech Sounds Perception Test, and the Seashore Rhythm Test (Reitan & Wolfson, 1993).

Brain injury may affect motor abilities, even in the absence of an obvious physical impairment. Neuropsychological evaluation should include measures of fine motor speed, dexterity, and simple strength. This is particularly important in assessing head injury because of the typically diffuse nature of cerebral involvement. Impairment of simple motor and psychomotor abilities can have major implications for rehabilitation and job placement (Cullum et al., 1987). These abilities can be measured by the Finger Tapping Test (Reitan & Wolfson, 1993), Grip Strength (Reitan & Wolfson, 1993), and Grooved Pegboard Test (Matthews & Kløve, 1964).

VISUOSPATIAL, VISUOPERCEPTUAL, AND VISUOCONSTRUCTIONAL FUNCTIONS. This category includes a wide variety of nonverbal skills that may be affected in brain injury, particularly when there is damage to the right (or nondominant) hemisphere. Some of the more common functions in this area include constructional ability, visuospatial problem solving, visuoperceptual abilities, and drawing ability (Cullum et al., 1987), and these can be assessed by cancellation tasks, Hooper Visual Organization Test (Hooper, 1958), the Benton Visual Form Discrimination Test (Benton, Hamsher, Varney, & Spreen, 1983), and the Rey-Osterreith Complex Figure Test (Osterreith, 1944; Rey 1941). The visuoperceptual deficits that often follow brain injury have been implicated as factors in the ability of patients to return to work, even after extensive rehabilitation. In particular, left visual neglect is an especially disruptive sequela because of the implications for patients' safety in ambulation and independent functioning (Gouvier, 1987).

LANGUAGE FUNCTIONS. The basic language functions that should be assessed include comprehension of speech, spontaneous speech, reading comprehension, confrontation naming, verbal fluency, vocabulary skills, repetition, written language, and general conversational abilities. Impairment in basic abilities may affect performance on other neuropsychological tests, and therefore its evaluation is critical (Cullum et al., 1987). Difficulties in communication can be extremely frustrating to patients. The most commonly documented residual language deficit in closed head injury is a persistent word retrieval impairment (Mapou, 1992), which leads to difficulties in social interactions (Williams & Mateer, 1992). Examples of language measures are the Reitan–Indiana Aphasia Screening Test (Reitan & Wolfson, 1993), the Revised Token Test (McNeil & Prescott, 1978), the Boston Naming Test (Kaplan, Goodglass, & Weintraub, 1983), and the Boston Diagnostic Aphasia Exam (Goodglass & Kaplan, 1987). In any assessment of language competence, it is important to specifically contrast receptive comprehension such as that assessed by the Peabody Picture Vocabulary Test (Dunn & Dunn, 1981) with expressive abilities such as those tapped by the Controlled Oral Word Association Test (Benton & Hamsher, 1983).

EXECUTIVE FUNCTIONS. Executive functions are basic regulatory abilities that allow for the planning, initiating, and execution of goal directed behavior, especially in novel situations. At the most primitive level, this includes motor behavior; at the highest level, it includes the ability to establish, maintain, and shift response sets, monitor one's behavior, think flexibly, use feedback to correct one's errors, and modify emotional expression (Sohlberg, Mateer, & Stuss, 1993). These functions are typically associated with the frontal lobes and can be assessed by tests such as the Stroop Test (Stroop, 1935), Trail Making Test (Reitan & Wolfson, 1993), and Controlled Oral Word Association Test (Benton et al., 1983). Even mild impairment can substantially disrupt daily functioning (Mapou, 1992; Williams & Mateer, 1992). According to Mapou, impairments in attention and executive functioning account for most of the acquired behavioral difficulties following brain injury. For example, disruption of executive functions might alter response to sensory input or the basic knowledge base so that it is no longer possible for the individual to carry out even previously well-known or routinized behaviors effectively (e.g., dressing, grooming, eating, or driving) (Sohlberg et al., 1993). Components of behavior may be seen, but they may be sequenced poorly or they may not be triggered at the appropriate time and place (or both). The patient seems oblivious to a need to respond, initiate a response, or sequence steps to achieve a goal (Sohlberg et al., 1993). Defects in only one of these aspects of executive functioning is rare; they typically involve a cluster of deficiencies, with one or two especially prominent (Lezak, 1983).

INTELLIGENCE/INTELLECTUAL FUNCTIONING. The most popular measure of intellectual ability is the WAIS-R (Wechsler, 1981). This test assesses intelligence for patients between 16 and 74 years of age and yields information regarding global intellectual functioning (Full Scale IQ), verbal and nonverbal abilities (Verbal and Performance Scale IQs). Use of the WAIS-R with brain-injured individuals allows a valuable analysis of patient performance. These analyses include comparisons of individual performance with normative and clinical samples, comparisons of current performance with premorbid functioning, and analyses of intraindividual performance on subtests that assess different areas of cognitive functioning (Strub & Black, 1985).

It is critical that more emphasis be placed on interpreting the individual subtests rather than the overall IQ scores. For example, specific deficits such as visual field cuts or attention can influence individual subtest performance and result in an overall lower IQ score (Cullum et al., 1987). In this case, interpretation of the overall IQ as indicative of global intellectual decline would be a mistake. Furthermore, the verbal subtests as a group tend to be less sensitive to brain injury in adults. These tasks involve overlearned or "crystalized" cognitive abilities that are less susceptible to the effects of trauma. The performance subtests conversely involve more "fluid" abilities that are very sensitive to brain damage (Cullum et al., 1987).

ACADEMIC FUNCTIONING. Measures of academic functioning such as mathematics, reading, and writing skills are rarely diagnostic in terms of brain dysfunction, but they will often provide practical information regarding vocational options and independent living (Sohlberg & Mateer, 1989) and assist in selection of various treatment approaches (Sohlberg & Mateer, 1989). The Wide Range Achievement Test–Revised (Jastak & Wilkinson, 1984) and the Peabody Individual

Achievement Test–Revised (Markwardt, 1989) are frequently used academic assessment measures that provide a screening of abilities; however, the detailed examination procedures of the Woodcock Johnson (Woodcock & Mather, 1989) yields data more amenable to proceeding on to a Level 3 assessment.

LEARNING AND MEMORY. The tremendous complexity of memory functions requires a wide range of assessment measures. Evaluation should include tests of learning and recall (immediate and delayed) of verbal and nonverbal/visuospatial information, as well as tests of recognition memory (Sohlberg & Mateer, 1989). Testing of specific areas of memory can be done with the California Verbal Learning Test (Delis, Kramer, Ober, & Kaplan, 1987), Rey Auditory Verbal Learning Test (Lezak, 1983; Rey, 1964; Taylor, 1959), Benton Visual Retention Test-Revised (Benton, 1974), and the Warrington Recognition Memory Test (Warrington, 1984) while more comprehensive assessment can be accomplished with the Wechsler Memory Scale-Revised (Wechsler, 1987). Careful analysis of the pattern of performance may have further important implications for rehabilitation strategies (Cullum et al., 1987). Any deficits in new learning can have profound effects on progress (Mapou, 1992).

REASONING AND PROBLEM SOLVING. Reasoning and problem solving are often affected in brain injury. These functions generally involve the integration of a number of more basic cognitive abilities on which they depend, such as basic executive functions (Cullum et al., 1987). Disruption of any component of these processes may result in significant deficits. In addition, deficits in higher cognitive functions can occur concomitantly with relatively normal performance on other more basic tasks. These higher skills include insight, judgment, and abstraction (Mapou, 1992) and, without these skills, patients may be concrete in their thinking and limited in their ability to participate in treatment planning (Mapou, 1992). Commonly used assessment tools consist of the Wisconsin Card Sort Test (Heaton, 1981), Categories Test (Reitan & Wolfson, 1993), and Raven's Standard Progressive Matrices (Raven, 1960).

EMOTIONAL FUNCTIONING. Emotional disturbance following brain injury is a frequent sequela and may actually affect the patient and family more than the cognitive deficits present (Cullum et al, 1987). The Beck Depression Inventory (Beck, 1987) and the State-Trait Anxiety Inventory (Spielberger, Gorsuch, & Lushene, 1970) measure self-reported symptoms of emotional functioning, whereas the Minnesota Multiphasic Personality Inventory measures a wider range of psychopathology (MMPI: Hathaway & McKinley, 1943; MMPI-2: Butcher, Dahlstrom, Graham, Tellegen, & Kaemmer, 1989).

Following the completion of the neuropsychological battery, the pattern of deficits is identified. This pattern forms the basis for the Level 3 assessment.

LEVEL 3 ASSESSMENT: FRACTIONATING THE DEFICITS AND IDENTIFYING THE CAUSES OF TASK FAILURE

The practice of neuropsychology at Level 3 of assessment is the beginning of a diagnostic journey that picks up where traditional psychometrically oriented neuropsychological assessments end. So far we see that the output of a Level 1 assessment documents the patient's experiencing problems in everyday life. The output of the Level 2 assessment documents brain dysfunction that may be localized or lateralized in accordance with the type of lesion the patient has. Once the areas of deficit are identified, the really interesting diagnostic task begins. It is not sufficient to identify the impairment of functions; it is much more important to ascertain specifically why they are impaired. Level 3 assessments focus on identifying why specific tests are failed, to provide information used to design the treatment intervention that would be applied at Level 4.

The basic approach to conducting a Level 3 assessment is to identify first which main areas of deficit are present in brain function. This is done by identifying which specific tasks are failed, and then searching for commonalities among the failed tasks so that the clinician can test hypotheses to help develop efficient Level 3 assessment procedures that ultimately will lead to rehabilitation treatments. For example, a patient may demonstrate poor visual memory as assessed by tasks like the Memory for Designs (Graham & Kendall, 1960) or the Benton Visual Retention Test (Benton, 1974). Such a task requires patients to perceive the stimulus gestalt, to remember what they have perceived, and then to correctly draw from memory their reproduction of the stimulus gestalt. Failure on this task may occur for any of at least three different reasons. Perhaps patients do not have the perceptual abilities to accurately see and decipher the stimulus gestalt itself. On the other hand, perhaps they don't have the motor abilities to accurately reproduce the gestalt with paper and pencil. Or alternatively, they can see, perceive, draw, and reproduce adequately, but can't remember what they've seen. How does one go about differentiating these various components to identify the weak link in the processing chain? This is done by fractionating the task. If patients can't reproduce designs from memory, one should see whether they can reproduce similar designs from copy. If they can, then clearly the problem is not motor or perceptual in origin. Indeed, the inability to reproduce from memory in the presence of accurate reproduction from copy would suggest that memory itself is the source of the disorder. If, however, patients cannot copy, one cannot be certain whether the problem represents perception or motor defects, and, of course, memory also could be involved. But one can examine this further by seeing whether patients can match to sample, using, for example, a Benton Visual Form Discrimination Test (Benton et al., 1983) to assess the integrity of the perceptual system for decoding such designs. If patients can do that, one can then examine the role of memory by administering the same test in a memory format to see whether the patients can not only accurately perceive the stimulus designs, but also hold the designs in memory long enough to permit recognition of those designs when presented later for memory testing. Likewise, the role of motor dysfunction can be evaluated by examining whether patients can draw to command. If they have demonstrated an inability to copy, but can draw to command, some disruption of the connection between the visual analyzers and the motor programming centers of the brain is suspected.

Clearly then, the overall goal of neuropsychological assessment at Level 3 is to identify which specific links in the functional system are disrupted. By doing so, the neuropsychologist can begin to formulate hypotheses about a possible intervention strategy. This intervention strategy is derived directly from the results of Level 3 assessment. Once the disrupted links in the functional system are identified, the examiner's job is to test the limits to discover what steps are necessary to support the response. When support structures necessary to bolster a response that is missing from the subject's repertoire are identified, the assessment process can move into Level 4 assessment, which encompasses the day-to-day assessments conducted in the context of any cognitive intervention.

To summarize to this point, Level 1 assessment identifies problems in the real world that warrant further investigation. Level 2 assessment is the thorough neuropsychological battery that develops a tentative profile of neurocognitive strengths and weaknesses. Level 3 assessment is conducted to delineate more specifically the reasons for the weaknesses identified in the Level 2 assessment. The results of the Level 3 assessment lead directly to the formulation of hypotheses about possible support strategies that can compensate for missing or absent skills or abilities. Level 4 assessment involves monitoring the day-to-day impact of these strategies during a cognitive intervention.

LEVEL 4 ASSESSMENT: DAILY MONITORING OF TREATMENT EFFECTIVENESS

Level 4 assessments represent the daily monitoring of performance of subjects during any cognitive intervention. At a minimum such assessment should include periodic or perhaps daily recording of identified target behaviors or target performances. However, this approach to Level 4 assessments is less sophisticated than strategies that incorporate single subject experimental designs in the assessments (Webster, McCaffrey, & Scott, 1986). By conducting Level 4 assessments in the context of single subject research designs, one can have greater confidence that the observed treatment effects represent the influence of treatment itself and not merely the passage of time or the influence of some otherwise unidentified variable. An added benefit is accountability. In these days of cost containment and denial of coverage, treatment gains that can be causally related to the treatment intervention may be crucial in insuring that the professional activities of rehabilitation psychologists can be regarded as appropriately compensable.

Level 4 interventions involve daily monitoring of the patient's performance on identified targets of treatment. For example, if the patient is involved in visual scanning training for left-sided hemispatial neglect, the daily performance probes might include assessments of letter cancellation or light board scanning behavior (Gouvier, Bue, Blanton, & Urey, 1987). When the assessment focuses on memory, the daily performance probes might include memory for digit span or word span. In the context of single subject research designs, it may be desirable to include in the Level 4 assessment not only one or two measures of the target behavior but also one or two measures of behaviors that are not specifically the target of intervention so that treatment specificity and dissociation of performance can be identified. The inclusion of nontarget assessment foci also permits some estimation of the generalizability of the treatment intervention.

The entire process of neuropsychological assessment for rehabilitation planning in cognitive disordered patients involves working up the ladder through Level 1, Level 2, Level 3, and Level 4 assessments, and then back down the ladder to evaluate the day-to-day impact of treatment through continued Level 4 assessments, the generalizability of those findings with periodic probes at Level 3, occasional readministration of the neuropsychological battery for Level 2 assessment, and ultimately, confirmation of the impact of intervention strategies through Level 1 assessments involving interview and evaluation of the patient's performance as reported by family members.

APPLICATION OF THE LEVELS SYSTEM OF NEUROPSYCHOLOGICAL ASSESSMENT FOR TREATMENT PLANNING FOR COGNITIVE DISORDERS

Not all patients will require detailed analysis of their neuropsychological profile at each of the 4 levels, nor will each patient's treatment necessarily be conducted according to a rigorous single subject research methodology, but a case example presented later illustrates the approach outlined in this chapter (Webster & Gouvier, 1984).

An overview of the assessment process shows several specific areas of functioning that should be examined at each level. Some of these areas involve formal testing; some require careful interviewing and behavioral observation.

LEVEL 1. INTERVIEW THE FAMILY AND PATIENT

The interview process should attempt to cover the following points:

1. Review current symptom complaints, behavioral excesses, and behavioral deficits.
2. Establish a priority among the problems on the problem list.
3. Review events leading to the injury and establish the presence and significance of preexisting factors that might limit participation in treatment or recovery, such as educational, occupational, or psychosocial history factors.
4. Evaluate other concurrent or associated problems that might affect participation or recovery, such as financial constraints, ongoing medical problems, or reactive psychological disturbances.
5. Review current psychological stressors.

LEVEL 2. CONDUCT A NEUROPSYCHOLOGICAL EVALUATION TO QUANTIFY THE DEGREE OF DEFICITS AND THE EXTENT OF RESIDUAL STRENGTHS IN THE FOLLOWING DOMAINS OF NEUROCOGNITIVE FUNCTIONING

1. Arousal, attention, and concentration
2. Sensor, perceptual, and motor functioning
3. Visuospatial, visuoperceptual, and visuoconstructive functions
4. Language functions
5. Executive functions
6. Intellectual functions
7. Academic functions
8. Learning and memory abilities
9. Reasoning and problem solving
10. Personal and emotional functioning

LEVEL 3. CONDUCT A FUNCTIONAL ANALYSIS TO EXAMINE WHY TESTS ARE FAILED AND TO FRACTIONATE THE CAUSES OF THE DEFICITS

This will yield a detailed profile of strengths and weaknesses of underlying abilities within the content area of each failed task.

1. Compare testing results with original problem lists. If they agree, use a "testing the limits" procedure to tease out the core deficits and identify residual strengths to use in intervention development. The goal is to train functional behaviors that will support themselves in the criterion environment and not simply to train patients to pass tests. Keep foremost in mind the need to have the training results apply to their everyday performances in the criterion environment.
2. If testing results and problem list complaints do not coincide, investigate further as follows:
   a. Select different tests of the same or related functions, and reattempt to document deficit severity on par with symptom complaints.
   b. Simulate the problem situation in the laboratory by introducing factors to promote ecological validity of the assessment such as distractions, intrusive noise, heat, children, or other factors that are identified as relevant in the Level 1 assessment data.
   c. Have family and patient engage in monitoring at home to more clearly define the problem situation and factors that contribute to it. Look specifically for secondary gain, familial support of disabled behavior, or other nonorganic contributors to observed deficits in neurocognitive functions, and be aware that emotional changes in the family members may skew the data as well (Novack et al., 1991).
3. Attempt to assess problems and resources not accessible through standard assessment procedures. These would include issues such as anger control, eating disorders (Crane & Gouvier, 1987; Gouvier & Crane, 1987), social sensitivity, motivation for change, ability to sustain goal directed behavior, and awareness of the need for and consequences of behavioral changes.

LEVEL 4. PLAN AND IMPLEMENT REMEDIATION PROGRAM

At this stage, intervention work begins. There are a number of factors to consider before proceeding.

1. Identify the most important problem from the problem list, based on its degree of intrusiveness and level of effect on the patient's everyday functioning. This requires a careful definition of the problem and observation of the problem under simulated conditions in the clinic. It is also necessary to identify environmental factors that influence the problem, and this may require home observation or, at the very least, careful monitoring by the patient and his or her family.
2. Look at what has already been tried. By identifying previously attempted interventions and examining what went wrong with them, one can minimize errors.

Case Example

M. L. was a 59-year-old white male with a lifelong history of Type A behavior who in June of 1983 suffered a stroke which resulted in a dense left hemiplegia, left neglect, and severe affective liability. He was forced to take medical retirement from his position as the vice president of a large manufacturing corporation, and displayed substantial anger and reactive depression secondary to this retirement. Although separated in the year prior to the stroke, he had reunited with his spouse at the time of evaluation. M. L. was referred by his physical therapist in January of 1984 for treatment of left neglect.

Selected Assessment Findings

Level 1. The physical therapist reported that severe neglect and tendency to lean to the right was hampering M. L.'s ability to benefit from physical therapy, and limiting his ability to ambulate with sufficient competence, necessitating use of a wheelchair. Interview with his spouse supported these reports and provided details about his poststroke lifestyle of doing little except engaging in worry and anger. The patient interview suggested that he could accept his forced retirement better if it did not also involve being confined in his home.

Level 2. History: 59 y.o.w.m. suffered stroke 6/83 resulting in dense left hemiplegia, left neglect, and severe lability. Was forced to retire from V. P. position of large manufacturing industry. Significant depression and some agitation regarding this. Was separated at the time of stroke, but reunited with spouse at this time. Referred by P. T. six months postonset for treatment of left neglect. Has lifelong history of "Type A" behavior. Right MCA thrombotic stroke. Premorbid IQ (estimated) = 114.

Test	Score	Poor	Low	Average	High	Superior
I. Intellectual functions:						
A. Hold tests (verbal):						
1. Information	13				X	
2. Vocabulary	12				X	
3. Verbal IQ	110					
B. Hold tests (performance):						
1. Picture completion	9		X			
2. Performance IQ	81		X			
C. Full Scale IQ	95			X		
II. Motor tasks:						
A. Dominance RIGHT						
B. Strength:						
1. Right kgs.	38				X	
2. Left kgs.	0	X				

C. Speed:					
1. Right hand tapping	51		X		
2. Left hand tapping	0	X			
3. Digit symbol	6		X		
D. Agility:					
1. Grooved pegboard:					
a. Right hand (# dropped = 0, time = 93 sec)			X		
b. Left hand (# dropped =, time =)			X		
III. Verbal learning and recall:					
A. Immediate recall WMS paragraph	8		X		
B. Delayed recall WMS paragraph	10			X	
C. Percent recall	120				X
D. Paired associates WMS	11		X		
E. Paired associates delayed	11			X	
Nonverbal visual construction and memory:					
A. 1. Rev figure drawing	19	X			
2. Immediate recall	10	X			
B. WMS figure memory					
1. Immediate recall	6	X			
2. Delayed recall	6	X			
3. Percent retained	100%			X	
C. Block design	6	X			
D. Object assembly	7	X			
IV. Visual perception:					
A. Suppressions	2	X			
B. Benton facial recognition	33	X			
C. Visual form discrimination	24	X			
V. Language:					
A. Wepman					
1. Repetition	0		X		
2. Naming	0		X		
3. Read	0		X		
4. Math	0		X		
B. Information	13		X		
C. Comprehension	13		X		
D. Similarities	12		X		
VI. Attention/concentration:					
A. Digit span	8	X			
B. Arithmetic	8	X			
C. Mental control	8		X		
D. Trails A	73′	X			
VII. Problem solving:					
A. Serial processing:					
1. a. Trails A	73′	X			
b. Trails B	120′	X			
2. Picture arrangement	7	X			
B. Generative:					
1. Bicycle (__ pts.)					
2. Word fluency	52			X	
C. Complex:					
1. Wisconsin Card Sort					
a. Categories	6				
b. Perseverative responses	94	X			

(continued)

LEVEL 3

	7/83	1/84	3/8
WMS:			
MQ	99	101	137
% Total	40	75	110
Trails:			
A	180+	84	73
B	180+	208	120
Cancellation		95	7
Body awareness			
Left		1	5
Right		6	1
Visual RT			
Left		0.93	0.61
Center		0.62	0.48
Right		0.75	0.59
Visual search			
Left		45.2	15.8
Center		34.2	12.4
Right		42.4	12.4

LEVEL 4. Intervention program is designed and implemented.

Problem list:

1. Significant right hemisphere dysfunction
2. Adjustment to "forced" retirement
3. Increased dependency in marital relationship
4. Left neglect and body awareness
5. Visuospatial problems and dressing apraxia
6. Decreased serial processing ability
7. Emotional lability and secondary social withdrawal
8. No leisure skills
9. Significant tension and anxiety

REHABILITATION STRATEGY FOR M.L.

1. Visual scanning and body awareness training
2. Cognitive therapy for depression and adjustment to retirement
3. Exploration of leisure skills and development of a sense of purpose
4. Blocking strategy for emotional lability
5. Relaxation training
6. Checklists for dressing apraxia on bedroom mirror
7. Track increases in independence via conjoint sessions with wife

M. L. was able to benefit from the intervention program conducted in 16 sessions over a 2-month period. The visual scanning training with the lightboard (which followed the procedures outlined in Webster and colleagues (1984) and Gouvier and colleagues (1984) led to generalized improvements in

> his scanning behavior as reflected in improved Trails performance, improved letter cancellation performance, and improved visual search and reaction times. These gains allowed him to broaden his leisure repertoire by resuming reading as a hobby. Even more significant was the impact of the body awareness training, which was modeled after the seminal work conducted at New York University Rusk Institute for Rehabilitation Medicine (Diller et al., 1974). We modified this training program to include the use of a verbal mnemonic for fostering recalibration of his sense of body position in space: "If I feel like I'm standing up straight, I'm leaning to the right. If I feel I'm leaning to the left, I'm standing up straight." As M. L. became increasingly comfortable in the feeling of leaning to the left, and confident that he was not going to fall, his performance in physical therapy improved dramatically and his physical therapist discharged him from therapy at the same time he completed our program. Follow-up contacts with M. L. and his wife (Level 1 assessments) provided evidence of lasting intervention effectiveness; as M. L. came to view himself as prematurely retired rather than medically disabled, he was able to refocus his life activities onto the plans he had always held for retirement such as traveling and spending more time with his family and grandchildren.

3. Formulate a treatment plan based on the results of the Level 3 assessment. This must include consideration of other factors such as a cost-benefit analysis of its likelihood of success, evaluation of the acceptability of the intervention, and the degree to which the intervention will be useful or applicable in the various situations encountered in the criterion environment.
4. Obtain support from the patient and his or her family. Interventions that are offered without the enthusiastic support of the participants are doomed to failure. Because of this fact, at the beginning of training it is useful sometimes to assign a higher priority to a smaller intervention for a circumscribed problem when the chance for success is high. The experience of success on a smaller problem may help bolster motivation of the participants to work diligently on implementing a more involved intervention later on.
5. Implement the training program in a systematic fashion, and only after collecting sufficient baseline data to permit ongoing evaluation of the program effectiveness. This process involves the following steps:
   a. Shape skills from simple to complex.
   b. Once skills are in place, work to reduce execution time and prolong sustained performance times.
   c. Gradually fade initial props and cues.
   d. Once the response can stand unsupported, move to promote generalization by simulation of the natural setting for training exercises, and moving out into the criterion environment to conduct training in the environments in which the behavior will be required.
   e. Frequently evaluate training effects using continuous monitoring of performance on the training tasks and periodic probes of related performance abilities derived from the Level 3 assessment. Use the results of the daily monitoring assessments to make data-based decisions

about when to move on to the next topic of training. Our previous data on visual scanning training suggests that it may be better to train to a lower criterion and then move on to training in a related area sooner. Treatment gains are not readily lost when the focus of training shifts, and prolonged training on one task does not promote generalization of training to related tasks nearly as quickly as shifting the focus of training to the related activities sooner (Gouvier et al., 1987).

6. Provide support and counseling throughout. Don't underestimate the importance of the psychotherapeutic relationship that exists along with (and that probably predated) the rehabilitation/retraining relationship, and don't allow the patient to become dehumanized as a subject in a single subject experiment to promote improved brain functions.
7. As one problem is resolved on the list, go back to step one and reprioritize the remaining problems. Once the new priorities are identified, repeat the remediation process.

A Caveat to Clinicians (or Problems with Early Neuropsychological Strategies)

It is fairly straightforward and simple to train patients in strategies to improve their testing performance, and given sufficient numbers and breadth of training exemplars, training effects are seen to generalize to areas for which no specific training was provided. For example, DeBoskey, Dunse, and Morin (1986) offered a comprehensive neurorehabilitation program for children with head injuries of mild to moderate severity. The program is derived from the 37 variables of the Selz and Reitan (1979) adolescent scoring system and offers 35 skills modules for retraining. The modules are applied sequentially based of the severity of deficit noted on the neuropsychological testing, with the worst testing performances being addressed first.

The first module, entitled "Category Test (abstract thinking, categorization, problem solving, concept formation)," uses as retraining materials stimuli derived from the Cattell Culture Fair test, the IPAT Culture Fair Test (Scale 2), the Leiter International Performance Scale, Raven's Progressive Matrices, the Structure of Intellect, the Categories Test, a domino game involving the matching of Roman and Arabic numerals, and 3 books containing additional abstract reasoning exercises. The second module, entitled "TPT (kinesthetic spatial organization including a memory and localization component, tactile form discrimination)" uses as training materials a series of 21 wooden formboards arranged in increasing difficulty of 3 to 12 shapes, as well as tactile thickness matching activities, tactile identification of sandpaper letters of decreasing size, and a shopping bag containing foods, toys and common shapes for tactile object recognition practice. The third module, entitled "Trails A (dot to dot by number, recognizing the symbolic significance of numbers)" uses a number of dot-to-dot workbooks and wipe-off cards as the training stimuli. The remaining 32 modules are of similar construction.

This program represents an early example of cognitive interventions driven by Level 2 assessment results. No personal criticism of the authors or the program is intended, as this program was developed during the early period of cognitive rehabilitation. It must be noted, however, that little or no consideration is given to the issues of generalization to the criterion (Level 1) environment and, because of the breadth of training offered on the Level 2 assessment tasks and related activities, it is difficult to evaluate the significance of any posttraining improvements that are noted. It is difficult to envision any sort of Level 2 assessment battery that would not be directly confounded by this training program. This might stack the odds in favor of being able to document program gains, but the functional significance of such gains is dubious unless the patient is being trained for a professional career as a test taker.

Some Final Words

Ten and twenty years ago, cognitive remediation therapies were in the embryonic stage of development, and no one was sure of the outcome. Researchers and clinicians eagerly awaited the results of the controlled treatment outcome studies that were to shed light on the ultimate efficacy or failure of cognitive rehabilitation to promote meaningful changes in patients' ability to function in their everyday lives. As we approach the 21st century, the jury is still out, and the definitive data that all had hoped for have not been forthcoming. We know that we can train new skills in brain-damaged patients, and we know that those skills are not likely to show strong generalization unless we deliberately promote that generalization. The challenge to clinicians working in this area is to promote changes that affect the real-world abilities of the people we serve.

References

Beck, A. T. (1987). *Beck Depression Inventory: Manual.* San Antonio, TX: Psychological Corp.

Benton, A. L. (1974). *Revised Visual Retention Test* (4th ed.). San Antonio, TX: Psychological Corp.

Benton, A. L., & Hamsher, K. (1983). *Multilingual Aphasia Examination.* Iowa City, IA: AJA.

Benton, A. L., Hamsher, K. deS., Varney, N. R., & Spreen, O. (1983). *Contributions to neuropsychological assessment: A clinical manual.* New York: Oxford University Press.

Butcher, J. N., Dahlstrom, W. G., Graham, J. R., Tellegen, A. M., & Kaemmer, B. (1989). *MMPI-2: Minnesota Multiphasic Personality Inventory-2. Manual for administration and scoring.* Minneapolis: University of Minnesota Press.

Chelune, G. J., & Moehle, K. A. (1986). Neuropsychological assessment and everyday functioning. In D. Wedding, A. M. Horton, & J. S. Webster (Eds.), *The neuropsychological handbook* (pp. 489–525). New York: Springer.

Chelune, G. J., Heaton, R. K., & Lehman, R. A. (1986). Neuropsychological and personality correlates of patient's complaints of disability. In G. Goldstein & R. Tarter (Eds.), *Advances in clinical neuropsychology* (Vol 3, pp. 95–126). New York: Plenum.

Crane, M. U., & Gouvier, W. D. (1987). Feeding problems of head injured clients: Physiological factors. *Network: Dietetics in Physical Medicine and Rehabilitation, 6*(2), 1, 4.

Cullum, C. M., Kuck, J., & Ruff, R. M. (1987). Neuropsychological assessment of traumatic brain injury in adults. In E. D. Bigler (Ed.), *Traumatic brain injury: Mechanisms of damage, assessment, intervention, and outcome* (pp. 129–164). Austin, TX: Pro-Ed.

DeBoskey, D. S., Dunse, C., & Morin, K. (1986). *Educating the mild to moderately head injured child: A neuropsychology based program.* Tampa, FL: Tampa General Hospital.

Delis, D. C., Kramer, J., Ober, B., & Kaplan, E. (1987). *The California Verbal Learning Test: Administration and interpretation*. San Antonio, TX: Psychological Corp.

Diller, L., Ben-Yishay, Y., Gerstman, L., Goodkin, R., Gordon, W. A., & Weinberg, J. (1974). *Studies in cognition and hemiplegia* (Rehabilitation Monograph No. 50). New York: New York University Medical Center, Institute of Rehabilitation Medicine.

Dunn, L. M., & Dunn, L. M. (1981). *Peabody Picture Vocabulary Test–Revised manual*. Circle Pines, MN: American Guidance Service.

Golden, C. J. (1978). *Diagnosis and rehabilitation in clinical neuropsychology*. Springfield, IL: Thomas.

Goodglass, H., & Kaplan, E. F. (1987). *The assessment of aphasia and related disorders* (2nd ed.). Philadelphia: Lea & Febiger.

Gouvier, W. D. (1982). Using the digital alarm chronograph in memory retraining. *Behavioral Engineering, 7,* 134.

Gouvier, W. D. (1987). Assessment and treatment of cognitive deficits in brain-damaged individuals. *Behavior Modification, 11,* 312–328.

Gouvier, W. D. & Crane, M. U. (1987). Feeding problems of head injured clients: Psychological factors. *Network: Dietetics in Physical Medicine and Rehabilitation, 6*(3), 1–2, 4.

Gouvier, W. D., Cottam, G., Webster, J. S., Beissell, G. F., & Wofford, J. D. (1984). Behavioral interventions with stroke patients for improving wheelchair navigation. *International Journal of Clinical Neuropsychology, 6,* 186–190.

Gouvier, W. D., Webster, J. S., & Blanton, P. D. (1986). Cognitive retraining with brain-damaged patients. In D. Wedding, A. M. Horton, Jr., & J. Webster (Eds.), *The neuropsychology handbook, behavioral and clinical perspective* (pp. 278–324). New York: Springer.

Gouvier, W. D., Bua, B. G., Blanton, P. D., & Urey, J. R. (1987). Behavioral changes following visual scanning training: Observations of five cases. *International Journal of Clinical Neuropsychology, 9,* 74–80.

Graham, F. K., & Kendall, B. S. (1960). Memory for Designs Tests: Revised general manual. *Perceptual and Motor Skills, 11,* (Monograph Suppl. No. 2-VIII), 147–188.

Gronwall, D. M., & Wrightson, P. (1974). Delayed recover of intellectual function after minor head injury. *Lancet, 2* (7874), 1452.

Hathaway, S. R., & McKinley, J. C. (1943). *Booklet for the Minnesota Multiphasic Inventory*. New York: Psychological Corp.

Heaton, R. K. (1981). *Wisconsin Card Sorting Test manual*. Odessa, FL: Psychological Assessment Resources.

Heilman, K. M., & Valenstein, E. (Eds.). (1979). *Clinical neuropsychology*. New York: Oxford University Press.

Hooper, H. E. (1958). *Hooper Visual Organization Test: Manual*. Los Angeles: Western Psychological Services.

Jastak, S., & Wilkinson, G. (1984). *The Wide Range Achievement Test-Revised: Administration manual*. Wilmington, DE: Jastak.

Kaplan, E. F., Goodglass, H., & Weintraub, S. (1983). *Boston Naming Test* (2nd ed.). Philadelphia: Lea & Febiger.

Katz, M. K., & Lyerly, S. B. (1963). Methods for measuring adjustment and social behavior in the community: I. Rationale, description, discriminative validity, and scale development. *Psychological Reports, 13,* 503–535.

Keenan, J. S., & Brassell, E. G. (1974). A study of factors related to prognosis for individual aphasic patients. *Journal of Speech and Hearing Disorders, 39,* 257–269.

Lezak, M. D. (1978). Living with the characterologically altered brain injured patient. *Journal of Clinical Psychiatry,* July, 592–598.

Lezak, M. D. (1983). *Neuropsychological assessment*. New York: Oxford University Press.

Lezak, M. D. (1987). Relationships between personality disorders, social disturbances, and physical disability following traumatic brain injury. *Journal of Head Trauma Rehabilitation, 2,* 57–69.

Mapou, R. L. (1992). Neuropathology and neuropsychology of behavioral disturbances following traumatic brain injury. In C. J. Long & L. K. Ross (Eds.), *Handbook of head trauma: Acute care to recovery* (pp. 75–89). New York: Plenum.

Markwardt, F. C. (1989). *Peabody Individual Achievement Test-Revised*. Circle Pines, MN: American Guidance Service.

Matthews, C. G. & Kløve, H. (1964). *Instruction manual for the Adult Neuropsychology Test Battery*. Madison: University of Wisconsin Medical School.

McKinlay, W., & Brooks, D. N. (1984). Methodological problems in assessing psychosocial recovery following head injury. *International Journal of Clinical Neuropsychology, 6*, 87-89.

McNeil, M. R., & Prescott, T. E. (1978). *Revised Token Test*. Baltimore: University Park Press.

Novack, T. A. Bergquist, T. F., Bennett, G., & Gouvier, W. D. (1991). Primary caregiver distress following severe head injury. *Journal of Head Trauma Rehabilitation, 6*(4), 69-77.

Osterreith, P. A. (1944). Le test de copie d'une figure complex: Contribution a l'étude de la perception et de la mémoire. *Archives de Psychologie, 30*, 286-356.

Raven, J. C. (1960). *Guide to the Standard Progressive Matrices*. London: Lewis.

Reitan, R. M., & Wolfson, D. (1993). *The Halstead-Reitan Neuropsychological Test Battery*. South Tucson, AZ: Neuropsychology Press.

Rey, A. (1941). L'Examen psychologique dans les cas d'encéphalopathie traumatique. *Archives de Psychologie, 28*, 286-340.

Rey, A. (1964). *L'Examen clinique en psychologie*. Paris: Press Universitaire de France.

Rourke, B. P. (1982). Central processing deficiencies in children: Toward a developmental neuropsychological model. *Journal of Clinical Neuropsychology, 4*, 1-18.

Sbordonne, R., & Long, C. J. (Eds.), (1996). *The ecological validity of neuropsychological tests*. Orlando, FL: Deutch.

Selz, M., & Reitan, R. M. (1979). Rules for neuropsychological diagnosis: Classifications of brain function in older children. *Journal of Consulting and Clinical Psychology, 47*, 258-264.

Smith, A. (1973). *Symbol Digit Modalities Test. Manual*. Los Angeles: Western Psychological Services.

Sohlberg, M. M., & Mateer, C. A. (1989). *Introduction to cognitive rehabilitation: Theory and practice*. New York: Guilford.

Sohlberg, M. M., Mateer, C. A., & Stuss, D. T. (1993). Contemporary approaches to the management of executive control dysfunction. *Journal of Head Trauma Rehabilitation, 8*, 45-58.

Spielberger, C. D., Gorsuch, R. L., & Lushene, R. E. (1970). *Manual for the State-Trait Anxiety Inventory*. Palo Alto, CA: Consulting Psychologists Press.

Stroop, J. R. (1935). Studies of interference in serial verbal reaction. *Journal of Experimental Psychology, 18*, 643-662.

Strub, R. L., & Black, F. W. (1985). *The mental status examination in neurology*. Philadelphia: Davis.

Taylor, E. M. (1959). *The appraisal of children with cerebral deficits*. Cambridge, MA: Harvard University Press.

Warrington, E. K. (1984). *Recognition Memory Test manual*. Windsor, England: Nfer-Nelson.

Webster, J. S., & Gouvier, W. D. (1984, August). *Assessing and treating the brain injured patient*. Continuing education workshop presented at the 92nd convention of the American Psychological Association, Toronto, Ontario, Canada.

Webster, J. S., Jones, S., Blanton, P. D., Gross, R., Beissel, G. F., and Wofford, J. D. (1984). Visual scanning training with stroke patients. *Behavior Therapy, 15*, 129-143.

Webster, J. S., McCaffrey, R., & Scott, R. R. (1986). Single case design for neuropsychology. In D. Wedding, A. M. Horton, & J. S. Webster (Eds.), *The Neuropsychology Handbook* (pp. 219-258). New York: Springer.

Wechsler, D. (1981). *WAIS-R manual*. New York: Psychological Corp.

Wechsler, D. (1987). *Wechsler Memory Scale-Revised*. New York: Psychological Corp.

Williams, D., & Mateer, C. A. (1992). Developmental impact of frontal lobe injury in middle childhood. *Brain and Cognition, 20*, 196-204.

Woodcock, R. W., & Mather, N. (1989). *Woodcock-Johnson Tests of Achievement*. Allen, TX: DLM Teaching Resources.

12

Rehabilitation Assessment and Planning for Children and Adults with Learning Disabilities

SUE R. BEERS

INTRODUCTION

The concept of learning disability, developed in the United States in the 1950s and 1960s, received widespread recognition with the passage of legislation mandating identification of children having such disability in the 1970s. This term generally describes the problems of children who are of at least average intelligence but demonstrate impaired perception, cognition, gross or fine motor skill, or language development and who function one to several years below grade level in at least one academic subject. Learning disability (LD) is found in 3% to 4% of the population, although some estimates are substantially higher (Hynd, Obrzut, Hayes, & Becker, 1986). Until recently it was generally believed that learning disabilities would be outgrown. Young adults with learning disabilities, however, are the single fastest-growing group of persons with disabilities on the college campus today (Notebook, 1996). In fact, recent research has found that individuals outside the formal educational setting continue to manifest learning disabilities that have lifelong ramifications for vocational success and psychological adjustment.

Along with the mandate to provide services for persons with LD has come a growing body of information that addresses assessment and planning for the rehabilitation of this diverse group of individuals. Today, as the needs of this

---

SUE R. BEERS   Western Psychiatric Institute and Clinic, 3811 O'Hara Street, Pittsburgh, Pennsylvania 15213.

Rehabilitation, edited by Goldstein and Beers. Plenum Press, New York, 1998.

population come to be better understood, there is a burgeoning literature not only in education but also in psychology, neuropsychology, and rehabilitation. To help understand the phenomenon or experience of LD, the words of individuals with the disorder provide the foundation for this chapter. The following statements are excerpted from material provided at a training program in specific developmental disabilities for rehabilitation psychologists (McCue, Katz, & Goldstein, unpublished manuscript). These quotations capture the varied aspects of the disorder and highlight problems outside the classroom setting.

> Although my reading level was adequate, it seemed as if I could not read slowly enough to absorb information. . . . My disability was at its worst when attempting to make friends. . . . Both people and books had to be studied and people were definitely more difficult.
>
> I truly believe we could cut half of the employment problems, half the drug addiction, half the alcoholics, if employers and the public knew and understood learning disabilities.
>
> I had tried to make it through college, but I couldn't make it except for lecture classes. I could read, but I could read it over and over again and couldn't tell you what it said.
>
> The experts have a tendency to expect learning disabled people to perform far below what we're likely to do. . . . You have to . . . give the individual the chance to survive and cope in their areas of deficits.

Today, professionals are using a model developed by the World Health Organization (WHO; World Health Organization, 1980) to conceptualize disorders that occur at the organ system level and affect specific abilities as impairments. Impairments, in turn, lead to disability in a more general area. Finally, when this disability actually prevents the person from functioning in society, it is considered to be a handicap (see Rogers & Holm, this volume, for further discussion). LD is often referred to as a "hidden" disability. Upon first consideration, this may be because the disability is manifested by problems requiring thinking and problem solving rather than by more easily observed activities. Applying the WHO model, however, one realizes that there are probably more basic reasons to consider LD a hidden disability. First, although the *disability* of the individual may be recognized by some (e.g., teachers, parents), the underlying *impairment* that led to the learning disability may never be identified. Moreover, because persons with learning disabilities frequently learn to avoid (and, indeed, are often advised to avoid) situations in which they cannot function, their *handicap* may never become apparent.

The goal of this chapter is to shed light on both the impairments and the handicaps associated with LD through a multidisciplinary approach to assessment. Recognizing the diversity of LD and the experiences of individuals with this disability, the following chapter discusses assessment and planning for children and adults. First, we present a working definition of LD and a brief discussion of the controversy that surrounds the definition. Second, we review recent research proposing that LD is a heterogeneous disorder and should be assessed as such. Next, we describe a comprehensive assessment strategy that utilizes the expertise of various disciplines, thereby enabling other professionals who serve this population (e.g., educators, vocational rehabilitation counselors, mental health clinicians) to better understand the scope of LD and to plan more effective interventions. Finally, we briefly discuss intervention strategies based on assessment results.

DIAGNOSTIC CONSIDERATIONS

Learning disability is defined by the National Joint Committee on Learning Disabilities (NJCLD) as follows:

> Learning Disabilities is a general term that refers to a heterogeneous group of disorders manifested by significant difficulties in the acquisition and use of listening, reading, writing, reasoning or mathematical abilities. These disorders are intrinsic to the individual, presumed to be due to central nervous system dysfunction, and may occur across the life span. Problems in self-regulatory behaviors, social perception, and social interaction may exist with learning disabilities but do not by themselves constitute a learning disability. (NJCLD, 1994, p. 65)

Based on the Education for All Handicapped Children Act (P.L. 94-142, 1975), Chalfant identified the major components of LD diagnosis: (1) failure to achieve, (2) disorder of basic psychological processes, (3) exclusionary provisions (i.e., the disorder is *not* caused by sensory incapacity, mental retardation, emotional disturbances, or educational or cultural disadvantage), (4) a medical and/or developmental history suggestive of etiology, and (5) severe discrepancy between IQ and achievement (Chalfant, 1984). Today, as professionals seek to identify and serve individuals with LD, the subjectivity of the phrase "severe discrepancy" has caused much debate. The concerns regarding the definition of LD noted by Adelman and Taylor in 1985 are still with us today. Controversy regarding the scope of the LD diagnosis or how broadly it is to be applied and the definition of severe discrepancy continues to be evident in many different ways. Although a poorly operationalized diagnosis limits progress in understanding the disorder, there are also more practical ramifications. LD services vary across the country, depending on the interpretation by different states and agencies. Indeed, a child who may be classified as LD in one state may be ineligible for learning support services in another.

Reynolds (1984; Reynolds & Kaufman, 1990) has devoted a great deal of attention to the "severe discrepancy" controversy, attempting to establish statistical criteria that can be uniformity applied across individuals and settings. Reynolds criticizes the most frequently applied models, the calculation of grade level or standard score discrepancy, as having the tendency to overidentify children with lower IQs and to deny services to those with higher IQs who are actually functioning well below their abilities. In 1984, Reynolds presented a statistical model to calculate the discrepancy between a student's achievement and the achievement of all other children with the same IQ. Although he has also developed a computer program that computes this discrepancy based on psychometric information (mean, standard deviation, standard error of measurement) from the various achievement tests (Reynolds & Stanton, 1988), this procedure is not uniformly applied. More recently, publishers of the Wechsler Individual Achievement Test included tables that provide an easy and accurate way to compare achievement results with any of the Wechsler IQ tests (Psychological Corporation, 1992).

Before 1970, psychologists and educators considered LD to be a homogeneous diagnostic entity. Based on this assumption, researchers employed comparative-population or contrasting-groups methodology to determine the specific strengths and weaknesses of children with LD. In the 70s, however, investigators began to develop both clinical-inferential and empirical models of learning disability subtypes (Petrauskas & Rourke, 1979). Their work led to attempts to characterize patterns of perceptual and cognitive abilities and deficits that might be amenable to differential treatment interventions.

Rourke and his colleagues spearheaded this research effort with a series of studies investigating the Verbal and Performance IQ discrepancies noted in children with learning disabilities (Rourke, 1985). Groups were formed on the basis of the relationship between Wechsler Intelligence Scale for Children (WISC) Verbal and Performance IQs: VIQ > PIQ, PIQ > VIQ, and PIQ = VIQ. Results suggested that there are subtypes of learning disabilities. More important, it became apparent that some of the patterns of cognitive abilities and deficits were associated with distinct patterns of sensorimotor abilities and deficits as well as with rather unexpected score patterns on academic achievement tests. That is, some children had poor arithmetic ability but performed better in reading.

Continuing their studies, Rourke and his colleagues investigated the neuropsychological significance of these unexpected patterns of academic achievement to determine whether children who exhibited particular patterns on measures of academic achievement also exhibited predictable patterns of neuropsychological abilities and deficits. Children were divided into groups based on patterns of performance in reading, spelling, and arithmetic as measured by the Wide Range Achievement Test (WRAT). The three groups were equal with respect to WISC Full Scale IQ, but demonstrated important differences. Group 1 (Global Deficit) was uniformly deficient in reading, spelling, and arithmetic. Although their arithmetic skills were below age expectation, children in Group 2 (Reading Deficit) exhibited significantly better arithmetic performance relative to their reading and spelling scores. The third group (Arithmetic Deficit) displayed normal reading and spelling skills but were markedly impaired in arithmetic. Although the three groups showed no differences with respect to WISC Full Scale IQ, analyses revealed interesting Verbal–Performance discrepancies. In the Global Deficit group, Verbal IQ was less than Performance IQ. The Reading Deficit group showed the same pattern. The Arithmetic Deficit group, however, exhibited Performance IQs less than Verbal IQs.

When the neuropsychological test performance of the three groups was compared, results suggested differences with respect to verbal and auditory-perceptual tasks, with the Arithmetic Deficit group performing significantly better than the Reading Deficit group on neuropsychological measures within those categories. There were also differences in visual-perceptual and visual-spatial abilities. The Reading Deficit group was significantly superior to the Arithmetic Deficit group on measures of visual-perceptual and visual-spatial abilities.

It is important to note that the differences between the Reading Deficit and the Arithmetic Deficit groups related to the *pattern* of reading, spelling, and arithmetic performance rather than to the *levels* of performance in arithmetic itself. That is, although the groups were almost equal in arithmetic performance, they exhibited different performances when verbal and visual-spatial tasks were compared. From these results, the investigators inferred that the neuropsychological bases for the impaired arithmetic performance of the two groups were quite different. The students in the Reading Deficit group had problems with arithmetic that arose because of verbal difficulties; students in the Arithmetic Deficit group experienced difficulty with arithmetic because of visual-spatial problems.

Rourke and colleagues studied these same groups of children to investigate whether they would differ in a predictable way on motor, psychomotor, and tactile-perceptual skills. Results suggested that there were no significant differences among groups on the simple motor measures. The Global Deficit and Reading Deficit groups, however, performed at a significantly superior level when compared with the Arithmetic Deficit group on two complex psychomotor tests, Mazes and Grooved Pegboard. Comparisons showed the Global Deficit and Reading Deficit groups also performed significantly better than the Arithmetic Deficit group on tactile-perceptual measures for both hands.

In summary, the work of Rourke and his colleagues offered major support for the hypothesis that the Reading Deficit group would perform better on measures of ability subsumed by the right hemisphere and that the Arithmetic Deficit group would perform more poorly on these tasks. Based on these studies and further work done across developmental levels, Rourke (1989) developed a dynamic picture of the neuropsychological strengths and weaknesses of children with what has come to be called nonverbal learning disability (NLD). Although the identification of NLD remains somewhat controversial, it is important for two reasons. First, Rourke and colleagues' work shows the importance of completing a neuropsychological evaluation for children with LD in order to gain more knowledge regarding the impairments (or the CNS deficits) causing the disability, thereby rendering it more visible. Table 1, based on the work of Rourke and colleagues, compares the neuropsychological deficits of the two groups. Second, the discrimination between verbal and nonverbal LD is important in planning the most appropriate and effective interventions that address weaknesses and capitalize on strengths.

TABLE 1. NEUROPSYCHOLOGICAL DEFICITS IN TWO SUBTYPES OF LEARNING DISABILITY

Reading/Spelling disability	Nonverbal learning disability
Auditory perception	Tactile and visual perception
Auditory and verbal attention	Tactile and visual attention
Auditory and verbal memory	Complex psychomotor skills
Phonology	Exploratory behavior
Verbal reception and repetition	Tactile and visual memory
Verbal storage	Concept formation and problem solving
Verbal association	
Verbal output	

Note. Adapted from "Arithmetic Disabilities, Specific and Otherwise: A Neuropsychological Perspective," by B. P. Rourke, 1993, *Journal of Learning Disabilities, 4.*

As the LD subtypes gained empirical validation in groups of children, researchers began to address the issue of their validity and stability in adult populations. Several groups have reported the results of descriptive studies of college students having LD (Mangrum & Strichart, 1988; Zvi & Axelrod, 1992). Cordoni, O'Donnell, Ramaniah, Kurtz, and Rosenshein (1981) compared cognitive profiles of heterogeneous groups of college students having LD with student controls. They found that these students exhibited patterns of ability seen in younger students.

Unlike the studies with children, there is little research that explores homogeneous groups of older persons with LD. Goldstein and his group have completed a series of investigations of adults with LD (McCue & Goldstein, 1991). Their work reports the analysis of the test profiles of 100 adults (mean education: 10.5 years) with identified LD to determine whether subtypes could be identified using rule-based methods similar to those of Rourke and colleagues. They found that only small proportions of their sample exhibited either Reading Deficit or Arithmetic Deficit comparable with Rourke's groups. Most of the adult sample, however, showed poor performance on both reading and arithmetic similar to Rourke's Global Deficit subtype. However, McCue and colleagues' sample was referred by the Office of Vocational Rehabilitation and may not be representative of the adult LD population as a whole.

Another study attempted to identify the Rourke subtypes in higher-functioning individuals. Walter (Morris & Walter, 1991) studied a group of 104 college student volunteers recruited from remedial classes in several subject areas. Subjects were grouped using the WRAT-R. Using this a priori classification method, Walter identified only the Arithmetic Deficit subtype in the sample and many of the students showed no learning disability at all. In order to increase sample size, the original score requirements were slightly relaxed (discrepancy between 11 and 15 points) and additional subjects were included in this group. The evaluation of this group revealed that 21 students could be clinically diagnosed as having an arithmetic disability.

In summary, both diagnostic issues and the variability of the presentation of LD across the lifespan have important implications for assessment. First, it is important to utilize psychometrically sound instruments and statistically rigorous techniques to define the IQ/achievement discrepancy. Second, because LD continues throughout the lifespan, assessment techniques are needed for both children and adults. Finally, a comprehensive neuropsychological evaluation will delineate the impairment or underlying mechanisms of the disability, suggest appropriate remedial and/or compensatory educational strategies, and provide a foundation for career counseling or vocational rehabilitation services.

ASSESSMENT

Assessment strategies in LD, then, are based on several considerations, including the "hidden" aspects of the disorder, the heterogeneity of the deficits, and the occurrence of LD across the lifespan. This section describes the assessment process and discusses, where possible, the reliability and validity of representative instruments frequently used in that process. The comprehensive assessment of LD includes the clinical interview, functional assessment, neurological examination, psychoeducational assessment, and neuropsychological testing.

CLINICAL INTERVIEW

Various clinical interviews have been developed by agencies serving clients with LD (McCue, 1994; Sparks Center, 1985; Szuhay & Newill, 1980; Vocational Rehabilitation, 1983). These interviews usually consist of demographic information, the individual's own view of the learning problems, a checklist for both reported and observed behavioral and physical problems, a complete medical history, and detailed information regarding school performance including reading, spelling, writing, mathematics, and test-taking skills. For adolescents and adults, information is obtained regarding employment history. Available information is reviewed regarding previous interventions, including special education, learning support, and tutors, as well as with respect to previous psychoeducational and neuropsychological evaluations.

When conducting the clinical interview it is important to be sensitive to signs and symptoms that raise the possibility of LD. Tables 2 and 3 summarize the changing physical and emotional manifestations that can be associated with LD across the lifespan.

FUNCTIONAL ASSESSMENT

An important aspect of the LD assessment is to discover how the individual manages "real-world" problems, such as reading a bus schedule or finding the bus station. As McCue states, "Functional assessment is undertaken to determine the *impact* of the . . . disability on behavior" (Chapter 7, p. 114, this volume). The reader is referred to Chapters 6 and 7 (this volume), which discuss functional assessment in greater detail. Although neuropsychological test performance is not

TABLE 2. COMMON PHYSICAL SIGNS AND SYMPTOMS OF LD ACROSS THE LIFE SPAN

Childhood	Adolescence	Adulthood
Digestive problems	Digestive problems	Digestive problems
Enuresis	Enuresis	Migraine headaches
Febrile seizures	Migraine headaches	Nervous energy
Hyperactivity	Restlessness	Lack of drive
Impulsivity	Impulsivity	Impulsivity
Clumsiness	Poor posture	Clumsiness, poor athletic skills
Self-stimulation	Incoordination	Driving problems
Poor vocal modulation	Playing with hair, picking skin	Playing with hair, picking skin
Perserveration	Problems controlling temper	Problems controlling temper
Inappropriate sense of size, weight, distance	Inappropriate voice tones	Flat tonal quality
Abnormal EEG	Abnormal EEG	Abnormal EEG

Note. From *A Training Program in Specific Learning Disability (SLD) for Rehabilitation Psychologists*, by M. McCue, L. Katz, and G. Goldstein, unpublished manuscript. Adapted with permission.

perfectly correlated with everyday functioning (Heaton & Pendleton, 1981), information provided during the functional interview can often be corroborated by the results of neuropsychological tests. In addition to the obvious questions regarding "reading, writing, and arithmetic," functional capacity is evaluated across the domains described in the following. Using the definitions as a guide, individuals are interviewed regarding functioning in a variety of areas. Examples are provided to illustrate the many everyday implications of LD (Brown, 1982).

PERCEPTION. Perception is the ability to process sensory information. Individuals with LD can have varied types of perceptual deficits that affect everyday functioning. A person is said to have a *proprioceptive deficit* when he or she cannot sense the position of limbs without benefit of sight. Although at first this might seem rather obscure, many jobs (e.g., auto mechanic) or even simple tasks (e.g., dressing) require manipulation of equipment without the advantage of full view.

Auditory perceptual deficits result in difficulties in understanding and processing auditory information. Although the sense of hearing is not impaired, the perception of what is heard is inaccurate (e.g., a similar sounding word may be replaced with another). On the other hand, persons with this type of deficit may be unable to discriminate an unaccented syllable—often the beginnings or endings of words or similar sounding letters, such as "m" and "n." An individual with an *auditory figure-ground deficit* has difficulty hearing sounds over background noise and may, for example, have difficulty keeping track of conversation when music is playing. Finally, an *auditory sequencing deficit* is a disruption of the order in which sounds are heard, resulting in misheard telephone numbers or garbled melodies as the notes of songs are perceived differently.

Tactile perceptual problems involve the sense of touch, are quite varied, and may account for some of the social problems experienced by individuals with LD. Sometimes the tactile system is immature and responsive only to a heavy touch (e.g., a very strong hug). On the other hand, immaturity may also cause an individual to avoid any tactile contact. Some persons with LD have trouble judging the appropriate amount of pressure needed to perform various motor acts. From a social stand-

TABLE 3. COMMON EMOTIONAL MANIFESTATIONS OF LD ACROSS THE LIFE SPAN

Childhood	Adolescence	Adulthood
Immaturity	Immaturity	Immaturity
Impulsivity	Impulsivity	Impulsivity
Problems delaying gratification	Difficulty shifting emotion when appropriate	Emotional rigidity
Susceptible to stress	Vulnerability to stress	Impatience
Increased anxiety	Increased suicide risk	Suicidality, decompensation
Bossy and demanding	Somatic complaints	Somatic complaints
Indiscriminate social behavior	Inability to assume responsibility	Selfishness, disregard for others
Increased need for structure	Poor capacity for empathy	Poor capacity for empathy
Rigid attachments	Few or shallow relationships	No relationships of depth
	Increased anxiety and worry	Increased anxiety and worry
Difficulty with uncertainty		

Note. From *A Training Program in Specific Learning Disability (SLD) for Rehabilitation Psychologists*, by M. McCue, L. Katz, and G. Goldstein, unpublished manuscript. Adapted with permission.

point, an individual with this problem may be unable to discriminate a playful tap on the shoulder from a forceful slap. The ability to make tactile discriminations also enables one to feel the difference between similar objects. A secretary with a tactile discrimination problem might have difficulty discriminating high-quality bond stationery from utilitarian typing paper when preparing a formal document.

Vestibular perceptual problems involve balance. Persons with deficits in this area may have trouble walking through uneven areas and may frequently trip on curbs or miss steps when descending stairs.

There are several forms of *visual perceptual problems* that can impact everyday functioning. Problems include discriminating a face in a crowd, finding keys on a cluttered desk, or locating an assignment on a blackboard that includes other information. Visual sequencing deficits not only impair reading ability (e.g., letters are seen as reversed) but also inhibit the ability to visually scan and locate an item from an array (e.g., a particular book in the library stacks). Similarly, some persons with LD have trouble seeing the difference between two very similar objects, such as letters or shades of colors. Difficulty with depth perception frequently makes it difficult to judge how near or far away an object might be. Of course, many sports activities (e.g., tennis, basketball) are exceedingly difficult for anyone with this deficit.

OTHER SENSORY DIFFICULTIES. Brown (1982) defines a catastrophic reaction as "an involuntary reaction to too many sights, sounds, extreme emotions or other strong stimuli" (p. 6). Persons who experience sensory overload may have problems with temper outbursts or become dazed and unaware of their surroundings. Others may have difficulty integrating information from two or more senses at one time. For example, reading problems result when an individual cannot form an association between spoken and written letters. Persons with this type of difficulty often have trouble doing two things at once, such as listening to a lecture and taking notes.

MOTOR FUNCTIONS. Motor problems are associated with awkward and/or inefficient movement, particularly with respect to goal-directed behavior. Deficits can occur within the perceptual, visual, or auditory motor system. An individual with a perceptual motor problem may have trouble with coordination because of inaccurate sensory information. This deficit usually results in difficulty with activities such as sports. An individual who has visual motor problems may be unable to imitate another person's behavior. Quite diverse activities, from copying an assignment from the blackboard to learning a new dance step by watching the teacher, may be disrupted by such deficits. In contrast, an auditory motor problem is indicated when a child or adult has trouble hearing something and then doing it. Diverse examples might include problems following verbal directions, responding to rhythm with movement as in dancing, or taking notes.

MEMORY. Memory problems are pervasive in LD and can involve both verbal and nonverbal material. They affect all aspects of a person's life—educational, social, and vocational.

HIGHER-LEVEL COGNITIVE ABILITIES. Individuals with LD often experience *cognitive disorganization,* or difficulty completing an orderly, logical problem-solving strategy. Instead, their problem-solving style is often either impulsive or rigid. For further details associated with these latter two categories, the reader is referred to the earlier volume in this series that discusses the various aspects of neuropsychological deficits in detail (cf. Goldstein, Nussbaum, & Beers, 1998).

Neurological Examination

An individual being evaluated for LD often is referred for a neurological examination, although frequently the results of the examination are completely normal. This procedure, usually completed by a neurologist, evaluates numerous aspects of behavior. In contrast to the psychoeducational or neuropsychological assessment, these procedures are not as structured and often involve improvisation and observation. By tradition, the examination consists of an evaluation of arousal, movement and posture, the function of the 12 cranial nerves and tests of sensation, muscle strength and tone, and reflexes (Berg, Franzen, & Wedding, 1994). Children are often evaluated with respect to "soft signs." Baron, Fennell, and Voeller (1995) define soft signs as a significant deviation from the normal pattern of motor development (e.g., stereognosis, graphesthesia, mirror movements). If the deviation occurs during a particular period of development, it may reflect neurodevelopmental changes that remain stable into adulthood. The Revised Neurological Examination of Subtle Signs (NESS) has been developed by Denckla (1985) to provide a standardized and reliable assessment of neurological soft signs. (For a comprehensive discussion of the neurological examination, the reader is referred to Stowe, 1998).

Psychoeducational Assessment

By definition, an evaluation for LD cannot take place without the measurement of intellectual ability and academic achievement. Although assessment of IQ and school achievement are almost always included in the neuropsychological battery, they are also often evaluated within the school setting. Two of the most frequently used measures of intellectual function are briefly described along with examples of both general achievement tests and measures to assess the specific abilities of reading, spelling, mathematics, and language ability. For a detailed discussion of the diagnostic and clinical applications of achievement tests, the reader is referred to Katz and Slomka (1990).

INTELLECTUAL ABILITY. The Wechsler intelligence tests are probably the most widely recognized measures of intellectual function. An extension of the original Wechsler–Bellevue Intelligence Scale, the three current versions, the Wechsler Preschool and Primary Scale of Intelligence-Revised (WPPSI-R; Wechsler, 1989), the Wechsler Intelligence Scale for Children–3rd Edition (WISC-3; Wechsler, 1991), and the Wechsler Adult Intelligence Scale–Revised (WAIS-R; Wechsler, 1981), offer the advantage of assessment from early childhood to advanced age. Each of the instruments includes verbal (the capacity for verbal learning and the ability to use verbal skills in reasoning and problem solving) and performance (the efficiency and integrity of perceptual organization) measures, grouped by Wechsler based on clinical information. The tests yield a global or Full Scale IQ

(FSIQ) as well as Verbal and Performance IQs (VIQ, PIQ) that are customarily compared when evaluating individuals for LD. In addition, factor analytic studies of these instruments have identified a relatively stable factor structure across age and population groups (Kaufman, 1979; Sattler, 1992). The Verbal Comprehension factor measures verbal knowledge and understanding, obtained informally and through formal education. Perceptual Organization is composed of subtests assessing the ability to interpret and organize visually perceived material within a time limit. The third factor, Freedom from Distractibility, is felt to provide a measure of attention, anxiety, symbol ability, sequential processing, or memory. The interested reader is referred to Reynolds and Kaufman (1990) for a detailed discussion of the development and application of the Wechsler intelligence tests.

The Stanford–Binet Intelligence Scale—4th Edition (SB-4; Thorndike, Hagen, & Sattler, 1986a) also assesses intellectual ability in toddlers through adults, and is composed of 15 subtests and four theoretically derived domain scores—Verbal Reasoning, Quantitative Reasoning, Abstract/Visual Reasoning, and Short-Term Memory. Comprehensive information regarding test reliability and validity are presented in a technical manual that was published shortly after the test itself was made available (Thorndike, Hagen, & Sattler, 1986b). Internal consistency (Kuder–Richardson 20 coefficient) for the global or Composite index ranges from .95 to .99 across the age levels. Internal consistency for the domains is also satisfactory, ranging from a median r of .86 for Short-Term Memory to .95 for both Verbal and Abstract/Visual Reasoning. The validity of the SB-4 has been extensively investigated, with studies addressing developmental changes, fairness across gender and ethnic groups, and correlations with other IQ measures. With respect to convergent validity, in studies of at least 75 participants the SB-4 showed a correlation of .80 to .89 with earlier versions, the WISC-R, the WPPSI, and the Kaufman Assessment Battery for Children (K-ABC). Of particular importance to the topic of this chapter are the studies that have compared the performance on the Stanford–Binet and the WISC-R or K-ABC in children with LD. In a sample of 90 children, correlations of the Composite and area standard age scores with the WISC-R FSIQ, VIQ, and PIQ ranged from .52 (Quantitative Reasoning/PIQ) to .87 (Composite/FSIQ). A smaller sample of students completed the SB-4 and the K-ABC scales. In that case, the overall pattern of correlations was substantially lower. The poorest association was found between Abstract Visual Reasoning and K-ABC Sequential Processing (.31); the highest occurred between the Composite and K-ABC Achievement (.74).

GENERAL ACHIEVEMENT TESTS. The Wide Range Achievement Test—3rd Edition (WRAT-3; Wilkinson, 1993) is the most recent version of the original test developed in 1936. Like earlier versions, the test measures the three academic skills of reading recognition, spelling, and arithmetic. The age level ranges from 5 to 74, with two alternate forms. Four reliability measures were computed for the various forms of the WRAT-3. Coefficient alpha, a measure of internal consistency, was computed over all age groups. The median coefficients ranged from .85 to .95 over the 9 WRAT-3 test versions (i.e., subtests of the 2 alternate forms and the alternate forms combined). Correlations between the two alternate forms were also high, with a median correlation across the age groups for reading at .92, spelling at .93, and arithmetic at .89. The stability of the WRAT-3, as indicated by test-retest, is excellent. Coefficients ranged from a low of .91 (Arithmetic, Tan version) to .98 (Reading, both Combined and Blue versions).

Content, construct, and discriminant validity are discussed in the test manual. Test content was developed over the previous editions of the test. The Rasch statistic measured the extent of item separation, or how well the items define the variable being measured. Each of the 9 tests had an item separation of 1.00 (the highest level possible). Construct validity of the WRAT-3 is supported in several ways. First, there is a positive correlation between test scores and age. Second, as would be expected, the test scores are positively correlated with measures of academic abilities and these academic scores are also positively correlated with intellectual ability. Third, the current version of the test is positively correlated with previous versions. Finally, the WRAT-3 discriminates between groups. Children in gifted, learning disabled, educable mentally retarded, and normal groups were given the WRAT-3. Test scores correctly predicted the group for 85% of the gifted children and 72% of the children with LD, a rate well above chance.

The Peabody Individual Achievement Test–Revised (PIAT-R; Markwardt, 1989) is another wide-range measure of achievement that assesses reading, spelling, mathematics, and general information. Grade and age equivalent scores as well as standard scores are computed. The standardization sample for this test is large and is controlled for gender and socioeconomic status. Extensive technical information is provided in the test manual. Reliability was investigated using split-half, Kuder–Richardson, test-retest, and item response theory. Split-half reliability coefficients are uniformly high, ranging from .84 for kindergarten children (Mathematics) to .98 for third graders (Reading Recognition). A comparison of split-half and Kuder–Richardson reliabilities indicated little difference, suggesting that the test domains are homogeneous in content. Test-retest reliabilities indicate that scores remain stable from one administration to the next. Coefficients for selected grades ranged from .78 for fourth graders (Mathematics) and sixth graders (Reading Comprehension) to .98 (Reading Recognition) for eighth graders. The Rasch method was used to compute reliability based on estimated raw scores. Coefficients were reported to be high, in the mid- to high 90s.

Content and construct validity information is also provided in the test manual. The extensive developmental process for each PIAT-R subtest is reviewed as evidence of the instrument's content validity. The PIAT-R correlates with the original instrument (PIAT) at an acceptable level, with lowest correlations found on Spelling at higher grade levels. The PIAT-R shows a moderate relationship with the Peabody Picture Vocabulary Test–Revised, with subtests that are most dependent on verbal abilities showing the strongest correlations.

The Wechsler Individual Achievement Test (WIAT; Psychological Corporation, 1992) is a comprehensive achievement battery appropriate for persons ranging in age from 5 to 19 years. The test measures eight academic areas: Basic Reading, Mathematics Reasoning, Spelling, Reading Comprehension, Numerical Operations, Listening Comprehension, Oral Expression, and Written Expression. These are the areas specified by the Education of All Handicapped Children Act as relevant to LD evaluation. In addition, the WIAT is the only achievement battery linked to the three Wechsler IQ scales (WPPSI-R, WISC-3, and WAIS-R), thus allowing for the direct calculation of psychometrically based IQ–achievement discrepancies.

Reliability of the WIAT is extensively discussed in terms of internal consistency, stability, and interrater agreement. To evaluate internal consistency, the test manual presents split-half reliability coefficients for the subtest and composite standard scores across the test age ranges. Calculations were completed at the beginning and end of the school year. As is usually the case, the composite scores, which sample the broad ranges of achievement, have the highest reliability. For these scores, the reliability coefficients, averaged across the ages, range from .90 for Language and Writing to .97 for the Total Composite Index. Subtest reliabilities, also averaged across the age span, range from .81 for Written Expression to .92 for Basic Reading. The stability of test scores over time was also evaluated by testing 367 children across five grade levels on two occasions. Testing intervals ranged from 12 to 52 days (median interval = 17 days). Scores were evaluated across ages and grades as well as across time. In all cases, the subtest, composite, and grade score differences were small, generally ranging from 1 to 3 standard score points between the first and second testing occasions. Interrater reliability was evaluated for Reading Comprehension, Listening Comprehension, Oral Expression, and Written Expression, all subtests that require scorer judgment. Two studies evaluated 50 protocols, randomly selected from the standardization sample. Each protocol was scored by four raters. Results showed high interrater reliability, with the average correlation between raters generally falling in the .9 range.

The WIAT manual also reports extensive information regarding content, construct, and criterion-related validity. The determination of content validity depended on both expert judgment and item analysis. Expert reviewers examined each test item with respect to content, format, and wording, as well as the congruence between each item and the curriculum objective it was designed to assess. Controversial items were either deleted or rewritten. In addition to this procedure, item analyses were completed during both the tryout and standardization phases of test development. Evidence of construct validity is provided by the confirmation of expected relationships between various WIAT subtests and by the correlations between the WIAT scores and the various Wechsler IQ scales. Further evidence is documented by the score differences between various clinical groups (e.g., gifted, children with mental retardation, children with learning disabilities) and the standardization sample as well as between the various age groups within the standardization sample. Finally, the relationship between the Stanford–Binet and other test scores, indexes, and diagnostic categories documented criterion-related validity. For individually administered achievement tests (e.g., WRAT-R, Kaufman Test of Educational Achievement, Peabody Picture Vocabulary Test, Differential Ability Scales [DAS]), there was consistency among tests of reading, mathematics, and spelling ability, with the majority of correlations approximately .80, ranging from a low of .42 (WIAT Reading Comprehension versus DAS Word Reading) to a high of .88 (WIAT Spelling versus Basic Achievement Skills Individual Screener Spelling). Group achievement test comparisons were somewhat lower, generally in the range of .72.

DIAGNOSTIC TESTS. Diagnostic tests are designed to identify the underlying cause of disability and to identify both deficits and strengths with the goal of providing an educational intervention. The Gates–MacGinitie Reading Tests (MacGinitie & MacGinitie, 1989) is a norm-referenced instrument measuring vocabulary and reading comprehension of students from grades K to 12. The comprehension subtests are graded with respect to difficulty and require the individual to read short passages and respond to questions regarding content and those requiring inference and abstraction. Higher test levels represent content found in most public school social and natural sciences and the arts. Vocabulary reliability coefficients range from .90 to .95 and Comprehension coefficients from .88 to .94. The major validity emphasis for this test appears to be content validity, which is discussed at length in the test manual. Although the test is useful to identify general reading problems, it lacks the diagnostic specificity to identify weaknesses.

The Boder Test of Reading-Spelling Patterns (Boder & Jarrico, 1982) was developed as a screening test to differentiate dyslexia from nonspecific reading disability and to classify dyslexic readers into one of three subtypes based on the analysis of their reading and spelling patterns. According to Boder, children who are dysphonetic readers display a strength in the visual gestalt function (i.e., how the word "looks"), but a weakness in the auditory analytic ability. On the other hand, the dyseidetic reader shows a strength in phonetic analysis, but cannot process the visual gestalt of particular words. The test is appropriate for children from kindergarten through high school and takes about 30 minutes to administer. It includes an oral reading test and a written test of spelling. Each subtest is completed under two conditions to allow for careful error analysis. Sight-word reading (i.e., known words) and the application of phonetic skills to unfamiliar words are evaluated. Spelling word lists are made up of known and unknown words in order to assess the child's ability to revisualize words and to apply phonic word-analysis skills. The reliability of the Boder Test is discussed in terms of interrater and test–retest reliability and internal consistency. Although sample size was small, reliability was sufficiently high. Interrater reliability compared the results of four examiners (three teachers and one psychologist) in rating good phonetic equivalents in the Unknown Words list. After the evaluation of test protocols from 30 disabled readers, interrater agreement was calculated at .99. Test–retest reliability considered both gender and age. Fifty children were tested at a 2-month interval, with the lowest reliability found in spelling (mean = .72) and the highest in reading level (mean = .97). Internal consistency was computed after comparing 46 test protocols with respect to scoring of the following components: percentage of words read correctly on Flash presentation; reading level; percentage of Known Words spelled correctly; and percentage of Unknown Words spelled with acceptable phonetic equivalents. After a modified split-half procedure and the application of the Spearman–Brown formula to estimate the reliability of the full subtests, reliabilities for the four tests were as follows: Flash presentation, .99; Reading Level, .97; Known Words spelling, .82; and Unknown Words phonetic spelling, .92. As in the reliability studies, sample size in the various investigations of test validity was small, ranging from 42 to 214 subjects. Citing several independent studies of construct validity, the test manual reviews extensive information suggesting that the Boder identifies diagnostic subtypes of dyslexia. In addition, the results of two studies reported in the test manual indicate that the Boder Reading Level score is highly correlated with the Reading score of the WRAT (at above the .90 level in both studies).

The Gray Oral Reading Tests–3rd Edition (Wiederholt & Bryant, 1992) was developed to assess oral reading and to aid in diagnosing reading difficulties in students from first grade through high school. The newest edition of this instrument includes rate and accuracy scores that are combined to produce a Passage Score as well as an optional error analysis procedure that allows for the more fine-grained evaluation of oral reading skills. Although earlier versions of this test were criticized because of a limited norm group, the current test is normed on a sample of 1,485 children across 18 states. Test–retest reliability was completed for both forms of the test for the entire age range. Coefficients assessing both internal stability and alternate-form reliability are reported in the test manual and range from .62 (Comprehension score, alternate-form reliability) to .92 (Accuracy score, stability coefficient). The test manual also provides an extensive discussion of test validity, presenting information regarding content validity, criterion-related validity, and construct validity. Construct validity in particular is discussed at length, with information included regarding relationship to age, intercorrelations of the individual scales, item discrimination, the test's ability to discriminate among groups of readers, and its relationship to other specific language tests, to general achievement tests, and to intelligence tests.

The Test of Written Language—Third Edition (TOWL; Hammill & Larsen, 1996) is a standardized test for students in grades 3 to 12 (ages 7 through 17) that attempts to identify skill deficits in the written language areas of thematic maturity, spelling, vocabulary, grammar, punctuation, capitalization, and handwriting. Two equivalent forms are available. Coefficient alpha was used to determine the extent to which test items correlate in order to provide an assessment of internal consistency over the 11 age intervals. A review of coefficients for the subtests and composite scores averaged across age levels indicates that both forms of this test are highly reliable, with all but one coefficient (Contextual Conventions) falling above the .80 range. Test–retest reliability is reported for Forms A and B using students in grades 2 and 12, with a testing interval of 2 weeks. Coefficients ranged from .72 to .94, with most in the range of .85. Because some subtests of the TOWL rely on subjective judgment, interrater reliability is an important consideration. As might be expected, interscorer agreement (averages for Forms A and B) for the 8 subtests was highest for Spelling (.97) and lowest for Story Construction (.83).

The KeyMath–Revised (Connolly, 1988) is individually administered and assesses understanding and application of math concepts and skills over three areas: basic concepts (numeration, rational numbers, and geometry), operations (addition, subtraction, multiplication, division, and mental computation), and applications (measurement, time and money, estimation, data interpretation, and problem solving). Two alternate forms are available. Although this test is most frequently used for students through the ninth grade, it is also applied clinically to evaluate the math skills of older individuals. The normative sample for this most recent revision was based on approximately 873 students in grades K through 7, with a proportional representation of African Americans and other minorities. Both alternate form and split-half reliability coefficients were computed for grade- and age-based scores. Coefficients measuring the reliability of the total test are generally high, ranging from .88 to .99. Validity is discussed with respect to content and construct validity. It is important to point out that this most recent revision contains content different from the original version. Content validity was addressed by identifying and carefully specifying the mathematical content of the instrument. Based on a curricular review, essential mathematics content was divided into relatively equal domains and test items were developed that accurately assess mastery with respect to each domain. Construct validity of the KeyMath-R is discussed in terms of developmental change, internal consistency, and correlation with other tests. As expected, students in higher grades achieved higher scores than those in lower grades and in the validity sample mean performance level increased with grade level. Internal consistency is indicated by the high correlations between tests measuring similar content and low correlations between tests with less similar content. For example, Numeration, Rational Numbers, and Geometry subtests correlated with Basic Concepts, ranging from .75 to .85 across selected grade levels (K, 2, 4, 6, 8); lower correlations were noted when these subtest scores were compared with Operations and Applications. Finally, construct validity was assessed by comparing scores between the KeyMath-R and the previous version of the test and the mathematics scores from two achievement tests—the Comprehensive Test of Basic Skills (CTBS) and the Iowa Tests of Basic Skills (ITBS). A review of data presented in the test manual indicates that although subtest comparisons show a moderate correlation (probably explained by changes in item content), the total score coefficients are high when this most recent form is compared to the original version. Moderate correlations (range .40–.50) are reported between the KeyMath-R and both the CTBS and ITBS. Test authors indicate that the lower correlation is accounted for by differences in item format and content.

NEUROPSYCHOLOGICAL TESTING

The study of the brain–behavior relationship in learning disability is a principal specialty of clinical neuropsychology. Referring to the WHO model, neuropsychological assessment provides an adjunct to traditional psychoeducational testing, with the goal of identifying the central nervous system impairment that underlies the disability identified by psychoeducational testing. In their training manual for rehabilitation psychologists, McCue and colleagues (unpublished manuscript) suggest that a comprehensive assessment for LD should include assessment over the domains of sensory-perceptual functioning, motor functions, and higher-level cognitive functioning (i.e., attention, memory, language, visual-spatial function, and problem solving). These test results are then integrated with the findings on measures of intellectual functioning, academic achievement, and personality. An evaluation of personality is conventionally included in most neuropsychological assessments. These batteries are most effective when used in conjunction with the clinical interview, a functional assessment, and, frequently, a neurological evaluation.

To illustrate testing across the life span, this section includes three different neuropsychological batteries. Batteries appropriate for children and adults are outlined first. Then a battery developed for assessing young adults is discussed in more detail as an illustration of assessment procedures for this often overlooked group of students with LD. As will be seen, the batteries share some common instruments.

CHILDREN. As discussed earlier, the work of Rourke has been highly influential in the assessment of children with LD. Strongly influenced by the Halstead-Reitan tradition, several of his books provide detailed information regarding the assessment of children (cf. Rourke, 1985, 1989, 1991, 1995; Rourke, Fisk, & Strang, 1986). Rourke's work consistently emphasizes the necessity of a comprehensive battery capable of identifying a child's strengths and weaknesses. Although this concept might be considered the keystone of Rourke's philosophy, one finds that in this age of managed care it is a struggle to maintain this standard. The comprehensive battery proposed by Rourke (see Table 4) is strengthened by the inclusion of testing procedures that are highly standardized and include appropriate normative information. Finally, the comprehensive battery allows the clinician to evaluate the child's functioning using methods of inference that include level of performance, pathognomonic signs (i.e., signs or symptoms that are distinctive and characteristic of a disease or condition), a differential score approach to evaluate strengths and weaknesses, and a systematic comparison of the functioning of the two sides of the body (Reitan & Wolfson, 1992).

ADULTS. As noted earlier, McCue and colleagues found that LD continues into adulthood. McCue and Goldstein (1991) and their collaborators have completed extensive investigations using the Halstead–Reitan Neuropsychological Battery (HRB). They note that this instrument, used in conjunction with the Russell, Neuringer, and Goldstein (1970) ratings procedure, is capable of identifying adults with LD, with test performance generally falling between the normal range and that seen in individuals with structural brain damage. As these investigators indicate, neuropsychological testing not only has diagnostic applications but also aids in identifying optimal learning and communication modalities and educational needs, planning realistic vocational goals, and providing information regarding prognosis with respect to rehabilitation goals. Tests included in the HRB are listed in Table 5. The reliability and validity of the battery is described in detail in another volume of this series (Goldstein, 1998). In addition, several of the individual HRB instruments are described in more detail in the next section of this chapter.

YOUNG ADULTS. As the increased accessibility of postsecondary education has resulted in a rise in the number of students who continue their academic efforts, colleges and universities are beginning to acknowledge the learning problems of an increasingly diverse student population. Although there are now more programs in place to provide services to students with previously diagnosed LD, these college students remain an underserved group. However, postsecondary students with identified LD may represent the tip of the iceberg. In our recent study of students receiving services at a campus learning center, we documented a surprising number of heretofore unidentified learning disabilities (Beers, Goldstein, & Katz, 1994). In fact, 15% of the students we evaluated using the comprehensive battery of neuropsychological tests presented in the following met stringent diagnostic criteria for LD. Testing procedures, as well as reliability and validity information, are included in the following section. In this battery, tests were chosen that were expected to best delineate the strengths and weaknesses of students who might meet the criteria for LD but who demonstrated the minimum academic skills to gain admission to college.

Symptoms Check List (SCL-90-R; Derogatis, 1983). The SCL-90 is a 90-item self-report symptom inventory, often used as a psychiatric screening device, designed

TABLE 4. Pediatric LD Battery

Children to 16 years	
Intellectual functioning Wechsler Intelligence Scale for Children —3rd ed. (WISC-3) Peabody Picture Vocabulary Test	*Academic achievement* Wide Range Achievement Test—3rd ed.
Personality functioning Personality Inventory for Children	

Children ages 5 to 9 years	

Perceptual functioning–tactile
Reitan—Kløve Tactile-Perceptual
  and Forms Recognition Test
	Perceptual functioning–visual
	Reitan–Kløve Visual-Perceptual Tests
	WISC-3 PC, PA, BD, OA
	Target Test
	Constructional Items (Aphasia Screening Test)
	Color Form Test
	Progressive Figures Test
	Individual Performance Tests
	  (Matching Figures and Vs, Star
	  Drawing, Concentric Square Drawing)

Perceptual functioning–auditory
Reitan—Kløve Auditory-Perceptual Test
Auditory Closure Test
Auditory Analysis Test
Peabody Picture Vocabulary Test
Speech-Sounds Perception Test
Sentence Memory Test
Verbal Fluency
WISC-3 Inf, Comp, Sim, Voc, DS
Aphasia Screening Test
	Motor and psychomotor function
	Reitan—Kløve Lateral Dominance
	Grip Strength
	Finger Tapping Test
	Kløve–Matthews Motor Steadiness
	  Maze Coordination Test
	  Static Steadiness
	  Grooved Pegboard

Concept formation problem solving Halstead Category Test Children's Word-Finding Test WISC-3 Arithmetic Matching Pictures Test	*Other* Underlining Test WISC-3 Coding Tactual Performance Test (6 blocks)

Children ages 9 to 16	
Perceptual functioning–tactile Reitan-Kløve Tactile-Perceptual and Forms Recognition Test	*Perceptual functioning–visual* Reitan-Kløve Visual-Perceptual Tests WISC-3 PC, PA, BD, OA Trail Making A (children's version)

Perceptual functioning–auditory
Reitan-Kløve Auditory-Perceptual Test
Seashore Rhythm Test
Auditory Closure Test
Auditory Analysis Test
Peabody Picture Vocabulary Test
Speech-Sounds Perception Test
Sentence Memory Test
Verbal Fluency Test
	Motor and psychomotor function
	Reitan-Kløve Lateral Dominance
	Grip Strength
	Finger Tapping Test
	Kløve–Matthews Motor Steadiness
	  Maze Coordination Test
	  Static Steadiness Test
	  Grooved Pegboard Test

TABLE 4. *(Continued)*

WISC-3 Inf, Com, Sim, Voc, DS Aphasia Screening Test	
Concept formation problem solving	*Other*
Halstead Category Test	Underlining Test
Children's Word-Finding Test	WISC-3 Coding
WISC-3 Arithmetic	Tactual Performance Test (6 blocks)
	Trail Making Test (children's version)

Note. PC = Picture Completion; PA = Picture Arrangement; BD = Block Design; OA = Object Assembly; Inf = Information; Com = Comprehension; Sim = Similarities; Voc = Vocabulary; DS = Digit Span. Adapted from *Nonverbal Learning Disabilities: The Syndrome and the Model*, by D. P. Rourke, 1989, New York: Guilford. Also from *Neuropsychological Assessment of Children*, by B. P. Rourke, J. L. Fisk, and J. D. Strang, 1986, New York: Guilford.

to reflect psychological symptoms in the following dimensions: somatization, obsessive–compulsive, interpersonal sensitivity, depression, anxiety, hostility, phobic anxiety, paranoid ideation, and psychoticism. The measure also provides three global indices of distress: a severity index, a positive symptom distress index, and a positive symptom total. Norms and T-score conversions are provided for both males and females based on samples of psychiatric inpatients and outpatients and adult and adolescent nonpatients. Subjects are asked to report how much discomfort each symptom has caused them over the last year through their response on a 5-point Likert scale. The score on each dimension or factor is the average rating given the items that comprise that factor.

Reliability of this short test is high. Measures of internal consistency range from .77 for Psychoticism to .90 for Depression. Test–retest correlations for the various dimensions range from .78 to .90. Convergent validity has been supported by significant correlation between the test and other measures of psychiatric symptomatology such as the Beck Depression Inventory and various scores of the

TABLE 5. ASSESSMENT OF LEARNING DISABILITY IN ADULTS: THE GOLDSTEIN ADAPTATION OF THE HALSTEAD–REITAN BATTERY

Halstead–Reitan Neuropsychological Battery
Aphasia Screening Test
Finger Tapping Test
Grip Strength
Sensory-Perceptual Examination
Bilateral Simultaneous Sensory Stimulation
Visual, Auditory, Tactile
Tactile Finger Recognition Test
Finger-tip Number Writing Perception Test
Speech-Sounds Perception Test
Rhythm Test
Trail Making Test
Tactual Performance Test
Category Test
Wechsler Adult Intelligence Scale-Revised
Minnesota Multiphasic Personality Evaluation

Middlesex Hospital Questionnaire. Although the authors suggest that the convergent and discriminant validity are supported by correlations with the MMPI score, others take issue with that conclusion (Mitchell, 1985).

Wechsler Adult Intelligence Scale-Revised (WAIS-R; Wechsler, 1981). The WAIS-R is discussed earlier in this chapter; in addition to intellectual ability, it provides for an evaluation of individual strengths and weaknesses.

Wide Range Achievement Test—3rd Edition (WRAT-3; Wilkinson, 1993). This test, described in an earlier section of this chapter, is included to provide a measure of academic achievement. Because some do not support the use of a reading recognition test to establish learning disability in older students, two additional reading tests are included in the battery (Conoley & Kramer, 1989b).

Comprehension: Nelson-Denny Reading Test (Brown, Bennett, & Hanna, 1981). This test was developed for use with high school and college students in order to provide a measure of achievement in reading comprehension. Comprehension consists of eight passages, each with a different content area, and 36 multiple-choice questions relating to specific facts, relationships, authors' purpose, generalizations, and other interpretations. Normative information is criticized because it is overrepresented by smaller colleges and includes no "major" institutions. Test–retest reliability for this subtest ranges from .75 to .82. Validity information, however, is limited to evidence that the subtest is context dependent; that is, correct answers depend on reading the passage rather than former knowledge (Conoley & Kramer, 1989a).

Word Attack: Woodcock Johnson Psychoeducational Battery-Revised (Woodcock & Johnson, 1989a, 1989b). This complex battery measures both cognitive ability and achievement of persons ages 3 to 80. It is used for the individual identification of special problems of disabilities, the diagnosis of specific weaknesses that interfere with development, the purpose of individual planning and guidance, and the evaluation of individual growth. Supplemental tests and clusters of tests within the battery provide a more comprehensive diagnostic assessment of a student's problems in a particular area. The Word Attack subtest measures the student's application of phonic and structural analysis skills to the pronunciation of unfamiliar printed words. Nonsense or low-frequency English words are read aloud. The reliability and validity aspects of this test battery are thoroughly covered in the technical manual. Subtest reliabilities over the various age groups produced median coefficients of .80 to .95. Concurrent validity appears to be better evaluated than predictive validity for this battery. All validity information is generally discussed, however, in terms of the broad clusters of the test rather than in terms of the individual supplementary subtests. One criticism of this battery is that, although well standardized in other areas, the entire adult sample included only 600 people (Conoley & Kramer, 1989c).

California Verbal Learning Test (CVLT; Delis, Kramer, Kaplan, & Ober, 1987). The CVLT assesses the multiple strategies and processes by which verbal material is learned and remembered. Incorporating the findings from memory research into clinical assessment, it measures both recall and recognition of word lists over a number of trials. First, the student is asked to recall a list of 16 words forming four semantic categories. Five trials are administered. Next, an interference list, also of 16 words, is presented once. Finally, the student is asked to remember the first list using both free and category-cued recall. After a 20-minute delay, free recall, cued recall, and recognition are measured. Five primary recall standard scores are based on List A performance.

The test manual reports measures of internal consistency and test–retest reliability. Three methods were used to evaluate the internal consistency on 133 subjects selected from a narrow age range (mean = 60.9 years) to reduce age effects. Using spilt-half or odd/even and coefficient alpha, reliabilities ranged from .69 (coefficient alpha) for 5-trial item totals in the nonclinical sample to .92 (odd/even) for the 5-trial scores in the nonclinical group. To examine the stability of the scores across repeated administration, 21 subjects were retested after a 1-year interval. As might be expected, 13 of the 18 test–retest correlations were statistically significant, with the average improvement across the 5 test trials of List A at about 2 words and the average improvement for recall and recognition trials at about 1 word.

Validity of the CVLT has been investigated using factor analysis, criterion-related measures, and by comparison across well-defined clinical groups. Factor analyses were completed using normal, neurological, and clinical groups. Results across these groups have been consistent with the known effects of the various disorders on memory and support the theoretical basis of the CVLT. Convergent validity was investigated by comparing the performance of various patient groups (mixed brain-damaged, substance abusers, psychiatric) on the CVLT and the Wechsler Memory Scale. Statistically significant correlations, ranging from .47 (Short-Delay Free Recall) to .66 (List A Total Recall) occurred when major scores of the CVLT were compared to the WMS Memory Quotient. Construct validity is also demonstrated by the test's ability to demonstrate distinct deficit patterns in different classes of patients. The CVLT has been used to study the memory impairment associated with Alzheimer's and Parkinson's diseases, chronic alcohol abuse, and multiple sclerosis.

Logical Memory: Wechsler Memory Scale-Revised (WMS-R; Wechsler, 1987). This subtest assesses short- and long-term recall of connected discourse. Two brief stories are read to the student, who attempts to recall them immediately and again after a delay of approximately 30 minutes. The score is based on the number of correct phrases reported. The test manual provides extensive standardization, reliability, and validity information. Although age-based norms and data relative to the effects of age and education on subtest performance are provided, developers are criticized for failure to include all age groups (e.g., 18–19-year-olds) in the standardization sample (Kane, 1991). A positive aspect of this test is that reliability and validity information are available for the subtests of the battery. Split-half reliability coefficients of the Logical Memory subtest are .68 for both immediate memory and the recall condition for the 20- to 24-year age group. Test–retest reliability for the same group was .67 for immediate memory and .72 for the recall condition. The test manual reports validity studies that have focused on the relationship of the Logical Memory subtest to conditions such as depression, Alzheimer's disease, and head injury.

Grooved Pegboard (Matthews & Kløve, 1964). This is a test of finger dexterity that requires the student to place notched pegs into a board having 23 holes as fast as possible. Each hole contains a groove. Therefore, there is only one way a given peg will fit into a hole. Dominant and nondominant hands are tested and time to completion is scored. The complexity of this test makes it highly sensitive to the cerebral hemispheric components of motor performance. It is frequently included as part of the Halstead–Reitan Neuropsychological Battery (Goldstein, 1990). Bornstein (1986) investigated difference scores on this test using 365 nonclinical subjects. He found a significant proportion of "atypical" patterns of performances and cautioned against overinterpretation of dominant versus nondominant differences as well as a gender effect.

Rey Complex Figure Test (Meyers & Meyers, 1995; Osterrieth, 1944). This test is widely used to assess perceptual organization and delayed visual memory in both adults and children. It permits assessment of a variety of cognitive processes, including planning, organization, and problem solving, as well as perceptual, motor, and memory functions. The examiner often tracks the student's responses during the copy of the design and during the immediate and delayed recall condition, administered after a 30-minute delay. Overall evaluations of test performance are based on a total accuracy score for each of the different areas or details of the figures developed by Taylor. Interrater reliability coefficients presented in the test manual ranged from .93 to .99, at a level consistent with other studies evaluating the same scoring system. Temporal stability was determined in a sample of normal subjects and testing was completed at an average interval of 184 days. Test–retest reliability assessed with the Pearson correlations was .76 and .89 for the immediate and delayed trial, respectively. Convergent and discriminant validity were examined in both normal subjects and in a heterogeneous sample of patients with documented brain dysfunction. In both samples the immediate and delayed recall scores were highly correlated, with more moderate correlation noted between the copy and recall scores. Construct validity was assessed by evaluation of the relationship between Rey Complex Figure scores and WAIS-R performance in the brain-damaged sample. As might be expected, Recall and Copy scores were most highly correlated with WAIS-R Performance subtests, particularly Block Design, Picture Arrangement, and Digit Symbol. A principal components factor analysis, completed on the same two groups (normal and brain-damaged), produced two quite similar solutions made up of five factors that accounted for more than 95% of the variance in both cases and included visuospatial recall, visuospatial recognition, response bias, processing speed, and visuoconstructional ability. Finally, three discriminant function analyses using copy, recall, and combined recall and recognition scores were used to predict membership in one of three groups: brain-injured, psychiatric, or normal. Copy variables correctly classified 57.8% of the subjects and recall variables correctly classified 61.1% of the subjects. When recognition scores were included with recall scores, classification rate for the total sample improved to 77.8%. However, it is worth noting that in discriminant analyses, normal and brain-damaged subjects were predicted more accurately than were psychiatric patients.

Controlled Oral Word Association Test (Benton & Hamsher, 1978). This test of verbal fluency assesses the ability of the student to generate spontaneously a list of words beginning with a designated letter. The test includes three, 60-second word-naming trials. Adjustments are made for sex, age, and education. Research has shown this test to be a valid and sensitive indicator of brain dysfunction. It is frequently included in neuropsychological test batteries (Lezak, 1995). Gender- and age-based normative information is provided from a study of persons with more than 12 years of education by Yeudall, Fromm, Reddon, and Stefanyk (1986).

The following tests are taken from the Halstead–Reitan Neuropsychological Test Battery, probably the most researched and widely used measure to assess brain damage in the United States (Reitan & Wolfson, 1993). Much of the initial validity information for this battery came from the blind interpretation of assessment information; that is, assessment with no prior knowledge of the subject's case history. This procedure makes it difficult to separate clinical expertise from test data. In his initial predictive validity study, Reitan demonstrated that the test data could be used to draw clinical inferences regarding brain dysfunction independent of external neurological criteria. Although many are trained in its use, there is not yet research that replicates the original diagnostic applicability of the battery. Validity research over the last 30 years has established that the battery is sensitive to the effects of brain damage in many neurologic disorders and can aid in diagnosing the location of brain lesions. Recent work centers on discovering the effects of age, education, and sex on the interpretation of test results (Heaton, Grant, & Matthews, 1991).

Trail Making Test, A and B (Reitan & Wolfson, 1993). This test, originally part of the Army Individual Test Battery, involves motor speed and attention and measures visual conceptual and visuomotor tracking. On Trails A, the student is given a sheet of paper on which circled numbers are randomly printed. The task is to connect the numbers in order as fast as possible. On Trails B, the sheet contains both numbers and letters. The task is to alternate between the two, maintaining correct sequence and working as rapidly as possible. The score is the time to complete each sequence.

This test is one of the most sensitive to the presence of brain damage. Visual scanning and tracking problems provide evidence of how the student responds to a visual array of any complexity, how well the individual performs when following a sequence mentally, and whether the student maintains the flexibility needed to deal with more than one stimulus at a time. The most commonly used scoring system requires that the examiner point out errors so that the test is always completed without errors. This procedure has resulted in a lower test reliability, especially for Part B. Reliability for Part A, as measured by the coefficient of concordance, remained high ($W = .78$) for three administrations at 6- and 12-month intervals. Reliability of Part B was lower, at $W = .67$. Research has demonstrated a significant practice effect for Part A, but not for Part B, of this test (Lezak, 1995).

Booklet Category Test (DeFilippis & McCampbell, 1979). This instrument, a portable version of the Category Test from the Halstead–Reitan battery, measures abstract reasoning as well as the ability to maintain attention to a lengthy task. The test also has a visuospatial component and is considered to require learning skills for effective performance. Stimulus figures that vary in size, location, shape, color, and intensity are grouped by abstract principles. The task is to deduce the organizing principle that relates the items in each subtest and to signal the answer by pointing to the appropriate number appearing on a strip below the booklet. As the test proceeds, the examiner provides only the feedback of "correct" or "incorrect." Evidence suggests that the booklet form is equivalent to the original Halstead version and a high correlation between the two forms is reported for a neurologically normal college student sample. Somewhat lower correlations were found, however, in an alcoholic (.80) and a mixed brain-injured, psychiatric sample (.76.) This version of the Category Test is criticized for its reliance on validity and reliability information previously presented with the original Category Test. The test manual lacks any standardization or normative data and, again, refers the reader to the literature on the Reitan version for interpretive information.

Tactual Performance Test (TPT; Reitan & Wolfson, 1993). This test requires the blindfolded student to place shapes in a form board using dominant, nondominant, and both hands. Successful completion requires both recall of the shapes and their location. The task measures tactual discrimination, manual dexterity, kinesthesis, incidental memory, and spatial memory. A total time score is recorded. Next, the blindfold is removed and the form board concealed. The student is asked to draw the board from memory, indicating the shapes and their relative placement. A memory and a location score are reported.

Speech-Sounds Perception Test (Reitan & Wolfson, 1993). This test is purported to be especially sensitive to damage to the left hemisphere of the brain. It is also thought to measure the student's capacity to attend to a mechanically administered, boring task. Sets of nonsense syllables beginning and ending with different consonants but based on the vowel sound "ee" comprise the items. The test is administered by tape recorder and the student responds on a multiple-choice answer sheet.

APPLICATION

Although one might consider the goal of the assessment process in LD to be the diagnosis of the disability, more careful consideration suggests that the comprehensive assessment also has important implications with respect to planning appropriate interventions. Levine discusses the importance of a developmental-phenomenological model of LD that moves beyond simple diagnosis or identification toward an understanding of how and where the learning process breaks down (Levine, 1987, 1994). Because the development of appropriate intervention is the final step in the assessment process, this chapter concludes with a brief introduction to "management by profile." According to Levine, this highly individualized process, like the comprehensive assessment, is multidisciplinary and includes teachers, administrators, parents, counselors, and physicians. The steps in the intervention process described here relate to the WHO model. That is, the information from the clinical interview, functional assessment, neurological evaluation, pyschoeducational assessment, and neuropsychological evaluation are integrated and used to develop an intervention strategy that addresses the handicap and supports the individual's strengths.

Demystification

Levine suggests that it is impossible for a student to address a problem that he or she cannot describe. He stresses the need to develop a vocabulary and conceptual framework as a way to discuss the specific deficits and strengths in either an individual or group setting. The discussion at this stage is extremely concrete, focusing on observable effects. Essentially this step opens the lines of communication between the student and teachers and/or rehabilitation specialists.

Bypass Strategies

Based on both the functional and the neuropsychological assessment, strategies are developed that circumvent cognitive deficits and weak or underdeveloped skills. These strategies are often applied while the child receives educational interventions to address deficit areas. Often students with LD benefit from adjustments with respect to rate and/or volume of learning, multiple examples of complex concepts, an emphasis on key points, use of devices to assist in learning, and alteration in standard classroom routines. Levine cautions that these modifications should be completed so as not to suggest "giving up" on the learning process.

Interventions at Breakdown Points

Breakdown points are specific weaknesses that impede the student's ability to learn. Neuropsychological assessment and task analysis can help identify the particular areas that should be the target of intervention. Typical interventions, applied both at home and at school, might include reteaching subskills, increased drill and practice, organizing tasks into manageable steps, applying devices such as computer software that provide external structure to the learning process, and practicing skills in various modalities.

Strengthening the Strengths

Throughout this chapter we have stressed that an important focus of assessment is the identification of strengths. In Levine's model, interventions provide individuals with LD an opportunity to exploit, enhance, and demonstrate their strengths. In addition, strengths can be applied to support learning in more problematic areas.

Summary

This chapter has described a multidisciplinary approach to the assessment of children, college students, and adults with learning disability. Clinical interviewing and functional assessment were discussed along with the reliability and validity of instruments included in the psychoeducational and neuropsychological assessment of LD. Emphasizing that the goal of assessment in LD is to provide information relevant to interventions as well as to diagnosis, this chapter concluded by describing a phenomenologically based model of care applicable across the range of learning disabilities.

REFERENCES

Adelman, H. S., & Taylor, L. (1985). The future of the LD field: A survey of fundamental concerns. *Journal of Learning Disabilities, 7,* 423-427.

Baron, I. S., Fennell, E. B., & Voeller, K. K. S. (1995). *Pediatric neuropsychology in the medical setting.* New York: Oxford University Press.

Beers, S. R., Goldstein, G., & Katz, L. J. (1994). Neuropsychological differences between college student with learning disabilities and those with mild head injury. *Journal of Learning Disability, 27,* 315-324.

Benton, A. L., & Hamsher, K. deS. (1978). *Manual: Multilingual aphasia examination.* Iowa City: University of Iowa.

Berg, R. A., Franzen, M., & Wedding, D. (1994). *Screening for brain impairment.* New York: Springer.

Boder, E., & Jarrico, S. (1982). *The Boder Test of Reading-Spelling Patterns.* New York: Grune & Stratton.

Bornstein, R. A. (1986). Normative data on intermanual differences on three tests of motor performance. *Journal of Clinical and Experimental Neuropsychology, 8,* 12-20.

Brown, D. (1982). Rehabilitating the learning disabled adult. *American Rehabilitation, 7,* 3-11.

Brown, J. I., Bennett, M., & Hanna, G. (1981). *Nelson-Denny Reading Test.* Chicago: Riverside.

Chalfant, J. C. (1984). *Identifying learning disabled students: Guidelines for decision making.* Burlington, VT: Northeast Regional Resource Center.

Connolly, A. J. (1988). *KeyMath-Revised: A diagnostic inventory of essential mathematics.* Circle Pines, MN: American Guidance Service.

Conoley, J. C., & Kramer, J. J. (1989a). Nelson-Denny Reading Test, Forms E and F. In *The tenth mental measurements yearbook* (Vol. 1, pp. 1035-1037). Lincoln: University of Nebraska Press.

Conoley, J. C., & Kramer, J. J. (1989b). Wide Range Achievement Test-Revised. In, *The tenth mental measurements yearbook* (Vol. 2, pp. 897-905). Lincoln: University of Nebraska Press.

Conoley, J. C., & Kramer, J. J. (1989c). Woodcock-Johnson Psycho-Educational Battery. In *The tenth mental measurements yearbook* (Vol. 2, pp. 1759-1765). Lincoln: University of Nebraska Press.

Cordoni, B. K., O'Donnell, J. P., Ramaniah, N. V., Kurtz, J., & Rosenshein, K. (1981). Wechsler adult intelligence score patterns for learning disabled young adults. *Journal of Learning Disabilities 14,* 404-407.

DeFilippis, N., & McCampbell, E. (1979). *Manual for the Booklet Category Test: Research and clinical form.* Odessa, FL: Psychological Assessment Resources.

Delis, D. C., Kramer, J. H., Kaplan, E., & Ober, B. A. (1987). *California Verbal Learning Test: Manual.* San Antonio, TX: Psychological Corp.

Denckla, M. B. (1985). Revised neurological examination for subtle signs. *Psychopharmacology Bulletin, 21,* 773-800.

Derogatis, L. R. (1983). *Symptoms Checklist-90-Revised.* Baltimore: Johns Hopkins University.

Goldstein, G. (1990). Comprehensive neuropsychological assessment batteries. In G. Goldstein & M. Herson (Eds.), *Handbook of psychological assessment* (2nd ed., pp. 197-227). New York: Plenum.

Hammill, D. D, & Larsen, S. C. (1996). *Test of Written Language* (3rd ed.). Austin, TX: PRO-ED.

Heaton, R. K., & Pendleton, M. G. (1981). Use of neuropsychological tests to predict adult patients' everyday functioning. *Journal of Consulting and Clinical Psychology, 49,* 807-821.

Heaton, R. K., Grant, I., & Matthews, C. G. (1991). *Comprehensive norms for an expanded Halstead-Reitan Battery.* Odessa, FL: Psychological Assessment Resources.

Hynd, G. W., Obrzut, J. E., Hayes, F., & Becker, M. T. (1986). Neuropsychology of childhood learning disabilities. In D. Wedding, A. M. Horton, & J. Webster (Eds.), *The neuropsychology handbook: Behavioral and clinical perspectives* (pp. 456-485). New York: Springer.

Kane, R. L. (1991). Standardized and flexible batteries in neuropsychology: An assessment update. *Neuropsychology Review, 2,* 281-339.

Katz, L. J., & Slomka, G. T. (1990). Achievement testing. In G. Goldstein & M. Herson (Eds.), *Handbook of psychological assessment* (2nd ed., pp. 123-147). New York: Plenum

Kauffman, A. S. (1979). *Intelligent testing with the WISC-R.* New York: Wiley.

Levine, M. D. (1987). *Developmental variation and learning disorders.* Cambridge, MA: Educators Publishing Service.
Levine, M. D. (1994). *Educational care: A system for understanding and managing learning disorders.* Cambridge, MA: Educators Publishing Service.
Lezak, M. D. (1995). *Neuropsychological assessment* (3rd ed.). New York: Oxford University Press.
MacGinitie, W. H., & MacGinitie, R. K. (1989). *Gates-MacGinitie Reading Tests–Third Edition.* Chicago: Riverside.
Mangrum, C. T., & Strichart, S. S. (1988). *College and the learning disabled student: Program development implementation and selection.* Philadelphia: Grune & Stratton.
Markwardt, F. C., Jr. (1989). *Peabody Individual Achievement Test–Revised: Manual.* Circle Pines, MN: American Guidance Services.
Matthews, C. G., & Kløve, H. (1964). *Instruction manual for the Adult Neuropsychological Test Battery.* Madison: University of Wisconsin.
McCue, M. (1994). *Neuropsychological diagnostic and functional interview.* Unpublished manuscript.
McCue, M., & Goldstein, G. (1991). Neuropsychological aspects of learning disability in adults. In B. P. Rourke (Ed.), *Neuropsychological validation of learning disability subtypes* (pp. 311–329). New York: Guilford.
McCue, M., Katz, L., & Goldstein, G. *A training program in specific learning disability (SLD) for rehabilitation psychologists.* Unpublished manuscript.
Meyers, J. E., & Meyers, K. R. (1995). *Rey Complex Figure Test and Recognition Trial: Professional manual.* Odessa, FL: Psychological Assessment Resources.
Mitchell, J. V., Jr. (1985). SCL-90. In *The ninth mental measurements yearbook* (pp. 1324–1929). Lincoln: University of Nebraska Press.
Morris, R. D., & Walter, L. W. (1991). Subtypes of arithmetic-disabled adults: Validating childhood findings. In B. P. Rourke (Ed.), *Neuropsychological validation of learning disability subtypes* (pp. 330–346). New York: Guilford.
National Joint Committee on Learning Disabilities. (1994). *Collective perspectives on issues affecting learning disabilities: Position papers and statement.* Austin, TX: PRO-ED.
Notebook. (1996, January 12). *Chronicle of Higher Education,* p. A33.
Osterrieth, P. A. (1944). Le test de copie d'une figure complese *Archives de Psychologie, 30,* 206–356. (1993. [J. Corwin & F. W. Bylsma, Trans.]. *The Clinical Neuropsychologist, 7,* 9–15).
Petrauskas, R., & Rourke, B. P. (1979). Identification of subgroups of retarded readers: A neuropsychological, multivariate approach. *Journal of Clinical Neuropsychology, 1,* 17–37.
Psychological Corporation. (1992). *Wechsler Individual Achievement Test manual.* San Antonio, TX: Author.
P.L. 94-142. (1975). *The Education for All Handicapped Children Act of 1975,* 20 U.S.C. SS 1401 et seq., 45 C.F.R. 121(a).
Reitan, R., & Wolfson, D. (1992). *Neuropsychological evaluation of older children.* Tucson, AZ: Neuropsychology Press.
Reitan, R., & Wolfson, D. (1993). *The Halstead-Reitan Neuropsychological Test Battery: Theory and clinical interpretation.* Tucson, AZ: Neuropsychology Press.
Reynolds, C. R. (1984). Critical measurement issues in learning disabilities. *Journal of Special Education, 18,* 451–487.
Reynolds, C. R., & Kaufman, A. S. (1990). Assessment of children's intelligence with the Wechsler Intelligence Scale for Children–Revised (WISC-R). In C. R. Reynolds & R. W. Kamphaus (Eds.), *Handbook of psychological and educational assessment of children: Intelligence and achievement* (pp. 127–165). New York: Guilford.
Reynolds, C. R., & Stanton, H. C. (1988). *Discrepancy determinator DDI: Technical and interpretive manual.* College Station, TX: A & M University: Train.
Rourke, B. P. (Ed.). (1985). *Neuropsychology of learning disabilities: Essentials of subtype analysis.* New York: Guilford.
Rourke, B. P. (1989). *Nonverbal learning disabilities: The syndrome and the model.* New York: Guilford.
Rourke, B. P. (1991). *Neuropsychological validation of learning disability subtypes.* New York: Guilford.
Rourke, B. P. (1993). Arithmetic disabilities, specific and otherwise: A neuropsychological perspective. *Journal of Learning Disabilities, 4,* 214–226.
Rourke, B. P. (1995). *Syndrome of nonverbal learning disability.* New York: Guilford.
Rourke, B. P., Fisk, J. L., & Strang, J. D. (1986). *Neuropsycholoeical assessment of children.* New York: Guilford.

Russell, E. W., Neuringer, C., & Goldstein, G. (1970). *Assessment of brain damage: A neuropsychological key approach.* New York: Wiley-Interscience.

Sattler, J. M. (1992). *Assessment of children: Revised and updated third edition.* San Diego, CA: Author.

Sparks Center for Developmental and Learning Disorders. (1985). *Case history, SLD characteristics observation checklist, and attention deficit disorder checklist.* Birmingham: University of Alabama.

Stowe, R. M. (in press). Assessment methods in behavioral neurology and neuropsychiatry. In G. Goldstein, P. Nussbaum, & S. R. Beers (Eds.), *Neuropsychology.* New York: Plenum.

Szuhay, J. A., & Newill, B. (1980). Learning disability history form. *Field investigation and evaluation of learning disabilities: Vol. 4. A proposed model for state vocational rehabilitation agencies service delivery to the learning disabled.* Scranton, PA: University of Scranton Press.

Thorndike, R. L., Hagen, E. P., & Sattler, J. M. (1986a). *Stanford-Binet Intelligence Scale (4th ed.): Guide for administration and scoring.* Chicago: Riverside.

Thorndike, R. L., Hagen, E. P., & Sattler, J. M. (1986b). *Stanford-Binct Intelligence Scale (4th ed.): Technical manual.* Chicago: Riverside.

Vocational Rehabilitation Center of Allegheny County. (1983). *Specific learning disabilities: A resource manual for vocational rehabilitation.* Pittsburgh, PA: Author.

Wechsler, D. (1981). *Manual for the Wechsler Adult Intelligence Scale-Revised.* New York: Psychological Corp.

Wechsler, D. (1987). *Manual for the Wechsler Memory Scale-Revised.* New York: Psychological Corp.

Wechsler, D. (1989). *Manual for the Wechsler Preschool and Primary Scale of Intelligence-Revised.* San Antonio, TX: Psychological Corp.

Wechsler, D. (1991). *Manual for the Wechsler Intelligence Scale for Children–Third Edition.* San Antonio, TX: Psychological Corp.

Wiederholt, J. L., & Bryant, B. R. (1992). *Gray Oral Reading Tests–3rd Edition: Examiners manual.* Austin, TX: PRO-ED.

Wilkinson, G. S. (1993). *The Wide Range Achievement Test administration manual.* Wilmington, DE: Wide Range.

Woodcock, R. W., & Johnson, M. B. (1989a). *Woodcock Johnson Psychoeducational Battery–Revised Edition: Examiner's manual.* Dallas, TX: DLM Teaching Resources.

Woodcock, R. W., & Johnson, M. B. (1989b). *Woodcock Johnson Psychoeducational Battery–Revised Edition: Norm tables.* Dallas, TX: DLM Teaching Resources.

World Health Organization. (1980). *The international classification of impairments, disability, and handicaps.* Geneva, Switzerland: Author.

Yeudall, L. T., Fromm, D., Reddon, J. R., & Stefanyk, W. O. (1986). Normative data stratified by age and sex for 12 neuropsychological tests. *Journal of Clinical Psychology, 42,* 918–946.

Zvi, J. C., & Axelrod, L. H. (1992). Learning disabled college students: An analysis of the factors emerging from initial assessment [Abstract]. *Journal of Clinical and Experimental Neuropsychology, 14,* 119.

13

Assessment and Planning for Memory Retraining

SUE R. BEERS AND GERALD GOLDSTEIN

INTRODUCTION

Impairment of memory is a ubiquitous symptom of neurobehavioral disorders, appearing most prominently in the amnesic disorders, but also in the dementias associated with head trauma and the degenerative diseases of the elderly. Over the past few years, designing, applying, and, to a lesser extent, evaluating various memory rehabilitation methods have become a growth industry. Memory disorders associated with different types of brain damage vary both qualitatively and quantitatively; therefore, rehabilitation strategies are frequently developed to address the memory deficits that are characteristics of a particular disorder. From a conceptual standpoint, the rehabilitation of memory might be compared to the rehabilitation of aphasia, a diverse condition requiring various rehabilitation techniques that consider both type and severity of language disruption. That is, an effective technique for one form of aphasia is unlikely to be appropriate for other forms.

Memory rehabilitation at our clinic has generally focused on individuals with acquired memory problems that remain permanent and stable, such as patients with closed head injury and those with Korsakoff's syndrome. Recently, however, guided by theoretical considerations as well as by technical advances, we have broadened our considerations to include patient groups with memory difficulties that are associated with various medical illnesses (e.g., systemic lupus erythematosus) as well as the healthy elderly with so-called age-associated memory impairment.

SUE R. BEERS Western Psychiatric Institute and Clinic, 3811 O'Hara Street, Pittsburgh, Pennsylvania 15213. GERALD GOLDSTEIN VA Pittsburgh Healthcare System, Highland Drive Division (151R), 7180 Highland Drive, Pittsburgh, Pennsylvania 15206-1297.

Rehabilitation, edited by Goldstein and Beers. Plenum Press, New York, 1998.

The following chapter provides a historical review of the various techniques adopted in the rehabilitation of memory and discusses the theoretical considerations that provided the foundation for current applications. Next, we describe rehabilitation procedures that have proven effective in our clinic. The final section discusses the future directions of memory rehabilitation and includes a discussion of our own theoretical synthesis of memory rehabilitation, including possible applications to a broad range of individuals.

History of Memory Rehabilitation

Comprehensive reviews of memory training studies with brain-damaged patients are contained in B. A. Wilson and Moffat (1984), B. Wilson (1986), Leng and Copello (1990), Franzen and Haut (1991), and Baddeley, Wilson, and Watts (1995). The field represents a convergence of three areas: experimental psychology, behavioral neurology, and the writings of mnemonists.

In the clinic, professionals working with brain-damaged patients wondered whether or not the memory difficulties they observed were remediable to any extent. Their early attempts to answer that question relied heavily on the popular literature. A widely cited paper by Lewinsohn, Danaher, and Kikel (1977) demonstrated the efficacy of imagery-oriented training for a heterogeneous group of brain-damaged patients, stimulating an attitude of optimism regarding the potential for remediation of memory disorders using techniques clearly borrowed from the mnemonists. Crovitz and collaborators also utilized imagery with severely amnesic patients (Crovitz, 1979; Crovitz, Harvey, & Horn, 1979). Later, a confluence developed among popular, basic science, and computer application approaches. An example of this synthesis is perhaps best represented in the work of Gianutsos and Gianutsos (1979), who applied sophisticated experimental design and computer technology to memory training. There is now an emerging field of prosthetic memory in which the environment takes over what was previously biologically mediated (Kapur, 1995). These developments are discussed in the last section of this chapter.

Neuropsychologists have applied basic concepts from the experimental psychology of memory to the study of amnesic patients, and this application has provided a highly viable and useful framework for conceptualizing the various types of amnesia. For example, experimental methods were applied to classify amnesia with respect to problems in the initial encoding of information, the storage of the information in the brain, or the retrieval of the stored information on demand (Butters & Cermak, 1980). Other investigators used similar concepts as the foundation for their memory training programs (Jaffe & Katz, 1975).

Theoretical Considerations

In developing our rehabilitation programs, we found it useful to distinguish among three types of memory impairment. Some patients, particularly those with unilateral brain damage, show modality-specific amnesia involving either verbal or nonverbal content. For example, the patient may remember a shopping list, but be unable to find the grocery store. The memory rehabilitation method found to work best for modality-specific memory difficulties is that of using the intact modality to support the impaired modality (Gasparrini & Satz, 1979).

The second type of memory impairment is often seen in patients who have experienced closed head injury. Although many individuals who incur a moderate to severe head injury report memory problems, usually at least some of the biological mechanisms for memory remain intact. That is, we typically see patients who have memory difficulty, or forgetfulness, rather than frank amnesia. In general, one can say that their capacity to remember or to encode, store, and retrieve new information is not as efficient as it was prior to injury. Rehabilitation efforts can thus focus on teaching the patient some method, or series of mnemonic techniques, that aids in remembering in a variety of situations or contexts. Imagery, organization, and elaboration seem to be major contributions of these mnemonic devices, but a large number of specific techniques have been developed that emphasize different cognitive processes.

The third category of memory impairment is severe and permanent amnesia, such as found in patients with Korsakoff's Syndrome and in some patients with severe trauma, aneurysms of the anterior communicating artery, cerebral infarctions, or cerebral anoxia. (Patients with progressive dementias such as Alzheimer's disease also become very amnesic, but the memory deficit is accompanied by communication and intellectual disorders, making rehabilitation a far more complex issue.) There is an active literature indicating that Korsakoff patients are capable of some forms of new learning. A paper by Brooks and Baddeley (1976) provided an early discussion of this matter, with more recent discussions provided by Butters (1984) and, most recently, by Cermak (1994). From this body of work, investigators concluded that even densely amnesic patients not only appear to benefit from cuing but also have some capability of forming new procedural memories. That is, they can demonstrate the acquisition of new verbal and nonverbal skills without the declarative memory capabilities to code the learning of these skills as episodes. Thus, the skill is learned without awareness of the circumstances under which the learning took place, or, in fact, that it even did take place. Classical conditioning is described by Squire as a form of procedural or "implicit" memory (Squire, 1992). The existence of this preserved memory system is now well established in severely amnesic patients.

Recently, investigators have also shown a great interest in priming, or the ability to enhance memory through prior exposure of stimulus material. Schacter, Cooper, Tharan, and Rubens (1991) produced impressive data suggesting that amnesic patients demonstrate priming effects approaching a level seen in healthy individuals. Noting that when cues are withdrawn, these patients lose the memories they previously recalled, Glisky coined the term "domain-specific knowledge" to characterize this type of learning without generalization that densely amnesic patients can achieve (Glisky, 1992, 1993; Glisky & Schacter, 1987). It is precisely because of this lack of generalization, however, that priming or cuing techniques do not provide an effective basis for effective memory training in densely amnesic patients.

From a rehabilitation standpoint, then, research suggests a strategy that involves the reliance on implicit memory may be most effective with Korsakoff patients. In normal individuals, new skills are typically learned by practice and rehearsal. Thus, one of our goals with this group has been to teach new items of information as one would teach new skills. In one of our earlier studies, we taught patients with Korsakoff or Korsakoff-type amnesias several items of discrete information (e.g., doctor's name, name of a place) using rote rehearsal (Goldstein & Malec, 1989). Although we did not inquire systematically, it was clear that learning took place outside of the patients' awareness, with patients having no recollection of previous sessions during the current session. In spite of this documentation of a preserved memory system in these severely amnesic patients, we identified several drawbacks that suggested this procedure would not be feasible as a rehabilitation procedure. First, the material learned was quite specific to the training situation. It was not demonstrated when questions were asked by a different clinician. Second, all patients remained amnesic and disoriented to material other than that used in the training. Noting these limitations, our current rehabilitation method involves using the relatively intact procedural learning facility for the purpose of recalling how to use an assistive device, which then provides specific information of the sort usually stored in memory.

Current Applications

Our laboratory has completed memory training with demented (Goldstein, Turner, et al., 1982), Korsakoff and Korsakoff-type (Goldstein, Beers, Shemansky, & Longmore, in press; Goldstein & Malec, 1989; Goldstein et al., 1985), and head-injured patients (Beers & Goldstein, 1995; Goldstein, Ryan, & Kanagy, 1982; Goldstein et al., 1988; Goldstein, Beers, Longmore, & McCue, 1996). The following section will present the assessment methods used to document the severity of memory deficits and the presence of additional cognitive deficits. The memory rehabilitation procedures we have found to be successful with head-injured patients and those with Korsakoff's Syndrome are described. Both studies describe rehabilitation applications made possible by advances in technology. Although computer-assisted techniques appear to be well accepted by patients, the validity and utility of computer applications in the rehabilitation of memory remains controversial (Gianutsos, 1992; Lynch, 1992; Speight, Laufer, & Klaus, 1992). In fact, the division of Clinical Neuropsychology of the American Psychological Association has issued a policy statement concerning ethical and professional considerations of computer-assisted rehabilitation (Matthews, Harley, & Malec, 1991). The work described here discusses the clinical utility of using computers or other devices in memory rehabilitation applications.

Assessment

Before the selection of a rehabilitation protocol, it is important to define the cognitive profile of each patient. One must consider not only the individual's memory function but also level of orientation, general intelligence, and language skills. Both of the training protocols discussed in this chapter require a generally adequate level of intellectual function as indicated by Wechsler Adult Intelligence Scale–Revised, Full Scale IQ $\geq$ 85 and a score on the Dementia Rating Scale of $\geq$ 110 (Mattis, 1988; Wechsler, 1981). In addition, training procedures preclude application for patients with aphasia or related difficulties in language comprehension. We have found that the following assessment provides the information necessary to determine which of our protocols is likely to be the most effective for a particular patient. The battery, along with a brief discussion of reliability and validity issues, is discussed in this section.

DEMENTIA RATING SCALE (DRS) (Mattis, 1988). This instrument provides a brief screening of attention, visual-spatial ability, abstract reasoning, and memory. The split-half reliability of this measure is high, at .90, with the test–retest at 1 week measuring somewhat lower. Studies in the test manual suggest that the test discriminates between mildly to moderately demented patients, correlating with the Wechsler Memory Scale-Revised, Memory Quotient and the Wechsler Adult Intelligence Scale-Revised, Full Scale IQ (FSIQ) at .59 and .70, respectively.

WECHSLER ADULT INTELLIGENCE SCALE—REVISED EDITION (WAIS-R) (Wechsler, 1981). The WAIS-R, a widely used test of intellectual functioning, provides for assessment of global level of intelligence. The test is composed of 11 subtests, 6 of which constitute the Verbal IQ (VIQ) and 5 of which comprise the Performance IQ (PIQ). The appropriate split-half or alternate-form reliability coefficients, computed within nine age levels for each of the subtests, FSIQ, VIQ, and PIQ, are high. Reliabilities for FSIQ ranged from .96 to .98. Verbal and Performance coefficients are comparable, ranging from .95 to .97 and .88 to .94, respectively. The shorter, individual subtests showed the expected lower reliabilities, ranging from .52 for Object Assembly at the youngest age level (16–17 years) to .96 for Vocabulary at six of the nine age levels. With respect to construct validity, individual subtest correlations and correlations between the Verbal and Performance Scales are fairly high, ranging from .33 to .81, suggesting that the two scales have much in common. Factor analysis supports this conclusion, identifying the presence of a single factor that accounted for about 50% of the total variance. Three major group factors—verbal comprehension, perceptual organization, and memory—were also identified (Anastasi, 1982, Mitchell, 1985).

WECHSLER MEMORY SCALE—REVISED EDITION (WMS-R) (Wechsler, 1987). The WMS-R includes measures of verbal and visual memory, using an immediate and delayed recall format. Composite indices include General Memory, Verbal and Visual Memory, Attention, and Delayed Recall. Stringer (1996) reported that test–retest and reliability coefficients for the various aspects of the test varied from .41 to .90, and the average reliability for the composite indices was the highest. Factor analysis indicates that the test reflects two factors: general memory and learning and attention and concentration. However, in a brain-damaged population the test attains a three-factor solution that is composed of verbal memory, nonverbal memory, and attention.

TOKEN TEST—REVISED EDITION (DeRenzi & Vignolo, 1962). This test of language comprehension measures the ability to understand and complete commands using increasingly complex grammatical structure. For example, the patient is asked to "place the large red circle under the small blue square." According to Franzen (1989), few studies have investigated the reliability of this test. In the several studies he discussed, test–retest reliability ranged from .90 to .98.

Concurrent validity information available for a sample of patients with aphasia suggests that the Token Test correlates with the Porch Index of Communicative Ability at .67. An investigation of concurrent validity with normal subjects reported a correlation of .71 between the Token Test and a measure of verbal intelligence, the Peabody Picture Vocabulary Test. This finding suggests that this test of language capacity is not independent of the effects of general intelligence.

BOSTON DIAGNOSTIC APHASIA EXAMINATION (BDAE) (Goodglass & Kaplan, 1983). Selected tests from the BDAE are included in the screening battery. Repeating Phrases consists of "low probability" and "high probability" sentences that differ with respect to vocabulary as well as in the predictability of verbal content. This distinction is based on the fact that patients with some types of aphasia are overly dependent on the predictability of content when they are asked to repeat phrases. In Responsive Naming, the patient is asked questions that require answers to be in the form of nouns, colors, verbs, or numbers. As the name implies, Visual Confrontation Naming asks the patient to look at a picture and respond with a descriptive response. Animal Naming is a fluency task that requires controlled word association. The patient is asked to name as many animals as possible in one minute.

The test manual reports that Kuder–Richardson split-half reliability of three of these four subtests ranges from .91 to .98. Factor analysis including all variables of the BDAE isolated 10 factors, with the phrase repetition tasks included in the Repetition/Recitation factor that involves voluntary speech production. On the other hand, visual confrontation and animal naming are included in the factor characterized by naming tasks. Franzen (1989) reported that the content and face validity of the BDAE are considerable; however, he noted basic validation work is lacking.

REHABILITATION PROCEDURES FOLLOWING HEAD INJURY

Persistent memory disorders are often associated with closed head injury. It is estimated that about 23% of closed head-injured patients experience an amnesic disorder after posttraumatic amnesia has resolved. This memory deficit may vary in severity, but it may be sufficiently disabling to impede the fulfillment of educational and vocational goals. The amnesia is frequently associated with an attention deficit: Patients often have difficulty concentrating, are distractible, and become somewhat impulsive. Sometimes the amnesia is an isolated, residual symptom. In other cases, however, it is associated with neuropsychological deficits such as dementia, aphasia, or visual-spatial disorders. Our rehabilitation procedures have focused on patients with relatively pure amnesia, who frequently report forgetfulness usually manifested by missing appointments, forgetting all or some components of instructions at school or work, and difficulty completing various activities of daily living that require memory (e.g., shopping, remembering names). Memory rehabilitation with head-injured patients generally has fairly ambitious goals that involve teaching the patient some method or series of methods to aid remembering in a variety of situations or contexts.

Since the publication of our earlier study (Goldstein et al., 1988) reporting successful application of two visual imagery-based techniques for improving memory in patients with closed head injury, we have developed computer-assisted technology for both training methods. There were several reasons for developing these procedures. Aside from cost-effectiveness, automation provides for a standardized application of the training across a variety of settings. In addition, the availability of a uniform rehabilitation procedure provides an opportunity for multicenter collaborative research. There are, however, potential disadvantages to computer-assisted rehabilitation procedures. Patients with severe impairments, either cognitive or physical, may be unable to operate the keyboard or understand screen displays. Training efforts may not be successful without the sustained and direct participation of an active therapist. Also, the reliability of the equipment itself may be a consideration (e.g., graphic quality, programming, and hardware problems).

To investigate these issues, we compared the results of training of the subjects included in the earlier study with the results of a new group of individuals who received highly automated training procedures in sessions held 2 to 3 times a week (Goldstein et al., 1996). With certain limitations, training procedures differed only in the application of the computer technology. Both protocols included the same ridiculously imaged story and face-name learning procedures described in the following section.

RIDICULOUSLY IMAGED STORY (RIS). In order to teach patients to learn a list of words, we adapted the ridiculously imaged story technique that had already been successfully applied with severely amnesic patients by Kovner and colleagues (Kovner, Mattis, & Goldmeier, 1983; Kovner, Mattis, & Pass, 1985). This method consists of embedding target words in a bizarre and humorous story. For example, suppose the first three target words were "shirt," "palace," and "grass." The first sentence of the ridiculous story might be: "The magic *shirt* sought by the prince was located in a *palace* made of *grass.*" Stories are presented on a computer screen with 20 target words printed in boldface type. In addition to reading the story out loud, patients are actively encouraged to use visual imagery techniques as an additional memory aid. The training involves three repetitions of the story at each session, with cues or hints provided. The first cue is the element of the story in which the word appeared (e.g., something about a prince). If the patient needs further help, we provide an appropriate category cue (e.g., an article of clothing). Because we wanted to build generalization into this training program, after several sessions using an identical story and target words, a new story and word list were given in subsequent sessions. In addition, in order to encourage patients to use this method spontaneously, we asked them to provide both the story and the word list. By the end of the training, all stories and lists were provided by the patient.

FACE-NAME LEARNING (FNL). At each training session, patients were also taught to use visual imagery to associate names with photographs of faces. A visual imagery technique was used to help form a link between the name and some specific or distinctive feature of the face (Lorayne & Lucas, 1974; Moffat, 1984). For example, a physical attribute (e.g., a great deal of hair) might be linked to the name of the individual (e.g., Harry). In other cases the picture might remind the person of a celebrity of the same name (e.g., "John looks like John Wayne").

The study used data from 10 patients from our original study (Goldstein et al., 1988) and 20 new patients. All had sustained serious head injury at least 1 year prior to assessment with significant impairment of memory and the absence of dementia documented by neuropsychological assessment (WAIS-R FSIQ ≥ 85). Training session data provided learning curves for both the RIS and FNL tasks. As in the earlier study, pretraining and follow-up statistical comparisons were completed. The variables included number (of words remembered on a selective reminding test, number of items recalled from a list of practical items, and correct responses on a face-name learning task.

TRAINING RESULTS. Inspection of the RIS learning curves indicated that the performance of the two groups over the 15 learning trials was quite comparable. No individual learning trial differences were statistically significant. A comparison of pretraining and follow-up selective reminding variables indicated that, as with the original sample, the computer-assisted patients showed significant differences before and after training; however, the amount of material learned was greater for the computer-assisted group. List-learning scores indicated that before and after training comparisons for the computer group were significant, but in this case the computer-assisted training showed no advantage over the original training procedures.

Computer-assisted training for the FNL procedure provided a clear advantage. The two training methods were compared over the 10 learning trials; 7 trials were statistically significant. In the original study, face recall score comparisons before and after training were not statistically significant. For the computer-assisted group, however, comparable comparisons were statistically significant. A comparison of the number of trials to learn a set before and after training for the original and the computer method indicated the differences were significant, but no clear training advantage was indicated by this outcome measure.

Our comparison of results of these two head-injured groups addressed the impact of computer technology on two methods of memory training previously demonstrated to be effective. In the case of the RIS method, the application of the computer-assisted training did not substantially change the original findings. It may well have been, however, that computer training could have adversely affected training results. This did not prove to be the case. For the FNL task there appeared to be a clear advantage for the computer-assisted training. It may be that the addition of the computer to the original training procedure increased the salience of the stimulus material and increased the structure of the cuing procedure (i.e., the trainer could not choose to ignore this part of the training). Successful replication with further cases will strengthen the recommendation that these computer-assisted procedures be applied in rehabilitation settings.

TRAINING PROCEDURES FOR PATIENTS WITH KORSAKOFF'S SYNDROME

Korsakoff and Korsakoff-like syndromes are characterized by the specificity and severity of impairment. They are relatively pure disorders, with minimal associated neuropsychological deficits. Thus, Korsakoff patients are typically neither demented nor do they have major disturbances of language, perception, or mobility. Instead, a remarkably dense amnesia is the most prominent and disabling condition. Patients with this disorder are typically disoriented and have no knowledge of recent personal or external events. They seem to be without recent memory, although they may have some recollection of remote events. The condition is irreversible, and is thought to be produced by hemorrhaging in subcortical, diencephalic structures, probably as a result of thiamine deficiency.

In our work with demented patients (Goldstein, Turner, et al., 1982), we used a reality orientation methodology, and showed that demented and amnesic patients learned and retained substantial amounts of the information taught during group sessions. However, there was no evidence of generalization to new material. A similar result was obtained for a patient with a Korsakoff-like syndrome resulting from cardiac arrest (Goldstein et al., 1985). This work led to the conclusion that rehearsal methods could be used to teach amnesic patients to learn specific information. Using a multiple baseline design, we taught densely amnesic Korsakoff patients to consistently recall a number of items of information (Goldstein & Malec, 1989). On the basis of this work, we concluded that some anmesic patients are capable of learning specific items of information that may be of adaptive significance.

These findings stimulated our interest in handheld assistive devices, as it was soon apparent that an appropriate device could contain the same information and probably more of it than we were training by rehearsal. Two problems emerged: the design of the device and the methodology needed to teach patients to use it consistently. We have designed, fabricated, and successfully field tested an assistive device as well as developed the methodology to teach patients to use the device when in need of information (Goldstein et al., in press).

In our current rehabilitation program, we evaluated this prototype using patients with Korsakoff-type amnesias. In order to be included in this group, patients met DSM-III-R criteria for organic amnestic disorder, as well as the Butters and Cermak (1980) psychometric criteria. That is, they did not meet criteria for dementia, but had relatively pure severe amnesia. To insure intact general intellectual function, a Wechsler IQ of 85 or higher was required, and the score on the Mattis Dementia Rating scale had to be at least 110. All patients had a Memory Index on the Wechsler Memory Scale that was at least 20 points lower than their WAIS-R FSIQ. Patients with aphasia or neurobehavioral disorders other than amnesia were not included in the training. Typically, Korsakoff and Korsakoff-type patients who meet these conditions are disoriented to time and place, and have minimal accurate recollection of recent events.

Based on the skill acquisition studies of patients with amnesia, we felt that rehearsal provided the most promising training strategy. After asking a question (e.g., What is your doctor's name?), we utilized a conditioning method in which a tone was paired with a prompt to use the assistive device. Next, the tone was gradually faded in an effort to elicit the desired motor response of using the device. This procedure can be considered a variant of classical conditioning, whereby the tone is viewed as the unconditioned stimulus and the question, the conditioned stimulus. Providing conditioning to use the device was successful; our next consideration was generalization. Because our review of the literature suggested the possibility that learning might be situation specific, we made the distinction between generalization across persons and settings. That is, the patient might make use of the device only when the training circumstances were exactly the same. We planned our training procedure to involve generalization sessions with different clinicians, different settings, and a combination of both.

TRAINING PROCEDURES. It cannot be overemphasized that while the technology of the assistive devices is of obvious significance, the essence of the rehabilitation method involves training the patient to productively use the device. Simply giving even a highly sophisticated device to a patient with memory failure, in our view, would not achieve any meaningful intervention.

Five patients who met criteria for substance-induced persistent amnestic disorder were trained to use a custom-built battery operated apparatus the size of a pocket computer with a window that showed two lines of LED display and had two buttons. The buttons turned on the apparatus and scrolled forward or backward through the information. The LED display was capable of displaying two lines of text (e.g., Nurse's name: Jane Jones). We customized the questions in the device for each patient by means of a receptacle that attached to a computer. The training was conducted over a course of daily sessions divided into an information gathering session and six training phases. Individual sessions were generally completed in 20 minutes or less.

INFORMATION GATHERING. This step involved developing a set of individualized questions to use in the training sessions. Staff members and the patients themselves identified information with relevance to daily functioning within the care setting. Typical information included the name of the ward, names of staff members, location of facilities within the hospital (e.g., canteen, library), and visiting hours. A question/answer format was developed for this information. The following procedure assured that the patients did not already know the answers to the questions developed for training purposes. Each patient was asked the list of questions daily for 5 consecutive days. Questions retained for the device training procedure were those that the patient answered incorrectly on at least 3 of the 5 days. A list of 6 questions was developed in this manner.

IMPLEMENTATION

Phase 1. After a single question was programmed into the device, it was introduced to the patient during the first day of this phase. The patient practiced turning the device on and reading the question and the answer from the LED screen. This procedure assured that the individual could physically manipulate the equipment used in the training. After this introduction, the patient was asked the question, a tone and a verbal prompt to use the device occurred immediately, and the patient's response was recorded. This pairing of the question and the tone/prompt continued for a maximum of 20 trials during each session. A correct response was scored when the patient picked up the device, turned it on, and read the answer out loud. Criterion for successful completion of Phase 1 was a correct response on 10 consecutive trials over 5 consecutive training sessions. After completion of this phase, the single question was removed and the remaining 5 questions were programmed into the device.

Phase 2. The training procedure remained the same, except that 5 questions were used during each session and the verbal prompt was given only as necessary. Responses were scored as correct when the patient picked up the device, turned it on, identified the correct question, and read the answer out loud. Each of the 5 questions was asked up to 20 times, or until the patient answered each question correctly for 10 consecutive repetitions. Criterion was met when each of the 5 question-tone pairs were responded to by the use of the device for 10 consecutive trials over 5 training sessions.

Phase 3. This phase was devoted to fading the tone. Thus, the same 5-question procedure as described previously was used but the tone was eliminated on a random basis. If the device was not used on a trial without the tone, the patient was prompted to respond as he had on those trials in which the tone was present. Again, each of the 5 questions was repeated up to 20 times. Criterion was reached when, for each of the 5 questions, the patient used the device without any cuing (tone or verbal prompt) for 10 consecutive trials over 5 consecutive training sessions.

Phases 4, 5, and 6. The questions in the device were not changed, but the tone was not used during any further training. The actual training procedures remained the same; however, these phases were devised to assess generalization across the trainers and settings. In Phase 4, a new trainer conducted the sessions; in Phase 5, the original trainer conducted the sessions in a new setting; and in Phase 6, both trainer and settings were different from the original training. The criterion for each of these phases was 10 consecutive correct trials for each of the 5 questions over three consecutive days.

All patients met criterion for each phase within eight days, with total training lasting approximately 26 sessions. Furthermore, the number of days needed to meet criterion diminished over time, and generalization across trainers and locations was achieved.

In essence, all five densely amnesic patients learned to use the device to access functionally relevant information in a short period of time. This finding has both theoretical and clinical significance. In addition to providing another demonstration of a procedural or implicit memory system in patients with Korsakoff's syndrome, we demonstrated that the learning was not situation specific. This finding suggests that overlearning of a skill may actually promote the transfer of that skill to other settings through the elaboration that comes with increased use. From a practical standpoint, then, rehabilitation efforts such as the type we used should be practiced to the point of overlearning rather than concluded upon the fulfillment of minimal criteria.

We are obviously quite encouraged by these results. From a practical standpoint, the relatively brief course of this training and the ability to generalize to other trainers and settings has important implications for directing rehabilitation interventions and improving the quality of life for severely amnesic patients. The remaining work in our rehabilitation program involves the assessment of outcome, which is discussed in the next section of this chapter.

CONTINUED CONSIDERATIONS

OUTCOME ASSESSMENT

As funding for rehabilitation procedures have come under tighter control, professionals are encouraged to develop methods to assess the functional outcome of the training process (Gordon & Hibbard, 1992; Levin, 1990). Ideally, objective measures of outcome would involve before and after rehabilitation assessments with instruments similar to the following.

Instruments

PATIENT ASSESSMENT OF FUNCTIONING INVENTORY (PAF) (Chelune, Heaton, & Lehman, 1986). This measure provides a standardized method with which to obtain information regarding the adequacy of basic adaptive abilities. The test is composed of 47 questions initially designed to assess patients' perceptions of their own ability to carry out everyday tasks and functions. There are now two additional forms of this scale for relatives and clinician caretakers. Items are grouped into eight scales according to the general nature of abilities (e.g., memory, language, communication). Five of these scales use a 6-point rating scale and three request a written paragraph that describes daily activities and problems. Validity studies completed by the test's authors indicated that the PAF is sensitive to the changes that persons with neurological deficits experience in their daily functioning. Furthermore, results of factor analyses indicated that the PAF scales represent meaningful clusters of everyday problems. However, predictive validity studies suggested that the presence or absence of complaints on the PAF are not sufficient to infer the presence of cognitive impairments.

SUNDERLAND SCALE (Sunderland, Harris, & Gleave, 1984). This questionnaire, completed by a relative or caretaker, attempts to provide for the systematic observation of memory functioning in everyday life. Questions are based on 27 behaviors found to discriminate severely head-injured patients from a control group with mild head injury (e.g., losing things around the house; doing some routine twice by mistake). Each behavior is rated with respect to frequency over the past 3 months on a 9-step scale that ranges from "Not at all" to "More than once a day." Preliminary studies of head-injured patients indicate that this instrument can provide a valid representation of everyday memory problems for at least several years after injury.

RIVERMEAD BEHAVIOURAL MEMORY TEST (B. Wilson, Cockburn, & Baddeley, 1985). This measure of "everyday memory" is published with four alternate forms, thereby lending itself to serial assessments. The test was developed, in part, to monitor change after treatment for memory problems and it attempts to fill the void between the clinical assessment of memory function and observation and rating scales. The test provides analogues to situations encountered in daily living, such as remembering a task or the name of a new acquaintance. At the beginning of the test, the individual gives a personal belonging to the examiner and is requested to ask that it be returned at the conclusion of the session.

Both reliability and validity of the Rivermead are high (B. Wilson, Cockburn, & Baddeley, 1989). Interrater reliability for the Screening Score was 1.00, while alternate form reliability of both the screening and profile measures among the four versions ranged from .67 to .88. The highest level of correlation for the Screening Score was between Forms A and B and A and D for the Profile Score. Test–retest reliability was .78 for the Screening Score and .85 for the Profile Score. Performance was slightly better on the second testing, primarily because on the second administration subjects were aware that they would be asked to remember a belonging given to the examiner. Validity was assessed by correlating test performance with performance on a variety of standardized memory instruments. Correlations between the Screening and Profile Scores ranged from approximately .20 to .60, with the highest correlations noted between the Profile Score and a word recognition and paired associate learning task. In order to assess the validity of the test as a reflection of everyday memory, therapists were asked to rate the patient's memory lapses during four daily sessions for a 2-week period. Test scores were negatively correlated with memory lapses at approximately .70.

PERFORMANCE ASSESSMENT OF SELF-CARE SKILLS (PASS) (Rogers & Holm, 1994). This instrument was developed to assess both the physical (e.g., making a bed) and cognitive (using a telephone) activities of daily living. It offers the advantage of utilizing observation of actual performances by trained clinicians rather than reports of caretakers or the patients themselves. Detailed information regarding the PASS can be found in Chapter 2 of this volume.

MULTIPLE ERRANDS TEST (Aitken, Chase, McCue, & Ratcliff, 1993). This test provides for a simulation of naturalistic problem solving and was developed to assess the ability of individuals to organize novel activities and complete purposeful behavior in a real-life setting. The individual must shop for six items, complete certain tasks (e.g., mail a postcard), and meet the examiner at a designated place and time. An examiner observes and records the individual's activity over 16 specific tasks or activities. Although some guidelines are provided, the requirements of some tasks are only implied. Preliminary information (McCue et al., 1995) indicates that this test is reliable; interrater reliability coefficients range from .99 (task failure) to .49 (misinterpretations of instructions). Test validity was more variable. Although the scores did not correlate with IQ or with the results of standardized neuropsychological instruments, scores did discriminate between normal and patient groups.

EVALUATION OF CURRENT REHABILITATION PROCEDURES

ALTERNATIVE METHODS OF MEMORY REHABILITATION

Despite advances in the field of memory rehabilitation and the development of a number of promising techniques for assisting amnesic patients, the field remains in some disarray. It appears that dissemination and application continue to outrun documentation of efficacy; there are many procedures, either reported in the literature or promulgated commercially, that purport to have at least some degree of efficacy with regard to improvement of memory in brain-damaged patients. Scientific demonstration of that efficacy, however, is often unavailable (Gordon & Hibbard, 1992; Levin, 1990; Lynch, 1992).

The thrust of efficacy studies reported in the scientific literature suggests that a modest degree of improvement of memory can be demonstrated in laboratory or clinical settings (Richardson, 1992; B. A. Wilson, 1995). Although much of this work has been positive in the sense that memory can be improved, practical consequences of such improvement are either negligible or understudied. For example, there have been no clinical trials evaluating long-term outcome. Studies have not evaluated how, or even whether, a patient's ability to function in day-to-day life is improved. The scant outcome evidence available includes before- and after-training comparisons on measures similar to those used in the training, with no evidence of generalization to the community. In summary, there is little evidence that extant memory rehabilitation techniques actually help patients with their memory in everyday life. This conclusion has led to some discouragement regarding the ultimate utility of laboratory- or clinic-based didactic rehabilitation methods.

In view of these findings, some have recommended the alternative strategy of environmental support for memory, utilizing such methods as supported work programs or technological innovations that substitute for memory (Jackson, Rogers & Kerstholt, 1988; Kapur, 1995). The model of training in the clinic with the hope of generalization to the environment appears to be in the process of being abandoned. Development and use of assistive devices is one aspect of this more recent thinking concerning rehabilitation of brain-damaged patients.

Evaluation of External Memory Aids

The use of external memory aids for encoding, decoding, reorganizing, rehearsing, and retrieving information, has been documented extensively in individuals with what might be termed subclinical memory problems (Harris, 1992; Rasking, 1993). Some of these aids are quite complex, with memory and comprehension demands that require one to remember the sequence of steps needed to program the device. However, the effectiveness of commercial external aids has not been explored to any great extent (Intons-Peterson & Newsone, 1992).

Unfortunately there is even less documentation of the usefulness of such aids with memory impaired persons. The most frequently described external memory aids or strategies used by a person after sustaining a severe head injury are various list- and note-taking methods, or dependence on others, with less frequent use of commercially available organizers or reminders (B. A. Wilson, 1992). While individuals with traumatic brain injury often rely heavily on external aids as reminders, they may lack the cognitive skills required to utilize many of the currently available commercial devices. Other, more general circumstances might also preclude successful use of currently available assistive devices. These include a lack of the finger dexterity or visual acuity necessary to program these somewhat complicated devices, resistance to reliance on any assistive device, or an avoidance of technology in general. In short, persuasive evidence of the conditions under which external memory aids do and do not facilitate memory is needed, since the situational specificity and complexity of external memory aids is so varied.

Future Directions

Although the efficacy of assistive devices in the cognitive rehabilitation methods with brain-damaged patients is under some question, the successful results obtained in our clinic have encouraged us to pursue the use of assistive devices in our future memory rehabilitation programs. Although it is clear that memory can be improved in the laboratory, questions remain regarding the application of that improvement to ordinary life. These considerations have stimulated a change in our own rehabilitation strategy. Previously, our theory was that memory impairment existed on a continuum of severity, and that training in mnemonic strategies was the optimal rehabilitation method for individuals with milder forms of disability (e.g., moderate head injury), whereas assistive devices were appropriate only for severely impaired individuals who had extremely limited declarative memory capacities.

We now are more impressed with the idea that assistive devices might aid in the rehabilitation of memory disorders across the full spectrum of severity. Recently, we proposed to broaden our work in prosthetic memory. Our initial investigations recognized the intrinsically severe and irreversible nature of the amnesia in patients with Korsakoff-type disorders and had limited goals in terms of improving the functional activities of daily living. We restricted the rehabilitation applications to densely amnesic patients because they represented an ideal experimental model for demonstrating efficacy. However, pure amnesia is a rare disorder. Our findings would be of broader applicability if it could be demonstrated that the procedures found to be effective with amnesics are also effective for patients with memory failure that does not constitute a pure syndrome, a far more common condition. We feel that research utilizing implicit memory has broad generalization to healthy elderly individuals as well as to neurologically impaired patients with mild amnesia.

Conclusion

This chapter has briefly reviewed the history of memory rehabilitation. The theoretical concepts that formed the basis for the memory rehabilitation of two patient groups, patients having head injury and those with Korsakoff's syndrome, were discussed and two distinct memory training procedures were outlined in some detail. Finally, we suggested several appropriate tests with which to assess the efficacy of memory rehabilitation outside the clinic and presented an evaluation of current trends in rehabilitation procedures.

In summary, an important goal of memory rehabilitation is to develop state-of-the-art technologies for the training programs described in this chapter. Future applications that integrate concepts derived from neuropsychologically oriented cognitive rehabilitation, artificial intelligence, and computer applications to education will provide powerful methods that move memory rehabilitation out of the clinic and that meaningfully enhance the everyday functioning of the patient.

References

Aitken, S., Chase, S., McCue, M., & Ratcliff, G. (1993). An American adaptation of the Multiple Errands Test: Assessment of executive abilities in everyday life [Abstract]. *Archives of Clinical Neuropsychology, 8,* 212.

Anastasi, A. (1982). *Psychological testing* (5th ed.). New York: Macmillan.

Baddeley, A. D., Wilson, B. A., & Watts, F. N. (Eds.). (1995). *Handbook of memory disorders.* New York: Wiley.

Beers, S. R., & Goldstein, G. (1995). Computer-assisted rehabilitation of memory in patients with closed head injury. *Journal of the International Neuropsychological Society, 4,* 327.

Brooks, D., & Baddeley, A. (1976). What can amnesic patients learn? *Neuropsychologia, 14,* 11-22.

Butters, N., & Cermak, L., (Eds.). (1980). *Alcoholic Korsakoff's syndrome.* New York: Academic.

Butters, N. (1984). The clinical aspects of memory disorders: Contributions from experimental studies of amnesia and dementia. *Journal of Clinical Neuropsychology, 6,* 17-36.

Cermak, L. S. (1994). Processing deficits of amnesic patients: Nearly full cycle? In L. S. Cermak (Ed.), *Neuropsychological explorations of memory and cognition: Essays in honor of Nelson Butters* (pp. 31-43). New York: Plenum.

Chelune, G. J., Heaton, R. K., & Lehman, R. A. (1986). Neuropsychological and personality correlates of patients' complaints of disability. In G. Goldstein & R. E. Tarter (Eds.), *Advances in clinical neuropsychology* (Vol. 3, pp. 95-126). New York: Plenum.

Crovitz, H. (1979). Memory retraining in brain-damaged patients: The airplane list. *Cortex, 15,* 131-134.

Crovitz, H., Harvey, M., & Horn, R. (1979). Problems in the acquisition of imagery mnemonics: Three brain damaged cases. *Cortex, 15,* 225-234.

DeRenzi, E., & Vignolo, L. A. (1962). The Token Test: A sensitive test to detect receptive disturbances in aphasics. *Brain, 85,* 665-678.

Franzen, M. D. (1989). *Reliability and validity in neuropsychological assessment.* New York: Plenum.

Franzen, M. D., & Haut, M. W. (1991). The psychological treatment of memory impairment: A review of empirical studies. *Neuropsychology Review, 2,* 29-63.

Gasparrini, B., & Satz, P. (1979). A treatment for memory problems in left-hemisphere CVA patients. *Journal of Clinical Neuropsychology, 1,* 137-150.

Gianutsos, R. (1992). The computer in cognitive rehabilitation: It's not just a tool anymore. *Journal of Head Trauma Rehabilitation, 7*(3), 26-35.

Gianutsos, R., & Gianutsos, J. (1979). Rehabilitating the verbal recall of brain-injured patients by mnemonic training: An experimental demonstration using single-case methodology. *Journal of Clinical Neuropsychology, 1,* 117-135.

Glisky, E. L. (1992). Acquisition and transfer of declarative and procedural knowledge by memory-impaired patients: A computer data-entry task. *Neuropsychologia, 30,* 899-910.

Glisky, E. L. (1993). Training persons with traumatic brain injury for complex computer jobs: The domain-specific learning approach. In D. F. Thomas, F. E. Menz, & D. C. McAlles (Eds.), *Community-based employment following traumatic brain injury* (pp. 3-27). Menomonie: University of Wisconsin-Stout Research and Training Center.

Glisky, E. L., & Schacter, D. L. (1987). Acquisition of domain-specific knowledge in organic amnesia: Training for computer-related work. *Neuropsychologia, 25,* 893-906.

Goldstein, G., & Malec, E. A. (1989). Memory training for severely amnesic patients. *Neuropsychology, 3,* 9-16.

Goldstein, G., Ryan, C., & Kanagy, M. (1982, June). *Neuropsychological assessment and retraining of an amnesic patient: A case report.* Paper presented at the 5th International Neuropsychological Society European conference, Deauville, France.

Goldstein, G., Turner, S., Holzman, A., Kanagy, M., Elmore, S., & Barry, K. (1982). An evaluation of reality orientation therapy. *Journal of Behavioral Assessment, 4,* 165-178.

Goldstein, G., Ryan, C., Turner, S., Kanagy, M., Barry, K., & Kelly, L. (1985). Three methods of memory training for severely amnesic patients. *Behavior Modification, 9,* 357-374.

Goldstein, G., McCue, M., Turner, S. M., Spanier, C., Malec, E. A., & Shelly, C. (1988). An efficacy study of memory training for patients with closed-head injury. *Clinical Neuropsychologist, 2,* 251-259.

Goldstein, G., Beers, S. R., Longmore, S., & McCue, M. (1996). The efficacy of memory training: A technological extension and replication. *Clinical Neuropsychologist, 10,* 1-11.

Goldstein, G., Beers, S. R., Shemansky, W. J., & Longmore, S. (in press). An assistive device for severely amnesic patients. *Journal of Rehabilitation Research and Development.*

Goodglass, H., & Kaplan, E. (1983). *The assessment of aphasia and related disorders* (2nd ed.). Philadelphia: Lea & Febiger.

Gordon, W. A., & Hibbard, M. R. (1992). Critical issues in cognitive remediation. *Neuropsychology, 6,* 361-370.

Harris, J. E. (1992). Ways to help memory. In B. A. Wilson & N. Moffat (Eds.), *Clinical management of memory problems* (pp. 59-85). London: Chapman & Hall.

Intons-Peterson, M. J., & Newsone, G. L. (1992). External memory aids: Effects and effectiveness. In D. J. Herrmann, J. Weingartner, A. Searleman, & C. McEnvoy. (Eds.), *Memory improvement: Implications for memory theory* (pp. 101–121). New York: Springer-Verlag.

Jackson, J. L., Rogers, H., & Kersholt, J. (1988). Do memory aids aid the elderly in their day to day remembering? In M. M. Gruneberg, P. E. Morris, & R. N. Sykes (Eds.), *Practical aspects of memory: Current research and issues: Vol. 2. Clinical and educational implications* (pp. 137–142). New York: Wiley.

Jaffe, P., & Katz, A. (1975). Attenuating anterograde amnesia in Korsakoff's psychosis. *Journal of Abnormal Psychology, 84,* 559–562.

Kapur, N. (1995). Memory aids in the rehabilitation of memory disordered patients. In A. D. Baddeley, B. A. Wilson, & F. N. Watts (Eds.), *Handbook of memory disorders* (pp. 553–556). New York: Wiley.

Kovner, R., Mattis, S., & Goldmeier, E. (1983). A technique for promoting robust free recall in chronic organic amnesia. *Journal of Clinical Neuropsychology, 5,* 65–71.

Kovner, R., Mattis, S., & Pass, R. (1985). Some amnesic patients can freely recall large amounts of information in new contexts. *Journal of Clinical and Experimental Neuropsychology, 7,* 395–411.

Leng, N. R. C., & Copello, A. G. (1990). Rehabilitation of memory after brain injury: Is there an effective technique? *Clinical Rehabilitation, 4,* 63–69.

Levin, H. S. (1990). Cognitive rehabilitation: Unproved but promising. *Archives of Neurology, 47,* 223–224.

Lewinsohn, P., Danaher, B., & Kikel, S. (1977). Visual imagery as a mnemonic aid for brain-injured persons. *Journal of Consulting and Clinical Psychology, 45,* 717–723.

Lorayne, H., & Lucas, J. (1974). *The memory book.* New York: Stein & Day.

Lynch, W. J. (1992). Ecological validity of cognitive rehabilitation software. *Journal of Head Trauma Rehabilitation, 7*(3), 36–45.

Matthews, C. G., Harley, J. P., & Malec, J. F. (1991). Guidelines for computer-assisted neuropsychological rehabilitation and cognitive remediation. *Clinical Neuropsychologist, 5,* 3–19.

Mattis, S. (1988). *DRS Dementia Rating Scale: Professional manual.* Odessa, FL: Psychological Assessment Resources.

McCue, M., Aitken, S., Chase, S. L., Petrick, J., Pramuka, M., & Ratcliff, G. (1995). Ecologically valid assessment of problem-solving ability: The American Multiple Errands Test. *Journal of the International Neuropsychological Society, 1,* 149.

Mitchell, J. V., Jr. (Ed.). (1985). Wechsler Adult Intelligence Scale-Revised. In *The ninth mental measurements yearbook* (pp. 1694–1705). Lincoln: University of Nebraska.

Moffat, N. (1984). Strategies of memory therapy. In B. A. Wilson & N. Moffat (Eds.), *Clinical management of memory problems* (pp. 63–88). Germantown, MD: Aspen Systems.

Rasking, M. (1993). Assistive technology and adults with learning disabilities: A blueprint for exploration and advancement. *Learning Disability Quarterly, 16,* 185–196.

Richardson, J. T. E. (1992). Imagery mnemonics and memory remediation. *Neurology, 42,* 283–285.

Rogers, J. C., & Holm, M. B. (1994). *The Performance Assessment of Self-Care Skills (PASS)* (version 3.1). Unpublished manuscript.

Schacter, D. L., Cooper, L. A., Tharan, M., & Rubens, A. R. (1991). Preserved priming of novel objects in patients with memory disorders. *Journal of Cognitive Neuroscience, 3,* 118–131.

Speight, I., Laufer, M. E., & Klaus, M. (1992). CIV (computer-aided interactive video): A novel application in neuropsychological rehabilitation. *Computers in Human Behavior, 9,* 95–104.

Squire, L. (1992). Memory and the hippocampus: A synthesis from findings with rats, monkeys, and humans. *Psychological Review, 99,* 195–231.

Stringer, A. Y. (1996). *A guide to adult neuropsychological diagnosis.* Philadelphia: Davis.

Sunderland, A., Harris, J. E., & Gleave, J. (1984). Memory failures in everyday life following severe head injury. *Journal of Clinical Neuropsychology, 6,* 127–142.

Wechsler, D. (1981). *Manual for the Wechsler Adult Intelligence Scale-Revised.* New York: Psychological Corp.

Wechsler, D. (1987). *Manual for the Wechsler Memory Scale-Revised.* New York: Psychological Corp.

Wilson, B. (1986). *Rehabilitation of memory.* New York: Guilford.

Wilson, B., Cockburn, J., & Baddeley, A. (1985). *The Rivermead Behavioural Memory Test manual.* Reading, England: Thames Valley Test Co.

Wilson, B., Cockburn, J., & Baddeley, A. (1989). The development and validation of a test battery for detecting and monitoring everyday memory problems. *Journal of Clinical and Experimental Neuropsychology, 11,* 855–870.

Wilson, B. A. (1992). Recovery and compensatory strategies in head injured memory impaired people several years after insults. *Journal of Neurology, Neurosurgery and Psychiatry, 55*(3), 177–180.

Wilson, B. A. (1995). Management and remediation of memory problems in brain-injured adults. In A. D. Baddeley, B. A. Wilson, & F. N. Watts (Eds.), *Handbook of memory disorders* (pp. 451–479). New York: Wiley.

Wilson, B. A., & Moffat, N. (Eds.). (1984). *Clinical management of memory problems*. Rockville, MD: Aspen.

14

Assessment and Planning for Psychosocial and Vocational Rehabilitation

Lynda J. Katz

Introduction

The rehabilitation of persons with severe and persistent mental disorders that falls under the rubric of psychosocial rehabilitation has historically involved the worlds of mental health, vocational rehabilitation, and community programs. Although these entities had distinct functions in the past, it has become apparent that their efficacy hinges on a collaborative process based upon a biopsychosocial model.

Psychiatry, through the mental health system, has typically focused on the biological aspects of mental disorders and their amelioration, at times without regard for the psychosocial consequences imposed by mental illness and/or consequent to a diagnosis of mental illness. While using the nomenclature of psychosocial interventions, studies emanating from a psychiatric framework have focused primarily on drug efficacy, family education, and day treatment programs which result in less frequent or less extensive periods of hospitalization, and greater periods of symptom remission and time in the community. The few exceptions to this overall focus can be found in the work done as part of the Fairweather Lodge movement (Fairweather, 1980) and the Assertive Community Treatment model (PACT) developed in Madison, Wisconsin (Stein & Test, 1980). However, a proper discussion of these two approaches is not in the purview of this paper. First, they have been previously documented in a most comprehensive manner. Second, both programs arose out of a concern primarily for residential and social support services for persons with long-term mental illness. Vocational

---

Lynda J. Katz   Landmark College, Rural Route 1, Box 1000, Putney, Vermont 05346.
Rehabilitation, edited by Goldstein and Beers. Plenum Press, New York, 1998.

interventions are valued mainly for their therapeutic effects (symptom remission and reduction of hospitalization), much as is the case with the majority of programs that emanated from a traditional mental health system approach. And third, assessment as practiced arose primarily out of the traditional psychiatric diagnostic interview process, that is, a concern for major psychiatric symptomatology as the basis for treatment interventions.

Vocational rehabilitation as a system, in its quest for employment outcomes has been primarily concerned with the "social" component of the model, one's place as a wage earner and tax payer in society. The federal legislation that lays the foundation for the State/Federal VR system makes it explicitly clear that employment is the ultimate outcome measure used to judge the system's viability for the taxpayer.

Psychosocial rehabilitation programs have downplayed the importance of the "bio" and "social" in favor of the "psycho," that is, the individual's sense of belonging and personal well-being. By definition psychosocial rehabilitation is the "process of facilitating an individual's restoration to an optimal level of independent functioning in the community" (Cnaan, Blankertz, Messinger & Gardner, 1988, p. 61). The process emphasizes the wholeness and wellness of the individual and encourages his or her active participation as a group member. It sees itself as the antithesis of the medical model, with a stress on skill building and social supports as opposed to symptom reduction, professional expertise, and concern with diagnosis.

It is our bias that the significance of an integrated model that includes an appreciation for the biological basis of most of the major psychiatric disorders can no longer be disregarded or demeaned. That having been said, our task is to examine the "psychosocial" legs of our triad, which must have equal footing if the combined effort is to succeed. It is hard to imagine the practicality of a three-legged stool with one leg more dominant than the other two.

The focus of this chapter is on those aspects of the rehabilitation process, specifically the functions of assessment and planning, that have fallen within the domains of psychosocial and vocational rehabilitation. Both systems have their own history, one emerging directly as a result of the deinstitutionalization of persons with mental illness from public institutions and the other (vocational rehabilitation) from work with persons with physical disabilities of war origin. Thus, the assessment and planning processes have traditionally been somewhat at odds. This was the case until the recent advent of supported employment, a concept that has bridged the gap between these somewhat disparate philosophies of rehabilitation.

In this chapter, we look briefly at those historical differences as they manifest themselves in assessment and planning efforts in particular. We then describe the various methodologies, examining their validity and reliability where possible. In detailing specific methods we offer specific strategies that appear to have clinical utility particularly in the assessment process and in the planning of rehabilitation programs that have employment as an outcome goal. Because research on efficacy is highly limited in the field of psychiatric rehabilitation, in most cases it will be possible only to address what appear to be essential ingredients if assessment and planning efforts are to yield beneficial results for those persons so engaged. What must be avoided, however, both in the present and into the future is the tendency to focus on programs rather than on the individuals whom they are intended to benefit. As we will discover, this can easily happen whether we are concerned with assessment or with planning. Thus, one of the final objectives of this chapter will be to address that very real limitation and to offer alternative visions for future work.

Psychosocial Rehabilitation

According to Bachrach (1992), psychosocial rehabilitation is a "therapeutic approach" that focuses on the development of persons with mental illness to their fullest capacities through learning procedures and environmental supports. She goes on to describe eight fundamental concepts that a variety of programs identified with the service modality of psychosocial rehabilitation have in common. These include enabling the individual to develop to the fullest extent of his or her capacities; stressing the importance of environmental factors in the process of rehabilitation; exploiting the individual's strengths; restoring hope in those who experience setbacks in their functional capacities and self-esteem; maintaining optimism regarding the vocational potential of the individual; emphasizing vocational as well as social and recreational pursuits; actively involving the individual in the design of his or her own rehabilitation protocol; and recognizing the necessity for ongoing, flexible, and planned interventions as opposed to a "one-time only kind of intervention" (p. 1458).

Clubhouse Model

These characteristics are often embodied in an approach that has been commonly known as the clubhouse model, stemming originally from the work of the 1950s Fountain House Program in New York City (Propst, 1992). Whether or not adhering strictly to this original model, psychosocial community programs have adopted general strategies for assessment and planning purposes. The context for both has been a "work-ordered" day (Propst, 1992, p. 27) for the most part. Activities necessary to maintain the clubhouse itself have formed the basis for vocational assessment in many instances. Thus, it was often felt unnecessary to have access to traditional paper and pencil or work sampling approaches to evaluation because real work activities themselves provided the context for evaluation to occur *in vivo*. Janitorial work, food preparation, filing and sorting, mailing, activity planning, and so on, are essential functions to the running of a clubhouse. In each instance members are assigned or volunteer for periods of time in each activity center. Assessment of work-related skills is made by staff assigned to a specific work unit. These observational reports have often been codified into checklists that reflect specific work functions as well as "worker attributes," that is, task completion, cooperation with coworkers, punctuality, dependability, and ability to deal with feedback from supervisory staff. Generally, the assessment process is a here-and-now approach. That is, there is little or no thought given to disability-specific concerns such as symptomatology and diagnostic information or to factors such as previous work history, years of education, prior professional training, or expressed vocational objective.

Focusing on the individual's strengths or skills has been viewed as the hallmark of the Boston University model of psychiatric rehabilitation (Cohen & Anthony, 1984). Psychiatric information (most particularly diagnoses or symptom pictures) has been viewed as either nonessential or stigmatizing for the individual (W. Anthony, 1980; W. A. Anthony & Jansen, 1984). W. A. Anthony, Cohen and Danley (1988), in justifying this view, cited an earlier review of literature on vocational outcomes and concluded:

1. Measures of psychiatric symptoms do not predict vocational rehabilitation outcome.
2. The psychiatric diagnosis does not predict vocational rehabilitation outcome.
3. Measures of psychiatric symptoms do not correlate with the psychiatrically disabled person's skills.
4. Measures of skills do predict vocational rehabilitation outcome.
5. Training in critical vocational skills improves vocational rehabilitation outcome. (p. 62)

Based on these conclusions, W. A. Anthony and colleagues established a specialty field that they named psychiatric rehabilitation. This new intervention model was based on a functional skills orientation with its own diagnostic and assessment procedures. Cohen and Anthony (1984) wrote: "Functional assessment . . . is a component of the rehabilitation diagnostic process . . . (and) serves to help clients understand what skills they need to develop in order to achieve their overall rehabilitation goal" (p. 82). The entire diagnostic process in rehabilitation is seen as composed of establishing an overall goal, developing a functional assessment, and developing a resource assessment; assessment occurs only after an overall rehabilitation goal is established and thus has validity only in light of that goal. Cohen and Anthony (1984) go on to list a series of principles that underlie the functional assessment process and then conclude:

> It is neither desirable nor necessary to rely solely on instruments when seeking a reliable and valid functional assessment. Practitioners have been trained to involve clients in the assessment process, to list skill strengths and deficits, and to define skills operationally, and to identify present and needed levels of skill functioning. (p. 95)

The validity of the methodology, however, has yet to be empirically demonstrated (Bond & Dietzen, 1993).

The view that the traditional psychiatric diagnostic approach is irrelevant to psychosocial rehabilitation has come to be accepted as a basic truth in the psychiatric rehabilitation community, which is composed largely of psychosocial programs. Few individuals have disagreed openly with this view, but as early as 1983 McCue and Katz stated:

> while there is considerable impetus in psychiatric rehabilitation to emphasize patient assets and strengths in the rehabilitation process, there is need to integrate information on diagnoses, symptoms, and weaknesses as well. With the severely psychiatrically disabled, to do otherwise would be counter-productive. To ignore significant behavioral difficulties which present numerous and, at times, insurmountable obstacles to functioning in favor of working with assets and strengths appears irresponsible, unfair to clients, and potentially dangerous. It is not suggested that psychiatric rehabilitation adopt the disease model and focus upon sickness, but rather, it is important to acknowledge symptomatology and deficits and work to overcome them. (p. 57)

In contrast to his earlier writings, W. Anthony's most recent publication (1994) speaks to an altered perspective with regard to symptomatology and its negative effect on vocational functioning.

> Furthermore, it is much more useful to draw programming implications when the sample is relevant to the question being asked. For example, the relationship between symptoms, work adjustment skills and employment outcome is much more important to know for those people who are attending a vocational program as opposed to a hospital based treatment program. What seems clear is that much less is known about the predictors of vocational outcome for the subpopulation of people receiving vocational interventions. . . . If a person seeks entrance and is engaged in a vocational program, then characteristics such as employment history, marital status, hospitalization history, and race are not predictive of outcome as was once thought. Intriguingly, these characteristics, as well as symptomatology, do predict work adjustment skills. (p. 10)

It is indeed unfortunate that the National Institute on Disability and Rehabilitation Research (NIDRR) did not access a comprehensive data base when it convened its Consensual Validation Conference in 1992 and published *Strategies to Secure and Maintain Employment for People with Long-Term Mental Illness* in which they state: "Research also shows that certain variables are *not* predictive of vocational success. . . . Neither particular patterns of symptomology, diagnosis, nor functioning in other life domains correlate strongly with positive employment outcomes" (NIDRR, 1993, p. 2, emphasis in original). Such categorical statements based on biased perceptions and preordained knowledge acquisition sources do not serve the field well nor, most importantly, do they benefit the persons for whom psychiatric rehabilitation as a legitimate intervention is intended.

As a consequence. therefore, the earlier and pervasive emphasis on skills or strengths with little if any regard for other salient factors critically colored perceptions of assessment and planning in psychosocial (psychiatric) rehabilitation. When individuals were referred to the state/federal VR system, many times their identified "strengths" did not manifest themselves once they were placed in competitive employment settings. What appears to have been left out of the assessment and planning processes were factors related to environmental specificity of behavior, responsivity to supports in the environment, vulnerability to stress (a consequence of the psychiatric illness itself), and the overarching reality of relapse. The nature of a psychiatric disorder forces the acknowledgment of its pervasive impact on the total life of the individual.

Transitional Employment: A Vocationally Based Psychosocial Model for Assessment and Planning

Returning then to an emphasis on strengths (skills) assessment and the resulting planning process, the psychosocial approach as exemplified by the clubhouse model initiated a vocationally based strategy to deal with the nontransferability of skills. This approach was to become known as transitional employment. Transitional employment (TE) programs were developed to bridge the gap between the somewhat sheltered work settings within the clubhouse itself and competitive and integrated work settings. In this model jobs, usually high-turnover, entry-level positions, were secured by the clubhouse or sponsoring agency. Clubhouse members were then assigned to jobs for less than 6-month periods on a part-time, 20-hour-per-week basis, the remainder of the time left for clubhouse participation and the peer-group support it provided. TE positions were seen as providing opportunity for *in vivo* assessment, as well as enabling the individual to acquire work experiences in a variety of settings, none of which were intended to become permanent careers. It was hypothesized that after a series of successful work experiences, the individual would be in a better position to pursue full-time competitive work if that were his or her desire or to return to an educational or training program to further enhance specific vocational skills (Bilby, 1992). In reality, only those individuals who possessed certain attributes relative to their premorbid history and their psychiatric illness itself were able to progress beyond unskilled or semiskilled positions in the job market. Thus, assessment based on situational observations alone did not and does not address certain other factors which appear to be essential to the psychiatric rehabilitation process. As a result, the model was to come under attack for a time, only to be "reborn" under the rubric of supported employment, another approach to assessment and planning that we address shortly.

Vocational Rehabilitation

Background

The state/federal system of vocational rehabilitation (VR) entered the world of persons with psychiatric disabilities through the Rehabilitation Act of 1943 but in reality VR staff did not begin to work with these individuals until well into the 1950s with an effort that was essentially institutionally based (Neff, 1988). State VR agency staff would be assigned to various state psychiatric facilities where their offices were located. They were asked to work with selected individuals who were thought by the mental health staff in charge to have potential to return to the workplace. Of course, in those early days those individuals selected by mental health staff in all likelihood could have found their own jobs if they were able to leave the institution in the first place. Individuals with severe and persistent mental disorders were, for all practical purposes, not discharged from residential settings, and if they were to work, their work was viewed as part of the therapeutic treatment approach. It had no real stand-alone intrinsic value nor was it deemed of particular value to the individual involved.

Thus, the early expectations of VR agencies were that individuals referred for services were work-ready and would benefit from the same kind of rehabilitation process that had been in place for more than 20 years for persons with physical disabilities. That process was the hallmark of vocational rehabilitation assessment and planning services until the most recent years when legislative mandate required that VR give priority to individuals with the most severely disabling conditions. The process is one of evaluation, restorative services, job training, education or work adjustment, job placement, and follow-up services. With persons with physical disabilities the focus is on evaluation and restorative services primarily. For example, given an individual with an amputated lower limb, evaluation serves to define specific functional limitations that are present regardless of the environment, for example, weight bearing, ambulation, physical stress tolerance. Restorative services focus on fitting the appropriate prosthetic device and prescribing physical therapy for proper care of the prosthesis and gait training. Once these objectives are accomplished, the process of returning to a similar or the same job setting as predisability is undertaken, with a goal of "closing" the case as successfully rehabilitated once the individual has completed 60 successive days on a given job. In some few instances the individual may have needed to be trained for another job, but in any case, once trained, placement occurs and the story ends with minor "postplacement" services required, for example, purchase of a second prosthesis.

How then did the model fit persons with psychiatric disabilities? Once the era of deinstitutionalization began, not well. For now those individuals who could have found their own jobs or returned to ones previously held were no longer in the caseload. The VR agency staff member had returned to the community and so had persons with severe and persistent mental illness. These were individuals who in many instances had never worked before, having been institutionalized in their late adolescence, or persons who could not return to an occupation they had held ten or fifteen years previously. Because the existing model of service could not be readily applied to this population, the implicit decision was made that real work was not a realistic goal. Instead, the idea of sheltered employment became a placement objective, bypassing the need for extensive evaluations, restorative services, or additional education or training. Because sheltered employment was not competitive employment. a long-revered goal of VR, the rationale for its existence focused around the work adjustment nature of such placements. VR funds were then allocated to workshops to provide work adjustment services for a limited period with the understanding that the workshop would then place the individual in competitive work.

SHELTERED WORKSHOP MODEL

In order to justify the workshop setting, VR required that vocational evaluations and work adjustment training take place. Vocational evaluations took on a special characteristic not unlike that in the clubhouse setting with the difference being in the definition of situational assessment. For the workshop, simulated job tasks were adopted; many of these were packaged and sold as comprehensive and state-of-the-art vocational evaluation systems. Instead of using transitional employment sites in the community, work stations were established through which the individual moved over the course of several weeks. Once specific skills were identified (in the case of persons with psychiatric disabilities, the usual procedure was to identify deficits/functional limitations), the goal was to place the individual in a job setting that most nearly matched identified limitations. VR staff then sought functional skills assessment measures that would purport to identify deficits and strengths and match these to specific occupations with minimal input from the individual consumer with respect to vocational interest or specific work values. Evaluation of aptitudes was seen as essential, as was the availability of job settings. If the individual's interests were in sync with the evaluation of abilities and the possibility of a job placement, that circumstance was the exception rather than the rule. One may have trained in food service with the desire to become a cook only to find out that the only option vocationally was to become a dishwasher at best. From a functional perspective, the individual had the necessary skills to perform the job regardless of whether it was ultimately his or her specific occupational goal.

In addition, since the workshop's survival depended upon contract work, focusing on the out-placement of its best employees was not particularly in the best interests of the workshop. So, once again, those individuals with the greatest need for work adjustment or specific skills training were placed on the bottom of the pile, as workshop staff and their VR counterparts viewed them as "too disabled to work." Those individuals who probably could have worked outside of the sheltered setting were never given that opportunity. Instead, they spent the next decade of their lives, in the majority of cases, working in a sheltered setting where their work met contract obligations while their own personal growth or vocational satisfaction was never a consideration. After all, they were working, and that was the ultimate goal of the VR counselor in charge of the case.

Supported Employment

As mentioned earlier, the practice of supported employment (SE) entered the world of persons with psychiatric disabilities almost as a template, lifted out of work done in the field of developmental disabilities. Disheartened by the multitude of shortcomings found in sheltered workshop programs, professionals primarily from the field of special education, not rehabilitation or mental health, established an alternative approach to the traditional assessment, work adjustment, and placement model espoused by the workshop industry. These individuals (Mank, Rhodes, & Bellamy, 1986; Wehman, 1986) looked upon assessment and work adjustment training as barriers to employment to a large extent. Instead they proposed a job placement process which matched an individual with a particular entry-level job. The individual was then trained on the job, with particular behaviors not conducive to successful employment targeted for modification through the efforts of a job coach. Once placed, the individual could remain in that position indefinitely rather than being subject to the time-limited constraints of the TE model.

Finally, with supported employment as an option, VR had an alternative to the dead-end workshop model, a model particularly unsuited in most instances to persons disabled by a mental disorder. "Functional limitations" for these people were generally not skill deficits but rather interpersonal communication and social skill problems imposed by their particular illness, in many cases, or part of the manifestation of that illness in others. The alliance with VR was so intense at one point that in order to apply for grants sponsored by the Rehabilitation Services Administration, a successful grantee had to specifically address how he or she would deal with the new "train/place" model.

Within a relatively short period, the model came under scrutiny particularly by individuals from the various clubhouse programs across the country who had seen some evidence of success with the TE model. As a result, TE became synonymous with SE for persons with severe and persistent mental disorders and thus both could qualify for VR service dollars as job placement strategies. The original SE model was also critiqued by Anthony's collaborators at Boston University who developed what has become known as the "Choose, Get, Keep" model

(Danley & Anthony, 1987). Assessment, still a relatively dirty word, became legitimate once again as part of the process of choosing a career direction or vocational option, but only if the individual could not benefit from an interpersonally processed "choosing" experience. Danley and Anthony wrote:

> The conceptual underpinnings of the *choose-get-keep* model of supported employment lead to certain programmatic emphases which may be more pronounced than in the *place-train* model of supported employment.... Programmatically, to ensure that clients are given the opportunity to choose the job they wish to try, the preemployment phase ... is typically longer than in the place-train supported employment model. (p. 28)

More recently, both of these programming models, TE and SE, have come to be accepted as intervention strategies for psychosocial and vocational rehabilitation programs (Marrone, 1993). Their limitations have been addressed by numerous writers as have their purported benefits (MacDonald-Wilson, Revell, Nguyen, & Peterson, 1991). Empirical data based on outcome measurements of efficacy are lacking to a large extent, although work by Bond and Boyer (1988) and Bond and McDonel (1991) is beginning to move the field in this direction.

Although Bond and Boyer's work (1988) does not critique supported employment, it does address the various models of vocational intervention in place until supported employment entered the scene. In reviewing studies over a 20-year period that claimed to espouse an experimental or quasiexperimental design, Bond and Boyer concluded that no one approach (TE, sheltered workshops, the Boston University model, hospital-based programs, the PACT model, work adjustment training program) outperformed the others with respect to job retention. That is, although individuals with psychiatric disabilities obtained employment equally well or less than well in each of these models, once the supports of the program were removed, the person no longer remained employed. Part of the assessment and planning process probably should have been to identify these necessary supports and set about implementing their accessibility. However, again as when the program takes priority over the individual, we may find that "we have won the battle but lost the war."

Predicted Validity and Vocational Assessment Practices

Prior to our discussion of two integrated model approaches to assessment and planning in the context of psychosocial and vocational rehabilitation practices and principles, it is necessary to review the work done by Bond and Dietzen (1993) on the predictive validity of vocational assessment as it relates directly to the assessment and planning processes that we address under the label of integrated models. Although the relationship is a negative one overall, the crux of Bond and Dietzen's work must be acknowledged by anyone wishing to establish the credibility of a particular methodology or practice in the field of psychosocial/vocational rehabilitation.

Bond and Dietzen (1993) describe and critique the assessment practices of interviewing, psychometric aptitude and interest testing, work sampling, and situational assessment procedures in terms of their predictive validity in particular. "Paradigmatically, predictive validity is measured by a correlation between a predictor variable (based on an assessment made before any intervention) and a criterion variable . . . , such as job placement and retention, or even more specifically, employment in a particular occupation" (p. 62). Although Bond and colleagues found limited evidence to suggest that systematic assessment is preferable to an unstructured approach (Potsubay, 1984), overall results from their extensive review might suggest that extended evaluations in artificial settings "appear to take on a life of their own, with sometimes tenuous connections to their stated purposes" (Bond & Dietzen, 1993, p. 78). The authors then reframe the question of vocational assessment in terms of where it occurs and its contribution to the goal of employment. They suggest that for many clients

> the optimal assessment approach includes only minimal prevocational training and assessment. The "place-train" paradigm is especially suited for entry-level jobs, which we assume will continue to be realistic job options for some clients. . . . There are other clients for whom careers requiring educational preparation (e. g., college preparation) are appropriate. The assessment paradigm is no doubt different for this latter group and may include aptitude and interest testing as is often currently used in industry. Focusing on clients with severe disabilities, we hypothesize the following guidelines for vocational assessments to be maximally valid:
>
>   1. Vocational assessments should occur in real work settings.
>   2. Assessments should be made repeatedly over time. . . .
>   3. The opportunity for job tryouts in more than one job will increase the likelihood of a job match.
>   4. Vocational assessment measures should include the following:
>   a. supervisor ratings of worker performance and attitudes,
>   b. client ratings of job satisfaction and perceptions of the work-place, and
>   c. objective ratings of the work environment. (p. 80)

Implicit but never articulated in Bond and Dietzen's work is their proposal of an individualized approach to assessment based to some extent on the client's current psychiatric status ("severe disabilities," for example). Inherent in these kinds of statements is a reliance on a biopsychosocial model that begins with some assessment of the nature of the psychiatric disorder and its impact on functioning in particular environments.

Integrated Models of Assessment and Planning

Two models, one that incorporates components of a psychosocial philosophy and a vocationally oriented programming approach and the other that espouses a developmental-vocational orientation, are addressed in this final section. They are discussed in some detail because of their integration of diverse methodologies as opposed to adherence to a specific model of assessment and planning and because of the efforts made to evaluate their validity.

New Hampshire IPS Model

Becker and Drake (1994) have drawn upon the Program of Assertive Community Treatment (PACT), the Boston University model, supported employment, and the empirical literature by Bond and his colleagues in the development of an approach to vocational rehabilitation they call Individual Placement and Support (IPS). The program emphasizes entering the work world as soon as possible (as opposed to seeking jobs after a lengthy period of vocational training and assessment) much like the supported employment model. Although vocational assessment is seen as an ongoing process continuing after the individual acquires the job rather than an extensive testing and evaluation period prior to the acquisition of employment (à la Wehman), the program does indeed conduct an initial vocational assessment. In the words of Becker and Drake (1994):

> The initial assessment includes the client, the CMHC staff, the case record, family members, and previous employers. Key data include work background (education and work history), current adjustment (physical health, endurance, grooming, interpersonal skills, medication management, symptomatology), work skills (job-seeking skills, job skills, aptitude, interests, motivation, work habits relating to attendance (dependability, stress tolerance), and other work-related factors (transportation, family support, substance use, expectations regarding personal, financial, and social benefits of working). (p. 199).

The authors go on to say, "The nature of the psychiatric disorder often determines the optimal type of job, the work environment, and needed job supports" (p. 200). And while the program utilizes the "choose, get, keep" approach, depending upon "levels of disability and verbal, cognitive, and interpersonal skills, clients need varying degree of staff guidance and support in finding, acquiring, and maintaining a job" (p. 197). Thus, we see in practice an example of adapting somewhat rigid models to meet the actual needs of real people. What is even more exciting, however, is the fact that these investigators have established a research protocol to measure the effectiveness of the IPS approach (Drake et al., 1994). Preliminary data appear to favor the IPS approach over a rehabilitative day treatment model, but the results must be seen as short-term successes for the present. Drake and colleagues (1994) comment that like the PACT model in which vocational adjustment attenuated over time (Test, 1992), their approach emphasizing rapid placement may also have greater initial than long-term success.

UNIVERSITY OF PITTSBURGH PROGRAM IN PSYCHIATRIC REHABILITATION

A similar model utilizing an eclectic approach but more firmly rooted in a developmental model of vocational maturity was placed in operation in the mid 1980s at the University of Pittsburgh's Department of Psychiatry Program in Psychiatric Rehabilitation. The program had incorporated a biopsychosocial approach since its inception and as a consequence had placed a heavy emphasis on comprehensive, ongoing, and individually based assessment and planning strategies. Built into the assessment process were a variety of psychometric and neuropsychological measures in addition to an extensive vocational evaluation that took into account interests, values, and aptitudes. The program's director and staff have published some of their empirical work with the California Occupational Preference System (COPS), in particular the Career Ability Placement Survey (CAPS), a measure of vocational aptitudes associated with domains of occupational clusters (Katz, Beers, Geckle, & Goldstein, 1989; Katz, Goldstein, Geckle, & Eichenmuller, 1991; Katz, Goldstein, Geckle, Morrissey, & Daily, 1991). In addition, the chapter by Slomka in *Psychiatric Rehabilitation: A Handbook for*

Practitioners (1992) details the roles that traditional psychometric and neuropsychological assessments play in operationalizing the meaning of "functional" in this developmentally oriented vocational model of assessment and planning with persons who have major disabling psychiatric disorders. Slomka writes:

> It appears . . . that the assimilation of a tremendous amount of information is required to conduct a comprehensive assessment. This is, indeed, true. There remain, however, the realities of time, resources and energy which limit the scope of the assessment process. The challenge in the development of assessment and data collection systems is the balance of accurate and reliable information. . . . In this regard, we must seriously entertain how additional assessment strategies might be modified to meet current needs. Rehabilitation, perhaps oversimplistically conceived, represents a learning process; an effort to assist clients toward a meaningful reintegration with their community. Our success in this regard is very much determined by our capacity to make valid, appropriate and reliable assessments of the variables which will influence this process. (p. 38)

As part of a Rehabilitation Services Administration (RSA) federally funded training grant, Katz and her staff at the University of Pittsburgh developed a series of data collection documents to facilitate the reliable transfer of information between mental health and vocational rehabilitation program staff working directly with individuals with severe and persistent mental disorders. Examples of the Psychosocial Summary, Critical Skills Checklist, and Interagency Service Plan can be found in the appendix. These documents were a direct result of the work program staff had undertaken in the area of deriving ecologically sound measures or indices of function in individuals variously impaired in the cognitive and affective domains by reason of their mental disorders.

Concomitant with formal assessment methods (measures of intelligence, achievement, neuropsychological functioning) that enhance and augment psychiatrically based evaluations, the program utilized numerous informal assessment procedures, all of which contribute to the planning process. Clients were placed in volunteer-work settings, given opportunities to shadow persons employed in various occupations, and given time to process what they would like to achieve in the vocational arena. It had been the experience of program personnel that "quick fixes" did not pay off in the long run. Rather, individuals returned to the program often disheartened and discouraged; although they wanted to work and in many cases pushed for placement, once on the job the reality of the everyday nature of the beast gave added meaning to the process of career exploration. Obviously, there are those individuals for whom career exploration is not a realistic goal; work is a financial necessity at best, a means to an improved sense of personal productivity for others, or the most that their disability can allow at a particular time. While acknowledging the work of Bond and others on accelerated placement and its merit (Bond & Dincin, 1986), as with all other programs, the bottom line is the individual and his or her vocational needs, not program availability.

The developmental approach recognizes the necessity for systematic and ongoing assessment strategies if the planning process is to be at all realistic. Individuals who become disabled by virtue of a psychiatric disorder are often in the later years of their adolescent and young adult development. They have limited experience with what it means to work, and like most of us at that stage in our lives, little real sense of what a career might look like. The necessity for years of education rather than weeks may seem an insurmountable obstacle. The expressed desire to become a nurse or primary caregiver or a psychologist may have its roots in the most recent exposure to work role models of any sort. The family's planned career direction for an adult child may remain a foremost concern for an individual who has had no opportunity to explore his or her own interests or to recognize his or her own values. The use of career exploration techniques such as that provided by the COPS system in conjunction with homework assignments that involve targeting pockets of the service, technology, or product industries for study have demonstrated their utility from a consumer perspective (Geckle & Katz, 1993). That the process takes time appears to be a given, but if long-term successful outcome results, the time will have been well spent. The jury is still out.

Another highly effective developmental-vocational strategy is the Job Club Model used in the planning process. Azrin and Philip's (1979) original work was adapted and has been an ongoing part of the program since its inception. However, the client population dictates the utility of the Job Club just as it does any other planned program. Knowledge of the particular group of individuals who will benefit from such an approach is obtained as the result of a comprehensive assessment process that takes into account aptitude, ability to process in a group setting, task organization skills, and the levels of guided support and structure necessary for a particular individual.

Overall efforts to evaluate the effectiveness of the developmental-vocational–based approach to assessment and planning have been minimal, with early efforts published in 1983 (Katz-Garris, McCue, Garris, & Herring). At that time these authors argued that definitions of success should move beyond job placement and include parameters that reflect quality of life as well. More recently, follow-up of clients referred to the VR agency for cost services (education and/or training) have demonstrated the efficacy of the developmental approach at least with respect to obtaining rehabilitation services that adhere to the client's expressed vocational goal (Katz & Geckle, 1993). This is in direct contrast to traditional practices of job placement activated because of a job opening on any given day or watered-down training opportunities because a particular career goal might require postsecondary education. This model has actualized comprehensive, systematic, ongoing, consumer driven, and ecologically valid assessment and planning and has been in operation in a consistent manner for nearly 10 years. The methodology should now be empirically evaluated in a way similar to the IPS model described previously to demonstrate unequivocally its predictive validity as a sound approach for particular individuals, in light of their particular attributes, limitations, and interests.

Future Directions

It should be clear to the reader by now that we must begin to address assessment from an ongoing, evolutionary perspective as opposed to a more traditional closed-ended entity if it is to have any meaningful relevance or utility in the psychosocial/vocational realm of planning and programming. But throwing the baby out with the bathwater is equally foolish. Particular formal evaluative procedures can be exquisitely sensitive to the collection of meaningful, individualistic data. Is predictive validity the gold standard that Bond and Dietzen (1993) suggest, or is it just one standard that loses some of its validity when one takes into account a broad and diverse goal such as successful vocational or rehabilitation outcome?

Assessment can be more broadly defined as a learning process; it is lifelong, open-ended, and inclusive. It may take dead-end streets and wrong turns along the way. Systematic approaches to assessment and planning minimize those risks and help put them in perspective. Specific learning along the way or as needed is a meaningful use of assessment but learning to learn is equally important and calls for a means of assessment as well. In other words, assessing specific skills or deficits in a specific job setting is an important way to establish predictive validity for a specific environment but the process tells us nothing about the individual's capacity to learn in a metacognitive sense, which for many individuals is the key to lifelong vocational success and personal satisfaction. These are overriding quality of life issues which will never be addressed as long as our planning for persons in psychosocial rehabilitation programs utilizes discrete, binary assessment measures; that is, completed training versus uncompleted training, placement versus nonplacement, employed versus not employed, retained in the job setting versus not retained.

Summary

In this chapter we have attempted to review the salient models employed in the assessment and planning processes of psychosocial and vocational rehabilitation programs. These models included the clubhouse approach, Boston University's Psychiatric Rehabilitation approach, sheltered workshops, transitional employment, supported employment, and integrated models of psychosocial/vocational rehabilitation. Where data were available, the reliability and validity of these models was addressed with reference to existing comprehensive reviews of outcome studies. The review has given rise to a discussion of the major limitations of each of the models and has outlined salient issues that will need to be addressed in the future. If assessment and planning strategies are to have a meaningful impact on the lives of individuals disabled by a major psychiatric disorder, lives that include the full range of human options from work to play and from learning to doing, work that addresses these issues is critical.

References

Anthony, W. (1980). *The principles of psychiatric rehabilitation.* Baltimore: University Park Press.
Anthony, W. (1994). Characteristics of people with psychiatric disabilities that are predictive of entry into the rehabilitation process and successful employment. *Psychosocial Rehabilitation Journal, 17*(3), 3–13.
Anthony, W. A., & Jansen, M. A. (1984). Predicting the vocational capacity of the chronically mentally ill. *American Psychologist, 39*(5), 537–544.
Anthony, W. A., Cohen, M. R., & Danley, K. S. (1988). The psychiatric rehabilitation model as applied to vocational rehabilitation. In J. A. Ciardiello & M. D. Bell (Eds.), *Vocational rehabilitation of persons with prolonged psychiatric disorders* (pp. 59–80). Baltimore: Johns Hopkins University Press.
Azrin, N. H., & Philip, R. A. (1979). The job-club method for job handicapped: A comparative outcome study. *Rehabilitation Counseling Bulletin, 17,* 144–155.

Bachrach, L. (1992). Psychosocial rehabilitation and psychiatry in the care of long-term patients. *American Journal of Psychiatry 149*(Special Articles), 1455-1463.

Becker, D. R., & Drake, R. E. (1994). Individual placement and support: A community mental health center approach to vocational rehabilitation. *Community Mental Health Journal, 30*(2), 193-203.

Bilby, R. (1992). A response to the criticisms of transitional employment. *Psychosocial Rehabilitation Journal, 16*(2), 69-82.

Bond, G. R. (1992). Vocational rehabilitation. In R. P. Liberman (Ed.), *Handbook of psychiatric rehabilitation* (pp. 244-275). Boston: Allyn & Bacon.

Bond, G. R., & Boyer, S. L. (1988). Rehabilitation programs and outcomes. In J. A. Ciardiello & M. D. Bell (Eds.), *Vocational rehabilitation of persons with prolonged psychiatric disorders* (pp. 231-263). Baltimore: Johns Hopkins University Press.

Bond, G. R., & Dietzen, L. L. (1993). Predictive validity and vocational assessment: Reframing the question. In R. L Glueckauf, L. B. Sechrest, G. R. Bond, & E. C. McDonel (Eds.), *Improving assessment in rehabilitation and health* (pp. 61-86). Newbury Park, CA: Sage.

Bond, G. R., & Dincin, J. (1986). Accelerating entry into transitional employment in a psychosocial rehabilitation agency. *Rehabilitation Psychology, 31,* 143-155.

Bond, G. R., & McDonel, E. C. (1991). Vocational rehabilitation outcomes for persons with psychiatric disabilities: An update. *Journal of Vocational Rehabilitation, 1*(3), 9-20.

Cnaan, R. A., Blankertz, L., Messinger, K. W., & Gardner, J. R. (1988). Psychosocial rehabilitation: Toward a definition. *Psychosocial Rehabilitation Journal, 11*(4), 61-77.

Cohen, B. A., & Anthony, W. A. (1984). Functional assessment in psychiatric rehabilitation. In A. Halpern & M. Furhrer (Eds.), *Functional assessment in rehabilitation* (pp. 79-96). Baltimore: Brookes.

Danley, K. S., Anthony, W. A. (1987). The choose-get-keep mode. *American Rehabilitation, 13*(4), 6-9, 27-79.

Drake, R. E., Becker, K. R., Biesanz, J. C., Torrey, W. C., McHugo, G. J., & Wyzik, P. F. (1994). Rehabilitative day treatment vs. supported employment: I. Vocational outcomes. *Community Mental Health Journal, 30*(5), 519-532.

Fairweather, G. W. (1980). *New directions for mental health services: The Fairweather Lodge: A twenty-five-year retrospective, No. 7.* San Francisco: Jossey-Bass.

Geckle, M., & Katz, L. J. (1993). [Neuropsychological assessment and rehabilitation services patient feedback questionnaire]. Unpublished raw data.

Katz, L. J., & Geckle, M. (1993). [Neuropsychological assessment and rehabilitation services quality assurance policy and procedure]. Unpublished raw data.

Katz, L. J., Beers, S., Geckle, M., & Goldstein, G. (1989). The clinical use of the career ability placement survey vs. the GATB. *Journal of Applied Rehabilitation Counseling, 20*(1), 13-19.

Katz, L. J., Goldstein, G., Geckle, M., & Eichenmuller, A. (1991). Can the performance of psychiatric patients on the general aptitude test battery be predicted from the career ability placement survey? *Journal of Applied Rehabilitation Counseling, 22*(1), 22-29.

Katz, L. J., Goldstein, G., Geckle, M., Morrissey, J. & Daily, R. (1991). Adult psychiatric norms for the career ability placement survey. *Journal of Job Placement, 7*(1), 12-17.

Katz-Garris, L. J., McCue, M., Garris, R., & Herring, J. (1983). Psychiatric rehabilitation: An outcome study. *Rehabilitation Counseling Bulletin, 26*(5), 329-335.

MacDonald-Wilson, K. L., Revell, W. G., Nguyen, N., & Peterson, M. E. (1991). Supported employment outcomes for people with psychiatric disability: A comparative analysis. *Journal of Vocational Rehabilitation, 1*(3), 30-44.

Mank, D. M., Rhodes, L. E. & Bellamy, G. T. (1986). In W. E. Kiernan & J. A. Stark (Eds.), *Pathways to employment for adults with developmental disabilities* (pp. 139-159). Baltimore: Brookes.

Marrone, J. (1993). Creating positive vocational outcomes for people with severe mental illness. *Psychosocial Rehabilitation Journal, 17*(2), 43-62.

McCue, M, & Katz, L. J. (1983). The severely disabled psychiatric patient and the adjustment to work. *Journal of Rehabilitation, 49,* 52-58.

National Institute on Disability and Rehabilitation Research (NIDRR), Office of Special Education and Rehabilitative Services. (1993). *Strategies to secure and maintain employment for people with long-term mental illness* (Vol. 15, No. 10). Washington, DC: Department of Education.

Neff, W. S. (1988). Vocational rehabilitation in perspective. In J. A. Ciardiello & M. D. Bell (Eds.), *Vocational rehabilitation of persons with prolonged psychiatric disorders* (pp. 5-18). Baltimore: Johns Hopkins University Press.

Potsubay, R. (1984). The impact of vocational assessment on occupational consistency and employment ability of rehabilitation clients. *Dissertation Abstracts International, 45*(6A), 1695.

Propst, R. (1992). Standards for clubhouse programs: Why and how they were developed. *Psychosocial Rehabilitation Journal, 16*(2), 25-30.

Slonika, G. T. (1992). Assessment in psychiatric rehabilitation. In L. J. Katz (Ed.), *Psychiatric rehabilitation: A handbook for practitioners* (pp. 33-82). St. Louis, MO: Green.

Stein, L., & Test, M. A. (1980). Alternative to mental hospital treatment I conceptual model: Treatment program, and clinical evaluation. *Archives of General Psychiatry, 37*, 392-397.

Test, M. A. (1992). Training in community living. In R. P. Liberman (Ed.), *Handbook of psychiatric rehabilitation* (pp. 153-170). Boston: Allyn & Bacon.

Wehman, P. (1986). Supported competitive employment for persons with severe disabilities. *Journal of Applied Rehabilitation Counseling, 17*, 24-29.

APPENDIX

Psychosocial Summary

PSYCHOSOCIAL AND VOCATIONAL REHABILITATION

Client Name: Evaluator:
ID Number: Agency: Date:

Basic Environmental Supports
Does client have an effective support system? (Circle Y or N)
    Family—Y N; Peer—Y N; Institutional—Y N; Does client maintain a stable residence? Y N; Is client active with an MH provider? Y N; Specify corrective action if necessary.

Physical Health
Are there significant health problems: Y N
Describe (i.e., well-controlled, may complicate programming):

If indicated, is follow-up provided: Y N
Describe (i.e., where, by whom, how often):

Psychiatric Status
Check for the presence of any major symptom cluster

If Present	Acute	Chronic/Residual	In Remission
____ Psychotic Experience	____	____	____
____ Depression	____	____	____
____ Manic Symptoms	____	____	____
____ Anxiety or Fears	____	____	____
____ Obsessions or Compulsions	____	____	____
____ Organic Dysfunction	____	____	____
____ Drug Abuse	____	____	____
____ ETOH Abuse	____	____	____
____ Suicide Attempts	____	____	____
____ Other Symptoms	____	____	____

Specify action necessary around any current symptoms manifestation:

Has the client demonstrated medication compliance? Y N;
For how long? _____

Intrapersonal Factors:
Can client readily discuss impact of his symptoms on behavior and lifestyle? Y N
If no, does he/she tend to: ____ Deny; ____ Avoid; ____ Use Sick Role
Client's personal rehabilitation aims/personal incentives toward change:

Noteworthy disincentives toward maintenance of employment:

CRITICAL SKILLS CHECKLIST

Client Name: Evaluator:
ID Number: Agency: Date:

Instructions: Circle the appropriate category indicating how well each of the following critical skills are performed according to the definition key below:

Very Poor (1)—skill performance is very poor or absent; *Poor* (2)—skill performance is poor or inconsistent at best; *Adequate* (3)—skill performance is adequate (meets current needs); *Good* (4)—skill performance is good (effective, consistent/efficient); *Superior* (5)—skill performance is highly satisfactory (particular strength) *No*—cannot be determined; skill performance in need of further evaluation.

The *Environmentally Specific Comments* section is used to *specify* environmental and behavioral parameters in summarizing skill performance, i.e. *quantity;* frequency, duration, consistency, reliability; *quality:* thoroughness, accuracy, desirability, expediency, efficacy; and *level of independence:* level/frequency of supervision, prompting. Comments here should further elucidate client strengths which may be capitalized upon, as well as deficit areas which may be targeted for intervention.

Specify the environment for which this assessment was completed:

A. *Language/Communication*
   —engages in every day conversation   1 2 3 4 5 N
   —understands verbal commands/directions   1 2 3 4 5 N
   —responds to nonverbal/contextual cues   1 2 3 4 5 N
   —expresses needs & feelings   1 2 3 4 5 N
   Environmentally Specific Comments:

B. *Memory and Learning*
   —recalls basic environmental information
     (names, address, telephone, simple
     directions, etc.)   1 2 3 4 5 N
   —remembers appointments and schedule
     of activities   1 2 3 4 5 N
   —recalls long-term information   1 2 3 4 5 N
   —comprehends written materials   1 2 3 4 5 N
   —learns from hands-on experience   1 2 3 4 5 N
   —learns from verbal instructions   1 2 3 4 5 N
   Environmentally Specific Comments:

C. *Attention/Concentration*
   —sustains attention to aurally presented information 1 2 3 4 5 N
   —sustains attention to task at hand   1 2 3 4 5 N
   Environmentally Specific Comments:

D. *Problem Solving and Conceptualization*
   —makes reasonable decisions   1 2 3 4 5 N
   —formulates a plan   1 2 3 4 5 N
   —carries out a plan   1 2 3 4 5 N

—generalizes a concept across settings	1	2	3	4	5	N
—responds to constructive criticism	1	2	3	4	5	N

   Environmentally Specific Comments:

E. *Daily Living*

—attends to personal hygiene	1	2	3	4	5	N
—uses a telephone	1	2	3	4	5	N
—engages in leisure time activities	1	2	3	4	5	N
—reads a newspaper	1	2	3	4	5	N
—budgets money	1	2	3	4	5	N
—plans a nutritionally balanced meal	1	2	3	4	5	N
—prepares meals adequately	1	2	3	4	5	N
—shops for personal items	1	2	3	4	5	N
—uses public transportation	1	2	3	4	5	N
—accesses necessary resources (Social Security, medical care, etc.)	1	2	3	4	5	N

   Environmentally Specific Comments:

F. *Self-Appraisal and Judgment*

—evaluates severity of external stressors	1	2	3	4	5	N
—copes with mild stress	1	2	3	4	5	N
—seeks support if unduly stressed	1	2	3	4	5	N
—verbalizes relevant and appropriate personal goals	1	2	3	4	5	N

   Environmentally Specific Comments:

G. *Stamina and Tolerance*

—tolerates a 6–8 hour workday	1	2	3	4	5	N
—lifts objects 35–50 lbs	1	2	3	4	5	N
—traverses 100 yards without fatigue	1	2	3	4	5	N

   Environmentally Specific Comments:

Key
1 = Very Poor
2 = Poor
3 = Adequate
4 = Good
5 = Superior
N = In need of further evaluation

FUNCTIONAL ASSESSMENT SUMMARY AND SERVICE PLAN
(Based on *DSM-IV* 5-Axial Diagnoses, formal testing,
and Critical Skills Checklist)

Client Name: _____ Case Manager: _____
ID Number: _____ Agency: _____
Date: _____

Instructions: Complete this functional assessment summary and service plan utilizing summary data from the three major data bases identified in the preceding.

Summary of personal resources, skills, strengths	Summary of missing or inadequate personal resources, deficits, limitations	Summary of resources needed to accommodate, ameliorate, or overcome problem areas	Responsible providers or of resource availability

15

Cognitive Remediation of Psychotic Patients

ROBERT S. KERN AND MICHAEL F. GREEN

INTRODUCTION

In the United States, recent estimates indicate that approximately $12 billion are spent annually to treat schizophrenia, and lost productivity as a result of the disease costs approximately $28 billion. It is well documented that schizophrenia also involves a host of cognitive, social, and occupational deficits that place limits on the ability to function independently. Approximately two-thirds of individuals diagnosed with schizophrenia are unable to maintain full independence and require some degree of supervised care. Because functional deficits are relatively refractory to traditional forms of treatment, it is imperative to develop innovative treatments that focus on improving social and occupational functioning.

Despite compelling data that indicate that schizophrenia is at least in part a neurobiological disorder with associated deficits in a number of cognitive domains (Bracha, 1989; Carlsson, 1988; Flor-Henry, 1976; Goldstein, 1986; Gur, 1978; Kovelman & Scheibel, 1984; Mednick, Parnas, & Schulsinger, 1987; Nuechterlein & Dawson, 1984), psychiatric rehabilitation programs have largely ignored these cognitive deficits. This failure is unfortunate when one considers the wealth of knowledge available about the information processing disturbances in schizophrenia. Over the past 30 years, studies of schizophrenia have led to the identification of deficits in a number of cognitive domains including attention, memory, executive functioning, language processing, motor speed, and others (Calev, Venables, & Monk, 1983; Goldberg, Weinberger, Berman, Pliskin, & Podd, 1987; King, 1991; Maher, 1991; Sacuzzo & Braff, 1981). A better understanding of these

---

ROBERT S. KERN   West Los Angeles VA Medical Center (B116AR), 11301 Wilshire Boulevard, Los Angeles, California 90073.   MICHAEL F. GREEN   UCLA–Neuropsychiatric Institute and Hospital, 760 Westwood Plaza (C9-420), Los Angeles, California 90024-4344.

Rehabilitation, edited by Goldstein and Beers. Plenum Press, New York, 1998.

core cognitive deficits has led to a better understanding of the etiology and course of the disorder; some investigators believe that emphasis should shift to remediation of these deficits (Green, 1993).

Perhaps efforts are already under way. Over the past five years, studies of cognitive remediation in schizophrenia have been found with greater frequency in the literature, and the level of sophistication of remediation techniques is advancing. The types of remediation studies range from investigations that attempt to modify discrete areas of cognitive dysfunction to more comprehensive programs that attempt to alter a wide range of cognitive and social functioning deficits.

What is cognitive remediation? The terms "remediation" and "rehabilitation" are often used interchangeably, even though they are not synonymous. For the purpose of this chapter, we shall use "cognitive remediation" to refer to those interventions designed to improve performance on a selected measure or a limited range of measures of cognitive processing. We will use "cognitive rehabilitation" to refer to general, broad-based efforts that address social and occupational deficits, as well as cognitive ones, and attempt to improve the overall functioning of the individual. Though the preponderance of findings presented in this chapter were collected on patients diagnosed with schizophrenia, it is believed that cognitive remediation may be applicable to psychotic patients in general.

At this point, cognitive remediation studies in schizophrenia have centered primarily around three areas: (a) testing the feasibility of modifying selected areas of cognitive dysfunction (Green, Satz, Ganzell, & Vaclav, 1992; Kern, Green, & Goldstein, 1995; Koh, Grinker, Marusarz:, & Forman, 1981), (b) testing the generalizability of treatment effects to other domains of functioning (e.g., cognitive, social, or occupational functioning) (Jaeger & Douglas, 1991; Kraemer, Zinner, & Moller, 1989; Olbrich & Mussgay, 1990; Wagner, 1968; Yozawitz, 1986), and (c) identifying "rate limiting" factors, or those cognitive deficits that limit the ability of schizophrenia patients to acquire new skills and abilities (Bowen et al., 1989; Corrigan, Wallace, Schade, & Green, 1994; Kern, Green, & Satz, 1992; Mueser, Bellack, Douglas, & Wade, 1991). It is beyond the scope of this chapter to address all three of these areas. Instead, we shall draw attention to issues related to testing the feasibility of cognitive remediation in schizophrenia. Feasibility studies, the term relates to cognitive remediation, refers to investigations that test whether or not a certain cognitive deficit can be modified or altered under controlled experimental conditions.

This chapter (a) discusses two models of information processing that have provided a conceptual framework for studies of cognitive remediation, (b) briefly describes selected clinical and laboratory-based measures of memory, abstraction/executive functioning, and early visual processing that have been used in studies of cognitive remediation, (c) reviews selected feasibility studies that have attempted to modify performance on measures of memory, abstraction/executive functioning, and early visual processing, (d) reviews selected comprehensive programs of cognitive remediation, and (e) presents preliminary conclusions about feasibility studies of cognitive remediation and discusses methodological considerations for future studies. Let us begin then with a look at two basic models of information processing that have been used to guide studies of cognitive remediation.

MODELS OF INFORMATION PROCESSING

In this section we discuss briefly our current understanding of two prevailing models of information processing and their relevance to the development of cognitive remediation interventions.

What cognitive model (or models) should we use to guide our cognitive interventions? Two separate, but overlapping, frameworks have been used to explain information processing in schizophrenia: capacity models and stage models (Nuechterlein & Asarnow, 1989). Capacity models emphasize an individual's overall processing capacity. Processing capacity is a limited resource that is used in the execution of various cognitive tasks (Kahneman, 1973). The amount of the resource, or the overall capacity, varies both between and within individuals. Differences in task performance may be explained by the assumption that certain individuals have less overall capacity (i.e., total amount of available resources) than others or are inefficient in the ability to access necessary resources required to perform certain cognitive tasks. In schizophrenia, the capacity model explains cognitive deficits as being the result of deficiencies in available processing resources (Nuechterlein & Dawson, 1984). These deficiencies may be the result of reduction in central capacity or the inefficient allocation of available processing resources. Factors such as arousal, motivation, and fatigue are considered insofar as they may influence overall processing capacity.

In contrast to capacity models of information processing, stage models emphasize a sequential series of processing stages. The stages range from elementary, basic cognitive processes to more complex, sophisticated processes. Information processed at earlier stages is fed to subsequent stages, where further processing takes place. At each succeeding stage, the output received from the previous stage is transformed or elaborated. Hence, at later stages, processing becomes increasingly complex. Cognitive deficits in schizophrenia are explained as being due to processing problems at a particular stage. Traditionally, the emphasis of these models is on finding the earliest stage of processing where dysfunction takes place. Information poorly processed at earlier stages leads to further processing disturbances in later stages through a cascading effect (Saccuzzo & Braff, 1981).

These two models overlap considerably, but have different emphases. Capacity models emphasize the function of a central processor. Interventions to ameliorate processing limitations focus on ways of increasing central capacity or bias allocation of processing resources. In contrast, stage models target their interventions for early processing stages in the hope of finding the earliest point of processing disturbance. Some theorists have proposed a model of information processing that incorporates features of both models in which capacity limitations are evaluated at differing stages of processing (Moray, 1969). In one such integrated model, the different stages are referred to as computational mechanisms and the capacity aspects are referred to as energetical mechanisms (Sanders, 1983). Cognitive measures vary by the degree of strain placed on computational and energetical mechanisms. All tasks involve computational mechanisms, that is, they consist of a series of basic processing stages that perform functions of sensory input, internal transformation or elaboration of information, and response output, but vary widely in their requirement of energetical mechanisms. For example, certain perceptual tasks place relatively small demands on energetical mechanisms; individuals can perform such tasks whether they are half awake, half asleep, or relatively unmotivated. Other cognitive tasks such as complicated problem-solving processes place high demands on energetical mechanisms, and typically require the full, undivided attention of the participant.

Assessment of Cognition and Information Processing

Let us now take a look at representative measures from selected cognitive and information processing domains that have been used in cognitive remediation studies. These measures include tests of memory (list learning tests), abstraction/problem-solving ability (Wisconsin Card Sorting Test—WCST; Heaton, 1981), and early visual processing (span of apprehension task; Estes & Taylor, 1966). Interest in these three measures stems from different reasons. Remediation of memory deficits is of interest because some investigators argue that memory may be disproportionately affected in schizophrenia (Saykin et al., 1991). In addition, several of the early feasibility studies focused on memory deficits (Bauman, 1971a; Koh, Kayton, & Berry, 1973). Recent interest in WCST performance has been based largely on the assumption of this test's sensitivity to deficits associated with the dorsolateral prefrontal cortex (Goldberg et al., 1987). And interest in the span of apprehension stems from the static, enduring course of deficit performance shown by schizophrenia patients on this test and the intriguing implications that modification of span deficits (a putative vulnerability indicator) might have for preventive treatment (Asarnow & MacCrimmon, 1978).

Memory—List Learning Measures

List learning measures are frequently used to assess verbal memory and learning in schizophrenia patients. One such measure, the California Verbal Learning Test (CVLT; Delis, Kramer, Kaplan, & Ober, 1987), includes a list of 16 items with four exemplars from four different taxonomic categories. The list is presented over a series of five learning trials, and is immediately followed by presentation of a distractor list. Free and cued recall are assessed immediately after the distractor list and again twenty minutes later. Recognition is assessed at the time of the 20-minute delayed free and cued recall.

Though list learning measures may seem dull (both to subject and examiner), they do provide certain advantages. One primary advantage is that list learning measures such as the CVLT provide a variety of indices related to memory and learning such as learning slope, free and cued recall scores, a discriminability index (recognition score), clustering scores, primacy and recency scores, and total number of intrusions. Clinically, the information obtained from such measures can be quite useful for making diagnostic distinctions between differing types of neurological syndromes and for characterizing specific processing deficits.

Schizophrenia patients show a shallower learning curve than normal controls on list learning tasks (Bauman, 1971a; Calev et al., 1983; Koh et al., 1973; Gold, Randolph, Carpenter, Goldberg, & Weinberger, 1992). Although the learning curve is shallow, it does reveal some learning of new information. The number of items freely recalled after a delay shows some decline, but does not reflect a complete loss of information. Schizophrenic patients have been reported to show a greater number of intrusions than psychiatric controls on list learning tasks (Nachmani & Cohen, 1969), although some investigators have failed to find a similar pattern of performance when comparing chronic schizophrenic inpatients and normal controls (Calev, Korin, Kugelmass, & Lerer, 1987). Although a number of early studies failed to show any significant differences between schizophrenic patients and normal controls on tasks of recognition (Bauman & Murray, 1968; Koh et al., 1973; Nachmani & Cohen, 1969), there is some evidence to suggest that recognition ability is also compromised, at least in chronic inpatients (Gold et al., 1992). In addition, Gold and colleagues showed that the discriminability index for schizophrenia patients on a word recognition task was worse than that of normals (i.e., included fewer number correct and a higher number of false identifications).

ABSTRACTION/PROBLEM-SOLVING ABILITY—WISCONSIN CARD SORTING TEST

The WCST has received a good deal of attention in the psychopathology literature in recent years. A traditional neuropsychological test that has been used for over 40 years (Milner, 1963), the WCST requires subjects to match a series of stimulus cards to one of four key cards. Detailed instructions are not provided to subjects except that they are told whether or not the match is correct. The four key cards and two decks of stimulus cards (128 cards) comprise the entire set of materials. The stimulus cards match the key cards by color, shape, or number of symbols. The subject must guess which of the three matching categories is correct. After the subject makes 10 consecutive correct matches, the correct matching category changes (e.g., from color to shape). However, the subject is not told about this change. The test is discontinued when the subject correctly completes six categories or when all of the stimulus cards are used.

In general, performance on this test is limited by (a) a failure to *attain* the correct matching category—the subject is unable to conceptualize that the stimulus cards can match the key cards by one or more of the sorting categories (i.e., color, shape, or number of symbols), (b) a failure to *maintain* set—the subject begins to match according to a particular category correctly, but fails to continue matching to the same correct category for 10 consecutive cards, (c) failure to *shift* set—the subject fails to shift to the new correct matching category following feedback of an incorrect response (Green et al., 1992). Even though the neuroanatomical specificity of this exercise to the prefrontal cortex has been questioned (Anderson, Damasio, Jones, & Tranel, 1991), it is generally agreed that the WCST is a valid measure of these cognitive abilities.

Chronic schizophrenia patients typically perform quite poorly relative to normal controls on this measure (Goldberg et al., 1987). In our lab, we have observed that chronic schizophrenia patients usually are only able to complete one or two categories, and they commit a high number of perseverative errors. A perseverative error occurs when the subject continues to match a stimulus card according to a previously employed principle (e.g., color) that the examiner now identifies as incorrect. A number of recent studies that have attempted to remediate performance on this task will be discussed in a later section.

The span of apprehension task has been used in a number of studies of schizophrenia (Asarnow, Granholm, & Sherman, 1991). One's span of apprehension refers to the number of items that can be attended to at one time (Woodworth, 1948). It can be measured by having individuals identify selected target stimuli (partial report) or entire stimulus arrays (full report). The Estes and Taylor detection task (Estes & Taylor, 1966) is a partial report visual span of apprehension test that has been used in a number of studies with schizophrenia patients. The task requires identification of tachistoscopically presented alphabetic stimuli of various array sizes (e.g., 3 or 12 letters). Under partial report procedures, subjects are required to identify one of two possible target letters from a multiletter array. Identification of the target stimuli may be done orally or by pushing a response button.

Some evidence supports the span of apprehension as a vulnerability indicator of schizophrenia. Deficits on this measure have been found in schizophrenia patients during states of remission (Asarnow & MacCrimmon, 1978, 1981; Nuechterlein et al., 1992) and deficit performance has also been noted in the offspring of schizophrenia mothers (Asarnow, Steffy, MacCrimmon, & Cleghorn, 1977). Hence, deficits on the span of apprehension seem to be associated with vulnerability to the disorder rather than the symptoms of the illness. These findings, although provocative, do not provide verifiable evidence for the span of apprehension as a vulnerability indicator. Still, remediation of deficit performance on the span of apprehension suggests intriguing clinical implications. If span of apprehension performance does reflect some cognitive processing limitation that predisposes individuals to developing schizophrenia, then improvement of those processes tapped by this measure could possibly serve a preventive function either by protecting individuals from initially developing the disorder or by preventing subsequent relapses.

FEASIBILITY STUDIES OF COGNITIVE REMEDIATION: "ATTACKING DISCRETE DEFICITS"

This section reviews selected feasibility studies that have attempted to remediate deficit performance on the measures just described: (a) memory (list learning tests), (b) abstraction/problem-solving abilities (WCST), and (c) early visual processing (span of apprehension task). Let us begin then with a look at a few representative studies that have tested the modifiability of memory functioning in schizophrenia.

REMEDIATION OF MEMORY DEFICITS: MODIFYING PERFORMANCE ON LIST LEARNING TASKS

In the 1970s Bauman and Koh and his colleagues conducted a series of studies that examined the performance of schizophrenia patients on tests of list learning. These studies sought to identify the underlying processing dysfunction, as well as test the modifiability of the deficit.

In one of the early remediation studies, Bauman (1971a) attempted to improve serial recall in a sample of adult inpatients with schizophrenia (the majority had been hospitalized less than one year). Patients were divided into two groups ($n = 12$ in each group) and presented 10 trigrams on a memory drum. Recall was assessed over three trials. One group received information about the trigram list (i.e., each trigram begins with a different letter of the alphabet) and specific instructions to "think of the first letter" of each trigram (an alphabetical encoding strategy). Patients in the other group were simply told to try their best. The results revealed no significant difference in the performance of the two groups, despite the fact that one group received specific instructions to organize the material alphabetically (perhaps because the subjects failed to employ the strategy). In an independent study Bauman (1971b) was able to demonstrate that schizophrenia patients' recall of list items was modifiable, but only under a condition where the list of words was preorganized, that is, arranged in clusters according to taxonomic category. The results of these studies suggested that the memory deficit in schizophrenia was due to a failure to initiate semantic categorizing and other clustering schemes in mnemonic organization of verbal material, a deficit that results in inefficient encoding.

A few years later, Koh and his colleagues reported that patients with schizophrenia were inferior to normal controls at serial recall on a variety of different type word lists that included unrelated (Koh & Kayton, 1974; Koh et al., 1973), categorically related (Koh et al., 1973), and affectively loaded words (Kayton & Koh, 1975). A series of investigations followed to see if the encoding difficulties could be remediated. These studies sought to clarify whether the impaired recall of schizophrenia patients was due to a permanent or temporary processing dysfunction. A number of procedures were introduced: (a) rating input words according to their pleasantness or unpleasantness, (b) rating the frequency of usage of words, and (c) classifying words according to their semantic meaning. These authors speculated that semantic elaboration of verbal information was central to remembering, and was a function performed routinely by most normal individuals.

In a seminal study, Koh, Kayton, and Peterson (1976) manipulated subjects' processing of input words. The study measured incidental and intentional free recall in a group of schizophrenia patients, nonschizophrenia psychiatric patients, and normal controls. In incidental learning paradigms, the subject is required to engage in an experimentally controlled task, but is not explicitly instructed to remember the material presented. An unexpected recall test then follows. Such procedures allow the experimenter to manipulate the subject's processing of the material during input. In Koh and colleagues' (1976) experiment, the subjects were required to rate a series of words according to their degree of pleasantness or unpleasantness. Then, without warning, the subjects were asked to recall the words they had just rated. Under these conditions, the performance of the schizophrenia and psychiatric patients was comparable to that of the normal controls. The same rating procedures were employed 1 week later for testing recall of a different list of words. This test was of intentional recall in that the subjects were informed beforehand that recall would later be assessed. Again, both patient groups performed comparably to the normal controls. An important factor in this study was that the subjects' mnemonic encoding strategy (i.e., rating words by degree of pleasantness or unpleasantness) was under the direct control of the experimenter and required little initiation or conscious effort on the part of the subject. The conclusion by Koh and his colleagues was that the semantic memory network of schizophrenia patients was probably intact, and that the disturbance observed in list learning was due to a failure to utilize those mechanisms involved in the elaboration of verbal information.

In an interesting follow-up study, Larsen and Fromholt (1976) tested the incidental recall of a group of schizophrenia patients and normal controls using a card sorting exercise as the experimental procedure. Instead of rating words as done in Koh and colleagues' (1976) study, subjects sorted cards with affectively loaded adjectives printed on them into piles labeled pleasant and unpleasant. After two sorting trials, an unexpected recall test was administered. The schizophrenia subjects performed comparably to the normal controls. The authors concluded that the effect was primarily due to a deeper level of processing. The sorting of cards into piles required the stimuli to be processed at a semantic level, resulting in improved recall for the schizophrenia group.

In conclusion, a few studies have reported success in modifying performance on list learning tasks. Those interventions that have been most successful have manipulated the semantic processing of the material. It appears that the recall of information in schizophrenia is largely affected by the way in which it is encoded and not by the intention to memorize per se.

Remediation of Abstraction/Problem-Solving Deficits: Modifying Performance on the Wisconsin Card Sorting Test

The following section reviews several studies that have tested the feasibility of remediating performance deficits on the WCST. This measure has received considerable attention in feasibility studies of cognitive remediation in recent years. We compare and contrast the findings from a few representative studies, pointing out possible explanations for the discrepancies between reports and presenting some general conclusions.

The crusade to remediate performance deficits of schizophrenia patients on the WCST followed a study conducted by Goldberg and colleagues (1987). This study tested the feasibility of modifying performance using detailed step-by-step instruction, and included three groups of schizophrenia inpatients described as having chronic, unremitting illnesses. All subjects received five serial administrations on the same day using an abbreviated version of the WCST, and a 2-week follow-up administration. The first group ($n = 15$) served as a control and received the WCST according to standard administration procedures (Heaton, 1981). The other two groups ($n = 15$ and $n = 14$, respectively) received (a) instructions about categories, (b) instructions about shifting of set, and (c) detailed step-by-step instructions on subsequent administrations after baseline. These latter two groups differed only by the order in which they received the separate interventions. Both experimental groups obtained significantly better scores than the control group under conditions that included detailed step-by-step instructions, but failed to maintain the advantage on subsequent follow-up administrations. Goldberg and colleagues (1987) concluded that teaching chronically ill schizophrenic inpatients through conventional techniques (i.e., providing rules and detailed step-by-step instructions) does not result in learning, and the disturbances of the dorsolateral prefrontal cortex are not responsive to rehabilitation.

After the report by Goldberg and colleagues (1987), a host of investigators designed studies to test the modifiability of deficit performance on the WCST in schizophrenia. Bellack, Mueser, Morrison, Tierney, & Podell (1990) tested schizophrenia inpatients who were being treated following an acute exacerbation of psychotic symptoms. Initially, these authors tested the hypothesis that poor performance exhibited by schizophrenia patients was due to a lack of motivation or inattention. To test this hypothesis a group of schizophrenia inpatients ($n = 16$) was administered the WCST four times and were sequentially assigned to receive either contingent or noncontingent positive reinforcement (i.e., a nickel). Neither procedure yielded a significant gain in performance. Next, the authors sought to examine whether the combination of instruction plus monetary reinforcement could improve WCST performance. A second group of subjects ($n = 12$) was administered the WCST four times and each subject received detailed instruction plus monetary reinforcement in the form of a nickel for each correct response. The instructions included general information about the task and specific instructions about matching categories and shifting set. The group that received the combination of instruction plus incentive performed better than the control group (monetary reinforcement only) at the time of the intervention and two subsequent follow-up administrations. This study yielded two important findings: (a) neither contingent nor noncontingent monetary reinforcement significantly improved performance from baseline, and (b) a brief training technique that provided information about the task and incentive dramatically improved performance from baseline and the performance gains were maintained at a 1-day follow-up.

Green and colleagues (1992) attempted to remediate WCST performance by using procedures that incorporated incentive and instruction, interventions that had been used in the previous two studies. Two groups of subjects were studied: a schizophrenia group ($n = 46$) and a mixed psychotic disorder group ($n = 20$). Both groups received four serial administrations of a computerized version of the WCST. At baseline (first administration), the two groups received a computerized version of the WCST using traditional administration procedures. During the second administration, contingent reinforcement was used (i.e., two cents for every correct response). For the third administration, both groups received detailed step-by-step instruction plus contingent reinforcement. For the fourth administration, both groups received contingent reinforcement only. As illustrated in Figure 1, the schizophrenia inpatient group showed significant improvement from baseline at the third and fourth administrations. No significant difference was noted between performance obtained at the second administration (contingent reinforcement only) and baseline. The conclusions drawn from Green and colleagues' (1992) and Bellack and colleagues' (1990) studies suggested the importance of motivational factors in conjunction with instructional training for remediating WCST deficits.

One study examined the effects of contingency interventions alone on WCST performance. Summerfelt and colleagues (1991) trained a group of schizophrenia and schizoaffective disorder inpatients and outpatients on the WCST. They found that training procedures that included monetary reinforcement and response cost yielded improved performance on the WCST, as demonstrated by fewer perseverative errors. The experimental design was a two-period crossover in which two administration conditions were counterbalanced over two periods of testing. In the one condition, subjects received the WCST under traditional administration procedures. In the second condition, subjects were initially provided a stake of $7.50 and could earn 10 cents for every correct response, but would lose five cents for every incorrect response. Their results indicated that subjects who received monetary reinforcement plus response cost performed significantly better than subjects who received traditional administration. In contrast to the findings by Bellack and colleagues (1990) and Green and colleagues (1992), the results of Summerfelt and colleagues' (1991) study suggest the importance of motivational factors *alone* on WCST performance.

The level of monetary reinforcement in Summerfelt and colleagues' (1991) study (10 cents per correct response) was higher than that used by Green and colleagues (1992) (two cents per correct response) and Bellack and colleagues (1990) (five cents per correct response). Therefore, it might be reasonable to suspect that differences in the findings of these studies may be due to value differences in the reinforcer. To address this question, Hellman, Green, Kern, and Christenson (1992) conducted a study in which patients were assigned to groups to receive either low (2 cents) or high (10 cents) levels of reinforcement per correct response. The study used a 2 × 2 between-group design with subjects as-

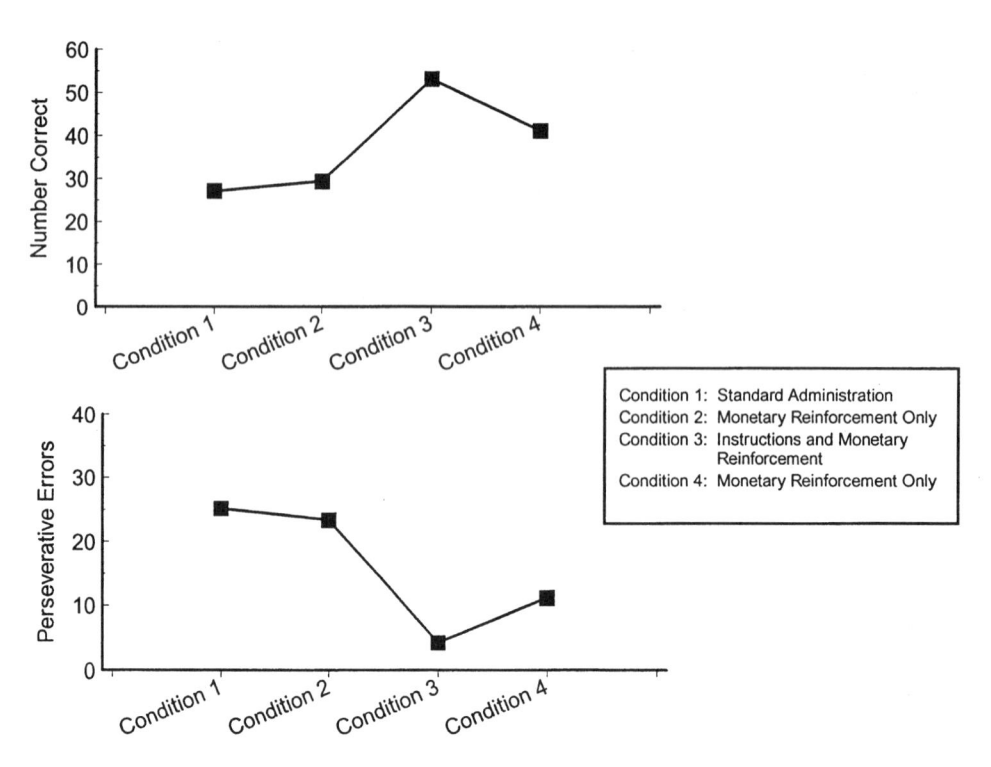

Figure 1. Schizophrenia patients' mean number correct and mean perseverative errors on the Wisconsin Card Sorting Test under four administration conditions.

signed to groups that were distinguished by level of monetary reinforcement (high versus low) and instructional training (instruction versus no instruction). The results showed that instruction, but not level of contingent reinforcement, improved performance on the WCST (i.e., fewer perseverative errors). Interestingly, the results of the monetary reinforcement were in the direction opposite to that predicted. Patients receiving low levels of monetary reinforcement showed a slight, nonsignificant tendency to perform better than patients receiving high levels of monetary reinforcement. Hence, these findings for differing levels of monetary reinforcement stand in contrast to Summerfelt and colleagues' study. However, Hellman and colleagues' study did not employ response cost procedures.

In general, the studies discussed suggest the need for incentive and instruction in training psychotic patients on the WCST. These studies tested the feasibility of modifying deficit performance on the WCST, and questions about durability were either not addressed or addressed in a limited fashion (e.g., 1 day). One approach that has only recently received investigative attention is errorless learning. Versions of errorless discrimination training, initially proposed by Terrace (1963), have been successfully employed with developmentally disabled individuals where learning is often impaired. Autistic children have been shown to acquire new skills following training, and subsequently maintain high levels of performance for prolonged periods of time. Our lab recently developed a training procedure based on errorless learning principles for the WCST (Kern, Wallace, Hellman, Womack, & Green, 1996). The training procedures, which dissected the task into its component parts, included a series of exercises on each component. Training began on simpler, more basic elements of the test (e.g., card identification) and progressed through increasingly more complex and cognitively challenging exercises (e.g., shifting set). Performance criteria were established for each stage, and subjects advanced to more difficult levels only after criteria for mastery of earlier stages were met. The level of difficulty was purposely manipulated so as to maximize successful performance. It was predicted that subjects who successfully completed this type of training (committing very few errors) would show improved levels of performance on the WCST, and that these levels would be maintained over a substantial period of time (up to 1 month posttraining). The preliminary results support these hypotheses.

REMEDIATION OF EARLY VISUAL PROCESSING DEFICITS: MODIFYING PERFORMANCE ON THE SPAN OF APPREHENSION TASK

Clearly, much effort has gone into modifying performance on list learning and the WCST. What about the feasibility of remediating performance deficits in other areas of cognition, for example, one in which there is no obvious strategy? Recently, we tested the feasibility of remediating performance on the span of apprehension task, a putative vulnerability indicator of schizophrenia (Kern et al., 1995). The version of the span of apprehension task that we employed used partial report procedures in which subjects are required to identify one of two target letters (Estes & Taylor, 1966). Arrays of either 3 or 12 letters were presented for 70 ms. Subjects were chronic schizophrenia inpatients who were sequentially assigned to one of four treatment groups. Each group received four administrations of the span of apprehension task: (a) at baseline, (b) during the intervention, (c) at immediate posttest, and (d) at a 1-week follow-up. The groups differed by the type of intervention received at the second administration. All of the remaining administrations followed traditional testing procedures. During intervention, the repeat administration group received traditional administration of the span of apprehension task without enhanced instruction or monetary reinforcement. The monetary reinforcement group received two cents for each correct response during the practice and test trials. The instruction group received readiness prompts that were designed to alert the subject about necessary task components. The fourth group received a combination of monetary reinforcement and enhanced instruction as described above.

The results of this study indicated a significant interaction between monetary reinforcement and instruction. As shown in Figure 2, the group that received both monetary reinforcement and instruction showed a substantial improvement in percent accuracy from baseline, and this level of performance was then maintained over two subsequent follow-up administrations (immediate and 1 week later). Of clinical importance, the improved performance of the group that received both interventions approached levels typically observed in normals on the span of apprehension task. These results illustrate that performance on a relatively simple measure of early visual processing, one with limited demands on higher cognitive processes, can be improved through extrinsic intervention. One avenue of hope for these and future investigations is that remediation of deficits on cognitive tests such as the span of apprehension, a putative vulnerability indicator of schizophrenia, may eventually serve as a protective factor against psychotic episodes.

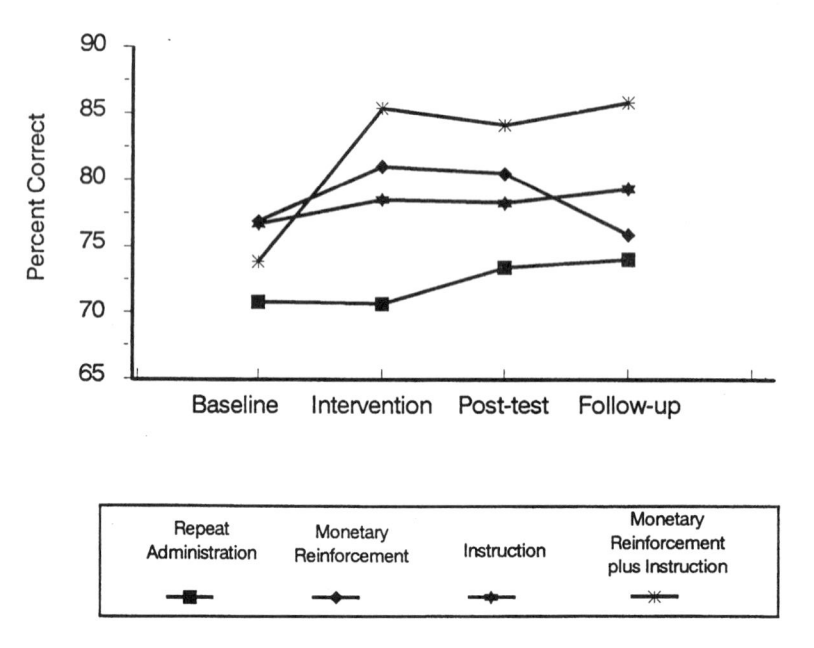

Figure 2. Mean percent accuracy scores on the span of apprehension task for schizophrenia patients receiving four differing interventions.

The studies surveyed thus far have emphasized remediation of discrete cognitive deficits. In contrast, a few investigators have been somewhat more ambitious in scope and have attempted to develop comprehensive training programs of cognitive remediation. These programs seek to remediate not only cognitive deficits but also, in some instances, social and occupational functioning. In the United States, Spaulding and Sullivan (1991), Yozawitz (1986), and Jaeger, Berns, Tigner, and Douglas (1992) deserve credit for developing such programs. The European contribution has had a particularly strong impact on the development of comprehensive programs. Two examples of such programs include Brenner's Integrated Psychological Therapy (IPT; Brenner, Hodel, Roder, & Corrigan, 1992) and van der Gaag's cognitive retraining program (1992).

Brenner's IPT is a systematic training program designed to improve the functioning of schizophrenia patients along multiple domains ranging from fairly elementary cognitive processes to more complex aspects of social behavior. Treatment consists of training in five subprograms: cognitive differentiation, social perception, verbal communication, social skills, and problem-solving skills. The subprograms are arranged hierarchically, so that initially the interventions target relatively basic cognitive skills, then proceed to verbal/social responses, and eventually to social problem-solving skills. The extended course of treatment consists of group sessions that are held five time per week in sessions lasting 60 to 75 minutes over approximately 3 months. The main evaluation of this program was conducted in Bern, Switzerland. Schizophrenia patients received standard treatment and were assigned either to an IPT group ($n = 14$), an attention-placebo group ($n = 15$), or a no-intervention group ($n = 14$) (Brenner, Kraemer, Hermanutz, & Hodel, 1990). The attention-placebo group participated in a variety of group activities such as games and discussions. The investigation sought to determine the efficacy of IPT at improving cognition and symptomatology in schizophrenia patients. The cognitive and psychopathology measures were administered pre- and posttreatment, and again 18 months later.

At the time of the first posttest, the IPT group showed significantly greater improvement than either comparison group on an attentional measure, self-assessment of cognitive dysfunction, and overall psychopathology (BPRS total). However, the groups did not differ on a measure of visual memory or the schizophrenia clinical scale from the MMPI. At the time of the 18-month follow-up, the IPT group continued to maintain significant improvement compared to the comparison groups on the measures mentioned, plus an additional psychosocial adjustment scale. In summary, IPT showed durable effects of training on selected measures of attention, subjective cognitive complaints, and overall psychopathology.

Another cognitive retraining program, developed by van der Gaag (1992), is based on a conceptual framework that integrates features of stage and capacity models of information processing. Like the program of Brenner and colleagues (1990), van der Gaag's training proceeds in steps from fairly elementary to complex cognitive exercises. Three main strategies were applied in training: self-instruction, mnemonics, and inductive reasoning. Sessions 1–12 included training on visual, auditory, tactile, and proprioceptive perception tasks, sessions 13–15 included training on tasks that integrated perceptual and language processing, and sessions 16–22 included training on more complex exercises of social perception. A study was conducted to examine the effects of this program on measures of attention, memory, (social) perception, and reasoning. Subjects participated in 22 training sessions over about a 3-month period. Sessions lasted 15 to 25 minutes, and were conducted two times per week. The experimental group was compared to an attention-placebo control group. The attention-placebo group received the same number of sessions, but instead of cognitive training, took part in a variety of recreational activities.

Pre-post differences were assessed in the following areas: (a) sustained attention and early visual processing (CPT, span of apprehension, Stroop test, Trail Making B); (b) social perception and reasoning (emotion matching and labeling tasks, WAIS Picture Arrangement subtest, a visual closure task); (c) verbal and nonverbal memory (Rey Auditory Verbal Learning Test, Rey–Osterreith Complex Figure Test with a delay, WAIS Digit Symbol subtest, a word fluency test); (d) concept formation and problem solving (modified WCST and WISC mazes subtest). The results indicated that the experimental group differed from the attention-placebo group on measures of social perception and memory but not attention or problem-solving ability. Given the nature of the training program, with a substantive emphasis on basic perceptual skills and social perception, these findings appear consistent with what might have been expected. Memory may have been improved through the activation of perceptual processes that may have allowed subjects to elaborate or process the material at a deeper level. Social perception may have been affected more directly through the training exercises.

One advantage of comprehensive remediation programs, besides the obvious possibility of impacting several different cognitive deficits, is that it allows testing of generalization effects. Specifically, such programs can test whether the effect of retraining in one area of cognition (e.g., attention) will yield improvement in another area of cognitive processing (e.g., memory).

Conclusions

At the beginning of this chapter we introduced three types of cognitive remediation studies: feasibility, generalization, and rate-limiting factors. Investigations logically would begin with studies of rate-limiting factors and move to studies of feasibility and then to studies of generalization. Studies of rate-limiting factors attempt to identify key cognitive deficits that limit or impede performance in other areas of functioning. For example, a number of independent studies suggest that the verbal memory deficits of psychotic patients act as rate-limiting factors for skill competence and acquisition in social skills training programs (Kern et al., 1992; Mueser et al., 1991). Identifying such key cognitive deficits can be seen as a logical prerequisite for planning feasibility studies. Feasibility studies may then be used to test whether the identified cognitive deficit(s) can be modified through intervention. The results from feasibility studies could then be applied to generalization studies. Improvement in certain areas of cognitive functioning might be predicted to impact related domains of behavior. For example, if vigilance is associated with social skill acquisition, then improving vigilance may have a beneficial effect on social skill acquisition.

In this chapter we reviewed several feasibility studies that targeted discrete deficits for remediation. One limitation common to these studies is the choice of a single dependent measure. The selection of one dependent measure makes it impossible to determine whether the intervention(s) altered the underlying cognitive construct or merely modified performance on that particular measure. The dependent measure is not distinguishable from the cognitive construct when only a single measure is used, thus limiting interpretation of results. One alternative method with more interpretable conclusions is to include several measures of the same construct, as well as several measures of different constructs (which are not expected to be affected by the intervention). In this way, the design tests the discriminant and convergent validity of the findings. For example, if we want to test the modifiability of early visual processing, we might include several measures of early visual processing as well as several measures of unrelated cognitive functions (e.g., language functioning). If the results reveal that the intervention modifies performance on several different measures of early visual processing, but not measures from unrelated cognitive domains, we have better support for the conclusion that the intervention modifies the construct of early visual processing.

Several challenges lie ahead for cognitive remediation efforts. First, future studies will need to address the issue of whether the interventions modify the cognitive construct or simply alter test performance. Second, studies of cognitive remediation will eventually need to move from inside to outside the laboratory. Brenner's IPT and similar programs have been useful in showing how improved cognitive functioning can generalize to more global aspects of behavior. Another way to move outside the lab would be to consider more ecologically valid dependent measures, as opposed to lab-based measures. For instance, instead of applying memory training strategies to lists of words or nonsense syllables, we could select lists of content material from a skills training program. The goal of these studies will be to develop cognitive interventions that address the processing demands of the real world. The bridging of laboratory findings to real-world problems will be one of the more pressing challenges for this area.

REFERENCES

Anderson, S. W., Damasio, H., Jones, D. R., & Tranel, D. (1991). Wisconsin Card Sorting Test performance as a measure of frontal lobe damage. *Journal of Clinical and Experimental Neuropsychology, 13,* 909–922.

Asarnow, R. F., & MacCrimmon, D. J. (1978). Residual performance deficit in clinically remitted schizophrenics: A marker of schizophrenia? *Journal of Abnormal Psychology, 87,* 597–608.

Asarnow, R. F., & MacCrimmon, D. J. (1981). Span of apprehension deficits during the postpsychotic stages of schizophrenia: A replication and extension. *Archives of General Psychiatry, 38,* 1006–1011.

Asarnow, R. F., Stetty, R. A., MacCrimmon, D. J., & Cleghorn, J. M. (1977). An attentional administration of foster children at risk for schizophrenia. *Journal of Abnormal Psychology, 86,* 267–275.

Asarnow, R. F., Granholm, E., & Sherman, T. (1991). Span of apprehension in schizophrenia. In S. R. Steinhauer, J. H. Gruzelier, & J. Zubin (Eds.), *Handbook of schizophrenia: Vol. 5. Neuropsychology, psychophysiology and information processing* (pp. 335–370). Amsterdam: Elsevier Science.

Bauman, E. (1971a). Schizophrenic short-term memory: A deficit in subjective organization. *Canadian Journal of Behavioural Science, 3,* 55–65.

Bauman, E. (1971b). Schizophrenic short-term memory: The role of organization at input. *Journal of Consulting and Clinical Psychology, 36,* 14–19.

Bauman, E., & Murray, D. J. (1968). Recognition versus recall in schizophrenia. *Canadian Journal of Psychology, 22,* 18-25.

Bellack, A. S., Mueser, K. T., Morrison, R. L., Tierney, A., & Podell, K. (1990). Remediation of cognitive deficits in schizophrenia. *American Journal of Psychiatry, 147,* 1650-1655.

Bowen, L., Wallace, C. J., Glynn, S. M., Nuechterlein, K. H., Lutzger, J. R., & Kuehnel, T. G. (1989). *Relationships among schizophrenic patients on attentional deficits, social problem-solving skills, and performance in psychoeducational rehabilitation tasks.* Paper presented at the 23rd annual convention of the Association for the Advancement of Behavior Therapy, Washington, DC.

Bracha, H. S. (1989). Is there a right hemi-hyper-dopaminergic psychosis? *Schizophrenia Research, 2,* 317-324.

Brenner, H. D., Kraemer, S., Hermanutz, M., & Hodel, B. (1990). Cognitive treatment in schizophrenia. In E. R. Straube & J. Hahleg (Eds.), *Schizophrenia: Concepts, vulnerability, and interventions* (pp. 161-192). New York: Springer-Verlag.

Brenner, H. D., Hodel, B., Roder, V., & Corrigan, P. (1992). Treatment of cognitive dysfunctions and behavioral deficits in schizophrenia: Integrated psychological therapy. *Schizophrenia Bulletin, 18,* 21-26.

Calev, A., Venables, P. H., & Monk, A. F. (1983). Evidence for distinct verbal memory pathologies in severely and mildly disturbed schizophrenics. *Schizophrenia Bulletin, 9,* 243-264.

Calev, A., Korin, Y., Kugelmass, S., & Lerer, B. (1987). Performance of chronic schizophrenics on matched word and design recall tasks. *Biological Psychiatry, 22,* 699-709.

Carlsson, A. (1988). The current state of the dopamine hypothesis of schizophrenia. *Neuropsychopharmacology, 1,* 179-186.

Corrigan, P. W., Wallace, C. J., Schade, M. L., & Green, M. F. (1994). Cognitive dysfunctions and psychosocial skill learning in schizophrenia *Behavior Therapy, 25,* 5-15.

Delis, D. C., Kramer, J. H., Kaplan, E., & Ober, B. A. (1987). *California Verbal Learning Test.* San Antonio, TX: Psychological Corp: Harcourt Brace Jovanovich.

Estes, W. K., & Taylor, H. A. (1966). Visual detection in relation to display size and redundancy of critical elements. *Perception and Psychophysics, 1,* 9-16.

Flor-Henry, P. (1976). Lateralized temporal-limbic dysfunction and psychopathology. *Annual New York Academy Sciences, 280,* 777-795.

Gaag, M. van der. (1992). *The results of cognitive training in schizophrenic patients.* Oegstgeest, The Netherlands: Eburon Delft.

Gold, J. M., Randolph C., Carpenter, C. J., Goldberg, T. E., & Weinberger, D. R. (1992). Forms of memory failure in schizophrenia. *Journal of Abnormal Psychology, 101,* 487-494.

Goldberg, T. E., Weinberger, D. R., Berman, K. F., Pliskin, N. H., & Podd, M. H. (1987). Further evidence for dementia of the prefrontal type in schizophrenia? A controlled study of teaching the Wisconsin Card Sorting Test. *Archives of General Psychiatry, 44,* 1008-1014.

Goldstein, G. (1986). The neuropsychology of schizophrenia. In I. Grant and K. M. Adams (Eds.), *Neuropsychological assessment of neuropsychiatric disorders* (pp. 147-171). New York: Oxford University Press.

Green, M. F. (1993). Cognitive remediation in schizophrenia: Is it time yet? *American Journal of Psychiatry, 150,* 178-187.

Green, M. F., Satz., P., Ganzell, S., & Vaclav, J. F. (1992). Wisconsin Card Sorting Test performance in schizophrenia: Remediation of a stubborn deficit. *American Journal of Psychiatry, 149,* 62-67.

Gur, R. E. (1978). Left hemisphere dysfunction and left hemisphere overactivation in schizophrenia. *Journal of Abnormal Psychology, 87,* 226-238.

Heaton, R. K. (1981). *Wisconsin Card Sorting Test Manual.* Odessa, FL: Psychological Assessment Resources.

Hellman, S. G., Green, M. F., Kern, R. S., & Christenson, C. D. (1992). The effects of instruction versus reinforcement on the Wisconsin Card Sorting Test [abstract]. *Journal of Clinical and Experimental Neuropsychology, 14,* 63.

Jaeger, J., & Douglas, E. (1991). Adjunctive neuropsychological remediation in psychiatric rehabilitation: Program description and preliminary data. *Schizophrenia Research, 4,* 304-305.

Jaeger, J., Berns, S., Tigner, A., & Douglas, E. (1992). Remediation of neuropsychological deficits in psychiatric populations: Rationale and methodological considerations. *Psychopharmacology Bulletin, 28,* 367-390.

Kahneman, D. (1973). *Attention and effort.* Englewood Cliffs, NJ: Prentice-Hall.

Kayton, L., & Koh, S. D. (1975). Hypohedonia in schizophrenia. *Journal of Nervous and Mental Disorders, 161,* 412-420.

Kern, R. S., Wallace, C. J., Hellman, S. G., Womack, L. M., & Green, M. F. (1996). A training procedure for remediating WCST deficits in psychotic patients: An adaptation of errorless learning principles. *Journal of Psychiatric Research, 30,* 283-294.

Kern, R. S., Wallace, C. J., Green, M. F., & Hellman, S. G. (1993). *The application of errorless learning on WCST performance in psychotic patients.* Paper presented at the eighth annual meeting of the Society for Research in Psychopathology. Chicago.

Kern, R. S., Green, M. F., & Goldstein, M. J. (1995). Modifiability of performance on the span of apprehension, a putative vulnerability indicator of schizophrenia. *Journal of Abnormal Psychology, 104,* 385-389.

King, H. E. (1991). Psychomotor dysfunction in schizophrenia. In S. R. Steinhauer, J. H. Gruzelier, & J. Zubin (Eds.), *Handbook of schizophrenia: Vol. 5. Neuropsychology, psychophysiology and information processing* (pp. 273-301). Amsterdam: Elsevier Science

Koh, S. D., & Kayton, L. (1974). Memorization of "unrelated" word strings by young nonpsychotic schizophrenics. *Journal of Abnormal Psychology, 33,* 14-22.

Koh, S. D., Kayton, L., & Berry, R. (1973). Mnemonic organization in young nonpsychotic schizophrenics. *Journal of Abnormal Psychology, 81,* 299-310.

Koh, S. D., Kayton, L, & Peterson, R. A. (1976). Affective encoding and consequent remembering in schizophrenic young adults. *Journal of Abnormal Psychology, 85,* 156-166.

Koh, S. D., Grinker, R. R., Sr., Marusarz, T. Z., & Forman, P. L. (1981). Affective memory and schizophrenic anhedonia. *Schizophrenia Bulletin, 7,* 292-307.

Kovelman, J. A., & Scheibel, A. B. (1984). A neurohistological correlate of schizophrenia. *Biological Psychiatry, 19,* 1601-1621.

Kraemer, S., Zinner, H. J., & Moller, H. J. (1989). *Cognitive training and social skills training in relation to basic disturbances in chronic schizophrenic patients.* Paper presented at the proceedings of the World Congress of Psychiatry, Amsterdam.

Larsen, S. F., & Fromholt, P. (1976). Mnemonic organization and free recall in schizophrenia. *Journal of Abnormal Psychology, 85,* 61-65.

Maher, B. A. (1991). Language and schizophrenia. In S. R. Steinhauer, J. H. Gruzelier, & J. Zubin (Eds.), *Handbook of schizophrenia: Vol. 5. Neuropsychology, psychophysiology and information processing* (pp. 437-464). Amsterdam: Elsevier Science.

Mednick, S. A., Parnas, J., & Schulsinger, F. (1987). The Copenhagen high-risk project, 1962-1986. *Schizophrenia Bulletin, 13,* 485-495.

Milner, B. (1963). Effects of different brain lesions on card sorting. *Archives of Neurology, 9,* 90-100.

Moray, N. (1969). *Attention: Selective processes in vision and hearing.* London: Hutchinson.

Mueser, K. T., Bellack, A. S., Douglas, M. S., & Wade, J. H. (1991). Prediction of social skill acquisition in schizophrenic and major affective disorder patients from memory and symptomatology. *Psychiatry Research, 37,* 281-296.

Nachmani, G., & Cohen, B. D. (1969). Recall and recognition free learning in schizophrenics. *Journal of Abnormal Psychology, 74,* 511-516.

Nuechterlein, K. H., & Asarnow, R. F. (1989) Perception and cognition. In H. I. Kaplan & B. J. Sadock (Eds.), *Comprehensive textbook of psychiatry* (5th ed., Vol. 1, pp. 241-256). Baltimore: Williams & Wilkins.

Nuechterlein, K. H., & Dawson, M. E. (1984). Information processing and attentional functioning in the developmental course of schizophrenia. *Schizophrenia Bulletin, 10,* 160-203.

Nuechterlein, K. H., Dawson, M. E., Gitlin, M., Ventura, J., Goldstein, M. J., Snyder, K. S., Yee, C. M., & Mintz, J. (1992). Developmental processes in schizophrenic disorders: Longitudinal studies of vulnerability and stress. *Schizophrenia Bulletin, 18,* 387-425.

Olbrich, R., & Mussgay, L. (1990). Reduction of schizophrenic deficits by cognitive training: An evaluative study. *European Archives of Psychiatry and Neurological Science, 239,* 366-369.

Sacuzzo, D. P., & Braff, D. L. (1981). Early information processing deficit in schizophrenia. *Archives of General Psychiatry, 38,* 175-179.

Sanders, A. F. (1983). Towards a model of stress and human performance. *Acta Psychologica, 53,* 61-97.

Saykin, A. J., Gur, R. C., Gur, R. E., Mozley, P. D., Mozley, L. H., Resnick, S. M., Kester, D. B., & Stafiniak, P. (1991). Neuropsychological function in schizophrenia: Selective impairment in memory and learning. *Archives of General Psychiatry, 48,* 618-624.

Spaulding, W. D., & Sullivan, M. (1991). From the laboratory to the clinic: Psychological methods and principles in psychiatric rehabilitation. In R. P. Liberman (Ed.), *Handbook of psychiatric rehabilitation.* Elmsford, NY: Pergamon.

Summerfelt, A. T., Alphs, L. D., Wagman, A. M. I., Funderburk, F. R., Hierholzer, R. M., & Strauss, M. E. (1991). Reduction of perseverative errors in patients with schizophrenia using monetary feedback. *Journal of Abnormal Psychology, 100,* 613-616.

Terrace, H. S. (1963). Discrimination learning with and without errors. *Journal of the Experimental Analysis of Behavior, 6,* 1-27.

Wagner, B. R. (1968). The training of attending and abstracting responses in chronic schizophrenics. *Journal of Experimental Research in Personality, 3,* 77-88.

Woodworth, R. S. (1948). *Experimental psychology.* New York: Cambridge University Press.

Yozawitz, A. (1986). Applied neuropsychology in a psychiatric center. In I. Grant & K. M. Adams (Eds.), *Neuropsychological assessment of neuropsychiatric disorders.* New York: Oxford University Press.

Index

Ability, vs. function, 4
Achievement tests
　in brain injury, 186-187
　in learning disability, 210, 211-213, 220
Activities of daily living: *see* ADL assessment
ADL assessment, *see also* Functional assessment; IADL assessment
　of extended ADL, 39, 51
　geriatric, 97-98, 101, 102, 113
　by neuropsychological testing, 120
　by occupational therapist, 14, 16, 17, 19-20
　by physical therapist, 39, 50-51
　research on, 24, 25, 26
　in simulated environments, 116
Affective impairment, assessment of
　in brain injury, 187
　geriatric, 91-92, 93-94
Alberta Infant Motor Scale (AIM), 44
Allen Cognitive Levels, 21, 22
Alzheimer's-type dementia, *see also* Dementia
　functional assessment in, 100-101, 102
American Multiple Errands Test (AMET), 124-127, 241
Americans With Disabilities Act (ADA), 13-14
Amnesia, 230-231; *see also* Memory rehabilitation
Amputees: *see* Prostheses, external
Anosognosia, 183
Anxiety, assessment of, 187
Aphasia, 59-61
　assessment of, 65-67, 185, 234
　in head trauma rehabilitation program, 175, 177
Apraxia of speech, 59, 60-61
　assessment of, 62-65
Arizona Battery of Communication Disorders of Dementia, 66
Arnadóttir OT-ADL Neurobehavioral Evaluation (A-ONE), 19-20, 25, 26
Arousal, assessment of, 184
Arthritis, functional assessment in, 99-100
Ashworth Scale, 50

Assessment of Motor and Process Skills (AMPS), 20-21, 24-27
Assessment of Preterm Infant Behavior (APIB), 45
Assisted living, 79
Attention, assessment of, 184
Attention-Deficit/Hyperactivity Disorder, 113
Auditory deficits
　geriatric assessment of, 94
　in learning disabilities, 208, 209
Autistic children, errorless training of, 277

Babinski sign, 144, 150
Balance, assessment of, 49-50
　geriatric, 94-95
　in learning disabilities, 209
Barthel Index (BI), 50-51, 97, 104
Batelle Developmental Inventory Screening Test (BDIST), 46
Bay Area Functional Performance Evaluation, 17
Bayley Scales of Infant Development (BSID), 42-43, 44
Berg Balance Scale, 49-50
Berkeley Prosthetic Project, 156-159
Biopsychosocial model, 247, 248, 256, 257
Blood pressure measurement, 48
Boder Test of Reading-Spelling Patterns, 214
Body awareness training, 194, 195
Borg Perceived Exertion Scale, 48
Boston Diagnostic Aphasia Examination (BDAE), 234
Boston University psychiatric rehabilitation model, 250-251, 254-255, 256
Brain injury: *see* Head trauma; Stroke
Brazelton Neonatal Behavioral Assessment Scale (BNBAS), 45
Brief Cognitive Rating Scale (BCRS), 100
Bruininks-Oseretsky Test of Motor Proficiency (BOTMP), 43-44

California Verbal Learning Test (CVLT), 187, 220-221, 270

Canadian Occupational Performance Measure (COPM), 22-23, 25
Capacity, vs. function, 4
Cardiopulmonary assessment, 48-49
Career ability placement survey (CAPS), 257, 259
Case management, by nurses, 78, 79, 81-84
Catastrophic reaction to stimuli, 209
Category Test (Halstead-Reitan), 223-224
　vocational outcome and, 119-120
Central nervous system, 138
　sensory testing and, 141-142
　strength testing and, 144
　systems model of, 36-38
Cerebral palsy, assessment in, 44, 46-47
Children
　head injury rehabilitation program, 196-197
　physical therapy assessment, 40, 41-47, 54
"Choose, Get, Keep" model, 254-255, 257
Clinical pathways, 82-83, 85
Clubhouse model, 249, 251-252
Coarse coding, 133-135
Cognitive disability, *see also* Head trauma
　functional assessment of, 114, 115
　　by FIM, 51, 98-99
　　geriatric: *see* Dementia
　　by interviewing, 117-118
　　by Multiple Errands Test, 124-127, 241
　　by neuropsychological testing, 119-122
　　by observation, 115-116
　　by rating scales and questionnaires, 118-119
　　in simulated environments, 116
　　by task analysis, 122-124
　neuropsychological assessment of, 181-197
　　case example, 192-195
　　caveats, 196-197
　　historical development of, 181-182
　　Level 1 (environment), 182-183, 189, 190, 197
　　Level 2 (tests), 182, 184-187, 189, 190, 196-197
　　Level 3 (fractionation), 182, 187-189, 191
　　Level 4 (intervention), 189-190, 191, 195-196
　　predictive value of, 119-122, 124-127
Cognitive rehabilitation, *see also* Memory rehabilitation; Psychotic patients, cognitive remediation of
　in Danish program, 174, 175
　defined, 268
　efficacy of, 197
　generalizability of, 196-197
　neuropsychological assessment and, 181, 189-190, 191-196
Communication disorders, 59-67
　assessment
　　of aphasia, 65-67
　　in brain injury, 185
　　categories of data for, 62
　　of dysarthria and apraxia, 62-65
　　purposes of, 61-62

Communication disorders (*cont.*)
　causes, 59-61
　in Danish head trauma program, 175, 177
Communicative efficiency ratio, 64-65
Community-based rehabilitation, 79-81
　case management in, 82, 83-84
Community integration, of developmentally disabled, 13
Computerized assessment, of task performance
　EASE, 23-24, 25, 26
　Observe, 23, 25, 104
　OT-FACT, 24, 25, 26
Computerized training, of head trauma patients, 175
Concentration, assessment of, 184
Construct validity, 133-135
Contracting, in case management, 84
Controlled Oral Word Association Test, 222-223
Coordination, assessment of, 149
Critical pathways, 82-83, 85

Delirium
　vs. aphasia, 65-66
　screening for, 92
Dementia
　vs. aphasia, 65-66
　cognitive assessment in, 91-93, 232, 233
　vs. depression, 93, 94
　functional assessment in, 100-102, 104, 120
　memory rehabilitation in, 229, 231, 232, 237
　screening for, 92-93
Dementia Rating Scale (DRS), 232, 233
Depression, assessment of, 187
　geriatric, 93-94
DER foot designs, 165
Dermatomes, 138, 142
Developmental disabilities
　community integration with, 13
　habilitation and, 4
Dexterity
　Grooved Pegboard, 222
　Tactual Performance Test, 224
Diadochokinesis, 149
Direct Assessment of Functional Status (DAFS), 101, 104
Disability, *see also* Cognitive disability; Learning disabilities
　assessment of
　　geriatric, 90-91, 95-102
　　by occupational therapist, 11-13, 16-23
　defined, 4, 11, 89, 90
Discharge planning
　functional assessment in, 91
　　KELS for, 16
　　PASS for, 17
　vs. rehabilitation nursing process, 77, 85

Dorsal column function, 141, 142
Dynamic assessment, 26
Dynamometry, 47, 48, 148
Dysarthria, 59, 60
  assessment of, 61, 62-65
  in head trauma patients, 175
Dyseidetic readers, 214
Dyslexia, *see also* Learning disabilities
  Boder Test for, 214
Dysphonetic readers, 214

EASE computerized assessment system, 23-24, 25, 26
Ecological validity, of neuropsychological testing, 119-120
Education for All Handicapped Children Act, 203, 212
Electromyography, in gait analysis, 160, 162
Emotional functioning, assessment of
  in brain injury, 187
  geriatric, 91-92, 93-94
Endurance testing, 48-49
Errorless learning, 277
Executive functions, assessment of, 186, 187
  Multiple Errands Test and, 124-127
  task analysis and, 123
Exercise testing, for endurance, 48-49

Facilitated movement approach, 37
Fairweather Lodge movement, 247
Family of patient
  in Danish rehabilitation program, 176, 178
  in neuropsychological assessment, 182-183, 190, 191
Flex foot, 161, 162, 165
Flexibility testing, 48, 140-141
Fractionation of deficits, 182, 187-189, 191
Frankel Scale, 53
Frenchay Activities Index, 51
Fugl-Meyer Assessment of Sensorimotor Recovery after Stroke, 52
Function, vs. ability, 4
Functional assessment, *see also* ADL assessment; Task analysis
  of cognitive disability, 114, 115
    by FIM, 51, 98-99
    by interviewing, 117-118
    by Multiple Errands Test, 124-127, 241
    by neuropsychological testing, 119-122
    by observation, 115-116
    by rating scales and questionnaires, 118-119
    in simulated environments, 116
    by task analysis, 122-124
  contexts of, 113
  defined, 114
  geriatric, 89-105, 113
    of disability, 90-91, 95-102
    future prospects, 103-105

Functional assessment (*cont.*)
  geriatric (*cont.*)
    goals of, 89-90
    of handicap, 90-91, 102-103
    of impairment, 91-95
    neuropsychological testing and, 120
  goals of, 114
  in learning disabilities, 207-208
  by occupational therapist
    computerized tools for, 23-24, 25, 26, 104
    of disability, 11-13, 16-23
    of handicap, 11-16
    levels of, 10-13
    methods for, 12-13
    research in, 24-27
    task orientation of, 9-10
  by physical therapist, 50-53
  in psychosocial rehabilitation, 250-251
Functional Assessment Staging of Dementia (FAST), 100-101
Functional Autonomy Measurement System (SMAF), 51, 54
Functional components evaluation, 63
Functional health patterns model, 73-74, 86
Functional Independence Measure (FIM), 51, 98-99, 113
Functional mobility, *see also* Motor functions
  geriatric assessment, 96
  occupational therapy assessment, 16, 17, 19-20, 26
Functional Reach Test, 49, 94-95
Functional Status Index (FSI), 99-100

Gait analysis, 150-152
  in geriatric assessment, 94
  historical development of, 156, 157-158
  in prosthetic research, 159-161
Gates-MacGinitie Reading Tests, 213-214
Gender bias, in IADL assessment, 104
Geriatric assessment: *see* Dementia; Functional assessment, geriatric
Geriatric Depression Scale (GDS), 93-94
Global Assessment Scale, 17
Global Deterioration Scale, 100
Goniometry, 48; *see also* Range of motion
  in gait analysis, 159-160
Government regulations: *see* Legislation
Graphesthesia, 141, 142
Gray Oral Reading Tests, 214-215
Grooved Pegboard, 222
Gross Motor Function Measure (GMFM), 46-47
Gross Motor Performance Measure (GMPM), 47

Habilitation, defined, 4
Halstead-Reitan Neuropsychological Battery (HRB), 217, 223-224
  Grooved Pegboard in, 222
  vocational outcome and, 119-120

Handicap
  assessment of
    geriatric, 90-91, 102-103
    by occupational therapist, 11-16
  defined, 4, 11, 89-90
Head trauma, *see also* Cognitive disability
  functional assessment in, 52, 53, 98-99
  memory rehabilitation in, 229, 231, 232, 234-236, 240
    assistive devices, 242, 243
  neuropsychological assessment in, 173-174
  rehabilitation in
    Danish program, 174-178
    future prospects, 179
    pediatric program, 196-197
    potential for, 171-172
    severe, 179
Health care systems: *see* Managed care
Hearing: *see* Auditory deficits
Home health care, nurses in, 78-80

IADL assessment, 96; *see also* ADL assessment; Functional assessment
  in dementia, 101-102, 104
  IADL Scale for, 99
  Lawton instrument for, 51
  limitations of instruments, 103-104
  by occupational therapist, 14, 16, 17, 18
IADL Scale, 99
Impairment
  defined, 4, 10-11, 89, 90
  geriatric assessment of, 90-95
Implicit memory, 231-232, 239, 243
Independence, measurement of, 96
Independent living, *see also* ADL assessment
  for developmentally disabled, 13
  functional assessment for, 16, 24
  for head trauma patients, 176, 177, 178
Index of ADL, 97-98, 104
Individual Placement and Support (IPS), 256-257
Individuals with Disability Education Act (IDEA), 45
Information processing
  models of, 269-270
  in schizophrenia
    assessment of, 270-272
    deficits in, 267-268
Instrumental activities of daily living: *see* IADL assessment
Insurance: *see* Managed care
Insurance nurses, 81, 82
Integrated Psychological Therapy (IPT), 279, 281
Intelligence testing
  for cognitive rehabilitation, 186
  in learning disabilities, 210-211, 220
  for memory rehabilitation, 232, 233
Intelligibility testing, 64-65
Interval scales, 91

Interviews, in assessment
  of cognitive disability, 117-118
  geriatric, 90, 91
  of learning disabilities, 207
  by occupational therapist, 12, 13
Isokinetic testing, 47, 148

Jacksonian sign, 136
Job analysis, 13-14; *see also* Task analysis; Vocational rehabilitation
Job Club Model, 259

Kaufman Assessment Battery for Children (K-ABC), 211
KeyMath-R, 215-216
Kinesthesia, assessment of, 141, 142
Kohlman Evaluation of Living Skills (KELS), 16-17, 25
Korsakoff's syndrome, memory rehabilitation in, 229, 231-232, 236-239, 243
Kurtzke Scales, 53

Language disorders: *see* Aphasia
Lawton IADLs, 51
Learning, *see also* Memory
  errorless training, 277
  neuropsychological testing of, 187
Learning disabilities, 201-226
  in adolescents, 207, 208
  in adults, 206
    neuropsychological testing, 217
    signs and symptoms, 207, 208
    young adults, 216, 217, 219-224
  assessment, 206-224
    achievement tests in, 210, 211-213, 220
    diagnostic tests in, 213-216
    functional assessment in, 207-208
    intelligence tests in, 210-211, 220
    interview in, 207
    of memory, 209, 220-221
    of motor functions, 209
    neurological examination in, 210
    neuropsychological tests in, 216-217, 218-219, 220-224
    of perception, 208-209
    of problem solving, 209-210
    psychiatric screening in, 217, 219-220
  vs. brain injury, 125, 126
  in children
    neuropsychological testing, 216-217, 218-219
    signs and symptoms, 207, 208
  defined, 203
  diagnosis of, 203
  heterogeneity of, 203, 204-205, 206
  interventions for, 224-225
  nonverbal, 205
  overview of, 201-203
  signs and symptoms, 207, 208
  in young adults, 216, 217, 219-224

Legislation
  Americans With Disabilities Act, 13–14
  Education for All Handicapped Children Act, 203, 212
  Individuals with Disability Education Act, 45
Life care plan, 82
Luria Neuropsychological Investigation, 172, 173, 174

Managed care
  learning disabilities in, 217
  neuropsychologists in, 189
  physical therapy assessment in, 38, 39–40, 51
  rehabilitation nursing in, 80, 82, 85–86
Manual muscle testing (MMT), 47–48, 144–149
Maryland Disability Index: see Barthel Index
Memory
  assessment of
    in brain injury, 187, 188
    in dementia, 92–93, 232, 233
    in learning disabilities, 209, 220–221
    for memory rehabilitation, 232, 233
    in schizophrenia, 270–271
  remediation of, in schizophrenia, 270, 272–274, 279, 280
Memory rehabilitation, 229–243
  assessment for, 232–234
  assistive devices for, 230, 237–239, 242–243
  computer-assisted, 230, 232, 235–236
  in dementia, 229, 231, 232, 237
  in head injury, 229, 231, 232, 234–236, 240
    assistive devices, 242, 243
  history of, 230
  imagery in, 230, 231, 234–236
  in Korsakoff's syndrome, 229, 231–232, 236–239, 243
  outcomes, 239–242
  theoretical basis of, 230–232, 243
Mental illness, see also Psychiatric screening tests; Psychosocial rehabilitation; Psychotic patients
  assessment of disability, 16–19, 20–22
  assessment of handicap, 14–16
Milwaukee Evaluation of Daily Living Skills (MEDLS), 18–19, 25, 26
Mini-Mental State Examination (MMSE), 92
Minimum Data Set for Nursing Home Resident Assessment and Care Screening (MDS), 102
Mobility: see Functional mobility
Motor functions, assessment of, see also Functional mobility; Neurological examination
  for cognitive rehabilitation, 185, 188
  geriatric, 91–92, 94–95
  in learning disabilities, 209
  pediatric, 42–45
Motor speech disorders, 60–65

Movement Assessment of Infants (MAI), 44
Multiple Errands Test, 124–127, 241
Multiple sclerosis, functional assessment in, 51, 52, 53
Muscle testing
  strength, 47–48, 143–149
  tone, 50, 139, 140–141

Neglect syndromes
  assessment for, 185
  case example, 192–195
Nelson-Denny Reading Test, 220
Nervous system, 137–138
  sensory testing and, 141–142
  strength testing and, 143–144
  systems model of, 36–38
Neurological Assessment of the Full and Preterm Infant, 45
Neurological examination, 131–152
  basic principles, 41, 131–138
  for cognitive rehabilitation, 184–185
  formal observation in, 139
  gait analysis in, 150–152
    geriatric, 94
  geriatric, 94–95
  history in, 138–139
  in learning disabilities, 210
  order for performance of, 135
  range of motion testing, 47, 48, 140–141
  sensory evaluation, 139, 141–142
  strength testing, 47–48, 143–149
    in cognitive assessment, 185
  validity issues, 133–135, 141
Neurological Examination of the Full-Term Infant, 45
Neuropsychological assessment
  of cognitive disabilities, 181–197
    case example, 192–195
    caveats, 196–197
    historical development of, 181–182
    Level 1 (environment), 182–183, 189, 190, 197
    Level 2 (tests), 182, 184–187, 189, 190, 196–197
    Level 3 (fractionation), 182, 187–189, 191
    Level 4 (intervention), 189–190, 191, 195–196
    predictive value of, 119–122, 124–127
  of learning disabilities, 216–217, 220–224
  Luria approach, 172, 173–174
  of psychotic patients, 270–272
  reimbursement under managed care, 189
  role of, in rehabilitation, 3, 4
Newborn assessments, 45
Nonverbal learning disability, 205
Normal function, as goal, 3–4
Nottingham Extended ADL Index, 51
Nursing: see Rehabilitation nursing
Nursing diagnoses, 74, 76–77

Nursing home residents, functional assessment in, 102
Nursing process, 74–78

OARS Multidimensional Functional Assessment Questionnaire, 92–93
Observation
 in functional assessment, 115–116
 in neurological examination, 139
 in occupational therapy assessment, 12
Observe computerized assessment system, 23, 25, 104
Occupational History, 14
Occupational Performance History Interview (OPHI), 14–15, 25
Occupational Role History, 14
Occupational therapy assessment, 9–27
 computerized tools for, 23–24, 25, 26, 104
 of disability, 11–13, 16–23
 of handicap, 11–16
 levels of, 10–13
 methods for, 12–13
 research in, 24–27
 task orientation of, 9–10
Ordinal measurement, 91
OT-FACT, 24, 25, 26

Pain
 referral patterns, 138
 sensory testing of, 141–142
Parkinson's disease, functional assessment in, 52, 53
Pathognomic tests, 133–135
Pathology, in WHO model, 10
Patient Assessment of Functioning Inventory (PAF), 240
Patient Competency Rating Scale, 126
Peabody Developmental Motor Scales (PDMS), 43
Peabody Individual Achievement Test–Revised (PIAT-R), 212
Pediatric assessment, in physical therapy, 40, 41–47, 54
Pediatric Evaluation of Disability Inventory (PEDI), 46, 54
Pediatric head injury rehabilitation, 196–197
Pegboard, 222
Perception: *see* Sensory evaluation
Performance Assessment of Self-Care Skills (PASS), 17–18, 24–27
 in memory training program, 241
Performance IQ, 211, 233
Performance-Oriented Assessment of Balance (POAB), 94
Performance-Oriented Assessment of Gait (POAG), 94
Peripheral nervous system, 137–138
 sensory testing and, 141–142
 strength testing and, 143–144
Person-task-context transaction, 9, 13, 24, 26

Physical therapy
 history of, 34–36
 neurology specialty in, 33
 nursing and, 80–81
 purpose of, 34
 training and licensure in, 33, 35–36
Physical therapy assessment, 36–55; *see also* Neurological examination
 adult, 47–53
 basic measurements, 41, 47–50
 diagnosis specific, 51–53
 functional assessment, 50–51
 future directions, 53–55
 pediatric, 40, 41–47
 sensitivity of tests in, 40, 54
 standardized, 38–40
 theoretical basis of, 36–38
 types of tests, 40
Physical therapy assistants, 35
"Place-train" model, 255, 256
Porch Index of Communicative Abilities, 67
Problem solving
 in brain injury, 187
 in learning disabilities, 209–210
 in psychotic patients, 271, 279, 280
Procedural memory, 231–232, 239
Program of Assertive Community Treatment (PACT), 247, 255, 256, 257
Proprioception
 assessment of, 141, 142
 deficit in, 208
Prostheses, external, 155–166
 clinical assessment of, 162–166
 historical development of, 155–156
 scientific analysis of, 156–162
  above-knee prostheses, 158
  below-knee prostheses, 158–159, 161–162
  gait analysis, 156–158, 159–161
Prosthetic memory, 230, 237–239, 242–243
Psychiatric rehabilitation model, 250–251; *see also* Psychosocial rehabilitation
Psychiatric screening tests
 for depression, 93–94, 187
 SCL-90-R
  in head trauma, 173
  in learning disabilities, 217, 219–220
Psychometric testing: *see* Neuropsychological assessment
Psychosocial rehabilitation
 basic concepts, 249
 Boston University model, 250–251, 254–255, 256
 clubhouse model, 249, 251–252
 defined, 248
 efficacy of, 248–249
 future directions, 259–260
 historical role, 247–248
 integrated models, 255, 256–259
 prediction of success in, 255–256, 260

Psychosocial rehabilitation (*cont.*)
  sheltered workshop model, 253–254, 255
  supported employment, 248, 252, 254–255, 256–257
  transitional employment, 251–252, 254–255
  vocational rehabilitation system and, 252–255
Psychotherapy, for head trauma patients, 172, 174, 175
Psychotic patients, *see also* Mental illness
  cognitive remediation of, 267–281
    assessment for, 270–272
    comprehensive programs, 279–280
    defined, 268
    feasibility studies, 268, 272–278, 280–281
    future directions, 280–281
    generalizability of, 268, 280
    goals of, 267–268
    information processing and, 267–268, 269–270
    rate-limiting factors in, 268, 280
  language impairments in, 65–66
  social skills training for, 279–280
Pulmonary assessment, 48–49

Questionnaires, as assessment methodology, 90, 91, 118–119

Range of motion, 47, 48, 140–141
Rappaport Disability Rating Scale, 53
Rasch analysis, 21, 24, 26–27, 99, 104
Reading comprehension, 220
Reasoning, assessment of, 187
Reflexes, assessment of, 140, 149–150
Rehabilitation, *see also* Cognitive rehabilitation; Head trauma, rehabilitation in; Memory rehabilitation
  defined, 3–4
Rehabilitation nursing, 71–86
  functional health patterns model, 73–74, 86
  future directions, 86
  goals of, 73
  history of, 71–73
  nursing process in, 74–78
  specialization in, 73
  subspecialties in, 78–86
    advocacy, 85–86
    case management, 78, 79, 81–84
    community-based care, 78–81
    education, 84–85
Reimbursement: *see* Managed care
Resource management, by nurses, 83–84
Restorative care, nurses in, 79–81
Rey Complex Figure Test, 222
Rivermead Behavioural Memory Test, 240–241
Role Checklist, 15–16, 25
Role performance: *see* Handicap
Routine Task Inventory-2 (RTI-2), 16, 21–22, 25

SACH foot, 159, 161, 162, 165
Schizophrenia: *see* Psychotic patients
Self-care: *see* ADL assessment
Self-Paced Walking Test, 48–49
Sensory evaluation
  clinical history, 139
  for cognitive rehabilitation, 184–185, 188
  geriatric, 94
  in learning disabilities, 208–209
  testing, 141–142
Sensory overload, 209
Sent-Ident, 94
Sheltered workshop model, 253–254, 255
Short Portable Mental Status Questionnaire (SPMSQ), 92–93
Single subject research designs, 189, 190, 196
Sit and reach test, 48
Skin, assessment of, 139; *see also* Touch
Social role: *see* Handicap
Soft signs, neurological, 210
Span of apprehension, in schizophrenia, 270, 272, 277–278, 280
Spasticity testing, 50
Speech disorders, 59–65
  vs. language disorders, 60
Speech-Sounds Perception Test, 224
Spinal cord injury
  functional assessment in, 51, 52, 53, 98–99
  motor presentations of, 144
Spinal nerve roots, 137–138, 141, 144
Spinothalamic tract function, 142
Stanford–Binet Intelligence Scale, 211
Stereognosis, 141, 142
Strength testing, 47–48, 143–149
  in cognitive assessment, 185
Stroke
  cognitive rehabilitation, 192–195
  functional assessment, 52
    Berg Balance Scale in, 50
    FIM in, 98–99
  speech-language therapy, 62
Structured Assessment of Independent Living Skills (SAILS), 101–102, 104
Subacute care, 79
Sunderland Scale, 240
Supported employment, 248, 252, 254–255, 256–257
Syme amputation, 159
Symptoms Check List (SCL-90-R)
  in head trauma, 173
  in learning disabilities, 217, 219–220

Tactile perceptual problems, 208–209, 224
Tactual Performance Test (TPT), 224
Task analysis, *see also* Functional assessment
  of cognitive disabilities, 122–124
  in geriatric assessment, 96–97
  of learning disabilities, 225

Task analysis (*cont.*)
    in neuropsychological assessment, 181–182, 187–189
    in occupational therapy assessment, 9–10, 11, 12–13, 27
    in physical therapy assessment, 39
Temperature sensibility, 142
Tensiometers, 148
Test of Infant Motor Performance (TIMP), 44–45
Test of Written Language (TOWL), 215
Timed Manual Performance Test (TMP), 95
Toddler and Infant Motor Evaluation (TIME), 45
Token Test–Revised, 233–234
Touch
    neurologic assessment of, 140, 141–142
    tactile perceptual problems, 208–209, 224
Trail Making Test, 223
    in stroke rehabilitation, 193, 194, 195
    vocational outcome and, 120
Transitional employment, 251–252, 254–255
Transportation, training for, 176
Traumatic brain injury: *see* Head trauma
Tremor, 139, 149

Vegetative state, rehabilitation in, 179
Verbal IQ, 211, 233
VF-14, 94
Visual functions, assessment of
    in cognitive disabilities, 185, 188
    geriatric, 94
    in learning disabilities, 209
    in schizophrenia, early processing, 270, 272, 277–278, 280, 281
Visual scanning, training in, 194–195, 196
Vocational rehabilitation
    federal legislation and, 13–14
    with head trauma, Danish program, 176, 177, 178

Vocational rehabilitation (*cont.*)
    neuropsychological assessment and, 119–120
    occupational therapy in, 13–14
    with physical disabilities, 252–253
    with psychiatric disabilities
        efficacy of, 248–249
        future directions, 259–260
        historical role of, 247–248, 252, 253
        integrated models, 255, 256–259
        prediction of success in, 255–256, 260
        psychiatric rehabilitation and, 249–251
        sheltered workshop model, 253–254, 255
        supported employment, 248, 252, 254–255, 256–257
        transitional employment, 251–252, 254–255

Wechsler Adult Intelligence Scale–Revised (WAIS-R), 186, 210–211, 220, 232, 233
Wechsler Individual Achievement Test (WIAT), 212–213
Wechsler Intelligence Scale for Children (WISC-3), 210–211
Wechsler Memory Scale–Revised (WMS-R), 221, 233
Wechsler Preschool and Primary Scale of Intelligence–Revised (WPPSI-R), 210–211
WHO model of function, 10–11
    geriatric assessment and, 89
    learning disabilities and, 202, 216, 224
    pediatric assessment and, 45–46
    physical therapy assessment and, 54
    sensory-motor evaluation and, 131, 152
Wide Range Achievement Test (WRAT-3), 211–212, 213, 220
Wisconsin Card Sorting Test (WCST), 270, 271, 274–277
Woodcock Johnson Psychoeducational Battery–Revised, 220
Word Association Test, 222–223
Work hardening, 14

ISBN 0-306-45662-1